The Regulation of Automated and Autonomous Transport

Kyriaki Noussia • Matthew Channon
Editors

The Regulation of Automated and Autonomous Transport

Springer

Editors
Kyriaki Noussia
School of Law
University of Reading
Reading, Berkshire, UK

Matthew Channon
Law School
University of Exeter
Exeter, UK

ISBN 978-3-031-32355-3 ISBN 978-3-031-32356-0 (eBook)
https://doi.org/10.1007/978-3-031-32356-0

This Springer imprint is published by the registered company Springer Nature Switzerland AG
The registered company address is: Gewerbestrasse 11, 6330 Cham, Switzerland

Preface

Automated vehicles, whether automated or autonomous vehicles, or connected and automated, or autonomous vehicles, autonomous drones and maritime autonomous surface ships as well as autonomous trains and rail systems are all part of the automated and autonomous transportation systems. Current questions about Automated Vehicles (AVs) and, generally, about autonomous transportation systems do not now revolve around whether such technologies should or should not be implemented; but the focus is shifting towards examining how such technologies will impact evolving automated and autonomous transportation systems. More importantly, how will mobility itself change as these independent evolving automated and autonomous transportation systems, first share and then dominate wide circulation pathways, be it roads, the skies, the seas or rail tracks? How will the public be kept informed and adapt towards such developments? These are all issues to be addressed, and this book on the law and regulation of automated and autonomous transportation systems is an attempt to shed light on these and other related regulatory legal questions. The book is addressed, in terms of primary audiences, to academic libraries and practitioners, and judges as well as policy-makers, to whom it will appeal as the law in this area is shaping and will continue to do so. Secondary audiences comprise engineers, sociologists, naval architects, all those involved in the automated industry, and AI people.

This book has contributions related to the law and regulation of automated and autonomous transportation systems in relation to cars/vehicle, drones and ships. A few important issues and challenges are addressed as well as international comparisons made. For example, the potential change in driving offences in both the UK and Australia in relation to automated vehicles. Moreover, following the latest consultation for a law reform in the area of the Automated and Electric Vehicles Act 2018; the way in which automation can reduce human errors, the impact of autonomous vehicular technologies and the ethical implications to be considered before autonomous vehicles can be safely deployed on the road. Further, the civil unmanned aerial vehicles (CUAVs) regulation in China, the legal provisions of the three regions in China for CUAVs and the need for effective supervision of UAV safe flight. Additionally, the social function and ethical issues relating to cyber risks and AVs, the digital shipping revolution that is coming, i.e. the development of unmanned maritime vessels, the legal and regulatory implications surrounding the usage of unmanned maritime vessels on civilian life and the ways in which the maritime industry is ready to accommodate widespread usage of unmanned vessels, is also discussed. Issues on effective maritime security are addressed as well as issues pertaining to salvage and autonomous maritime navigation. The legal issues which are expected to arise from the penetration of drones in passenger and cargo transports; the needed legal framework for drones to be used for delivery process, are also amidst the issues examined in this book. It is noted that the legal regulatory fragmentation observed worldwide in the legal regimes and law for automated and autonomous transportation systems is not a hinder but a challenge to better regulate as we enter the fully digitised section of the era of the Fourth Industrial Revolution.

London, UK & Athens, Greece
Exeter, UK
September 2022

Kyriaki Noussia
Matthew Channon

Contents

List of Contributors

Matthew Channon University of Exeter, Law School, Exeter, UK

Viktoria Chatzara Rokas Law Firm, Athens, Greece

Aybüke Naz Durmuş University of Reading, School of Law, Reading, UK

Maria Glynou Attorney at Law, Athens, Greece
LLM candidate and Research Assistant, LSE, London, UK

Zuhal Gocmen Attorney at Law, Istanbul, Turkey

Samantha Huneberg University of Johannesburg, Department of Mercantile Law, Johannesburg, South Africa

Francesco La Fata University of Florence, Department of Economics and Management, Florence, Italy

Callum Laffey University of Exeter, Law School, Exeter, UK

Sara Landini University of Florence, Department of Legal Sciences, Florence, Italy

Can Luo School of Artificial Intelligence and Law, Southwest University of Political Science and Law, Chongqing, China

Kyriaki Noussia University of Reading, School of Law, Reading, UK

Cecilia Severoni University of Udine, Udine, Italy

Andrea Signorino Barbat University of Montevideo, School of Law, Montevideo, Uruguay

Shanice N. Trowers Faculty of Law, University of Technology, Kingston, Jamaica

List of Contributors

Part I
Drones: Aviation/Aerial Automated, Autonomous Transportation Systems

Introduction

Kyriaki Noussia and Matthew Channon

Abstract Autonomous transport systems, for automated and autonomous transportation, in all modes of transport, i.e. road transport systems (automated and autonomous cars), aviation autonomous systems (automated/autonomous drones, unmanned aerial vehicles), shipping (maritime autonomous surface ships, underwater autonomous vehicles) and rail, are together with the proliferation of ICT and IoT will predominantly dominate our societies in the near future. In light of the above, this chapter summarises the main issues discussed in the various contributions related to the law and regulation of automated and autonomous transportation systems in relation to cars/vehicle, drones and ships. This introductory chapter also pertains to the challenge observed in this book, i.e. the need to have a regulation for AI and autonomous transport systems, as regulation will allow the creation of trust which will allow the massive deployment of AI and autonomous (transportation) systems. The chapters from invited authors address the law in the areas of automated and autonomous transportation systems in the categories of road vehicles, drones and vessels. This introductory chapter also states that in all of the contributions what is noted is the challenge of the shift of liability as manufacturers need to prepare for potential product liability risk from damages, as well as its effect on the way insurers will address insuring the operation of vehicles and drones or ships in an automated and autonomous transportation systems.

Keywords Automated transport · Autonomous transport · Regulation · Autonomous transport systems · Artificial intelligence · Drones · Autonomous vehicles

---

K. Noussia (✉)
School of Law, University of Reading, Reading, Berkshire, UK
e-mail: k.p.noussia@reading.ac.uk

M. Channon
Law School, University of Exeter, Exeter, UK
e-mail: m.channon2@exeter.ac.uk

K. Noussia, M. Channon (eds.), *The Regulation of Automated and Autonomous Transport*, https://doi.org/10.1007/978-3-031-32356-0_1

1 General

Autonomous transport systems, for automated and autonomous transportation, in all modes of transport, i.e. be it road automated and autonomous road transport systems (automated and autonomous cars), or aviation autonomous systems (automated/autonomous drones, unmanned aerial vehicles), or shipping (maritime autonomous surface ships, underwater autonomous vehicles) and rail, are going to dominate our societies and everyday life in the years ahead.

This book has various contributions related to the law and regulation of automated and autonomous transportation systems in relation to cars/vehicle, drones and ships. A challenge observed in this book, which needs to be addressed, is solving the legal regulatory fragmentation observed worldwide on legal regimes and law for automated and autonomous transportation systems.

When introducing new technology such as autonomous transportation systems, one of the basic challenges is to understand the risks that will emerge. Regulation should be considered and any relevant incidents should be gathered and disseminated. From the state's perspective, governments need to set minimum standards, establish responsibility, and follow up incidents to allow detailed and pragmatic approach in its regulation and rulemaking. As noted by Zwicker et al. 'when an emerging technology does not fit neatly within a pre-existing regulatory scheme, regulators have the difficult task of creating new rules that do not conflict with existing ones'.[1] Autonomous systems can create safer, more efficient and lower carbon transportation systems.[2] To do so we need to design and implement the future system in a way to embrace and respond to challenges. Some things to consider are (a) the expectations from such automated and autonomous transportation systems and the degree of public acceptability for the various modes of autonomous transportation systems; (b) the need for real world trials to assess the actual needs in practice; (c) the need to find a way to develop technologies and services that are trustworthy, ethical and inclusive, and this will only result after extensive consultation, and (d) as driverless transport evolves some common frameworks will be required to allow the design of algorithms to lead to ethical decisions.[3]

The chapters from invited authors address the law in the areas of automated and autonomous transportation systems in the categories of road vehicles (road automated autonomous transportation systems), drones (aviation, aerial automated autonomous transportation systems) and vessels (maritime automated autonomous transportation systems).

---

[1] Zwicker et al. (2018), pp. 109–124.

[2] On environmental effects of autonomous vehicles, e.g. see the review from Kopelias et al. (2020), pp. 1–6. On safety and autonomous ships, see de Vos et al. (2021), p. 1.

[3] Royal Academy of Engineering (2020), pp. 1–2.

2 Drones: Aviation/Aerial Automated Autonomous Transportation Systems

The chapter on drones and their function in the delivery service sector addresses this new frontier in the logistics area and the related benefits. The chapter discusses the need for adequate regulations to ensure the safe and effective use of these systems and the need for adequate insurance coverage via bespoke new products able to be drafted as suitable for this new service sector. It also addresses the need for regulators and insurers to cooperate to enable the legal and regulatory and insurance framework to get a vast future deployment of delivery drones, as this will allow their use to help reduce inequalities in access to quality products, particularly for elderly, disabled or disadvantaged people with special needs.

The chapter on unmanned air transport and drone usage and relevant legal issues examines prior authorisation and/or notification and operational requirements for drone activities under characteristic examples of existing legislations and highlights the needed legal revisions to allow drone cargo and passenger transports flourish. It is noted that legal issues on aviation safety, security of people, properties and the environment, privacy, e-communications, liability, insurance, risk sharing, oversight, regulatory/supervisory capacity and many more, will need to be duly addressed as well as the enhancement of further cooperation and coordination between the competent authorities and legislators, to ensure high level of aviation safety and the efficient integration of new technologies in air transports.

The chapter on the civil unmanned aerial vehicle (CUAV) law of China discusses the legal problems which have been encountered by China in introducing this law and discusses ways to enforce legislative supervision of UAVs as well as the attempts of the legislator to promote industrial development while ensuring safety of supervision, while posing the question of how in the future can the best balance between the two be achieved via a future law reform. Not least, it discusses the potential of establishing an insurance system for ‘civilian drones’, to better handle and share the risk of potential infringement liability of civilian drones. It is suggested that a good insurance scheme will also urge all parties to carefully produce and operate civil UAVs.

3 Ships: Maritime Automated/Autonomous Transportation Systems

The chapter on the intersection between law and technology in maritime law discusses autonomous marine vehicles (AMVs) and security issues as the former are to be used both in the private and state maritime sectors. The chapter analyses the intersection between law and technology in maritime law and proposes how AMVs correspond to the current legal framework in the light of recent developments. It is concluded that while AMVs present many opportunities for states to enhance their maritime security and law enforcement, at the same time they may pose threats to maritime security hence an international legal framework, to a more detailed extent than the current one is needed to regulate them effectively and address any existing gaps.

The chapter on salvage and autonomous maritime navigation also addresses the need for applicable legal framework, and the need to re-evaluate all obligations of conduct of the salvor given the use of AI to continue to have adequate insurance coverage. It also discusses the problem of the salvage reward if the activity is carried out by a maritime autonomous surface ships (MASS) and concludes that artificial intelligence intended for the management of the vessel will be increasingly based on the characteristics indicated by the European Parliament of the Autonomy, and on self-learning as the vessel will interact with the environment and will be able to learn from the surrounding and change the operational conduct if needed. The effect on the liability regime which is currently centred on the shipowner is also discussed and the need for operators of high-risk AI systems to hold liability insurance is proposed for the operators in AI dominated autonomous ships.

The chapter on unchartered legal waters addresses seaworthiness and good seamanship and their applicability to autonomous vessels and concludes that although traditional legal concepts they are still applicable to autonomous vessels; however, seaworthiness should no longer be viewed from the perspective of a competent master and crew to determine a vessel as seaworthy but the need for international regulations which specifically address autonomous vessels is needed alongside an amendment to the Convention on the International Regulations for Preventing Collisions at Sea (COLREGs) that would consider specific situations which would affect an autonomous vessel.

The chapter on reforming the law of the sea for the future of automated shipping signals to the reader's attention the most significant barriers to autonomous ship integration and pinpoints the priorities in the current regulatory framework that ought to be continued in the regime governing autonomous ships, as well as proposing to recommend solutions for the International Maritime Organization (IMO) to facilitate the adoption of autonomous ships in a merchant setting. It is suggested that for the amendment to the IMO's legal instruments, perhaps an implementation of an agreement or a Maritive Autonomous Vehicle (MAV) Code akin to the Polar Code could be used as a template to achieve the effect that occurs when amending simultaneously the terms of several conventions.

4 Road Vehicles: Road Automated/Autonomous Transportation Systems

The chapter on automated vehicles, liability, motor insurance production and distribution discusses the problems that algorithmic automation of vehicles poses in the areas of civil liability and liability insurance. The author highlights the need—when including AI in the systems relating to decisions for driving such vehicles—to focus on prevention rather than compensation and design algorithms to reduce risks rather than sanction harmful conduct. In this respect, the disadvantages of the strict liability system are presented and discussed alongside the new proposition for using technology to enhance the prevention of accidents era that arises with the vast use of AI which will allow the law to not only govern technology, but will also enhance the use of technology as a regulatory tool.

The chapter on the legal and ethical aspects of autonomous vehicles discusses the main legal and ethical issues deriving from the use of AVs to highlight the regulators' significant role in providing solutions to such issues and makes some propositions regarding liability issues and insurance implications, criminal and contractual liability, and data implications. It highlights that despite the hope or desire of the society and wider public for the wide public circulation of AVs sooner rather than later, it is pertinent that such technology is used and deployed massively in the wider public sector only when it becomes more utilisable and safer, as the public will only trust and use AVs when their benefits exceed their risks, and the trust needed is present for the allowance of massive deployment of AI and autonomous vehicles. Insurance can help instil this trust as it can be a solution, and a compulsory mutual insurance scheme and the use of telematics can also help. Nevertheless, the user's as well as the AV occupant's data privacy remains a question to be solved so that ultimate protection is offered.

The chapter on cyber risk discusses risks methodologically classified in three dimensions that the author called social, functional and ethical dimensions, and invites the reader to consider new technologies' risks and the ways that insurance can help address them. It is discernible that according to the author, accomplishing due balance in the design, application and use of technologies is the solution which would allow humans and society to adapt to a cybernetic world. Insurance also plays a role as it helps spread the risk and can act as a strategic aid for due risks management.

The chapter on automated vehicles and road traffic law compares the current UK and Victoria road traffic laws and application to automated vehicles. It further discusses the application of the function of road traffic law to automated vehicles. The chapter then discusses reform proposals from both the UK and Australia. The chapter notes that there are similarities between the proposals from the UK and Australia, with some differences. Finally, the chapter discusses whether legal personality should be provided to automated vehicles.

5 Conclusions

It is discernible that in all automated/autonomous transportation systems the need for regulation is apparent. Even in legal systems and jurisdiction where such regulation currently exists, it is not detailed enough to offer adequate compensation regime and liability coverage, as well as adequate insurance coverage and protection of the subjects using the automated/autonomous transportation systems in relation to their data. As our discussion will show, the regulation of automated/autonomous transportation systems is of pivotal importance to establish trust and allow the vast deployment of AI.

A challenge posed is the shift of liability as manufacturers need to prepare for potential product liability risk from damages in the ambit of operation of vehicles and drones or ships in an automated and autonomous transportation systems. This entails further challenges as the pool of companies potentially liable for accidents will deepen, and questions on whether the software or hardware caused a particular event will be around. This in turn will affect the way insurers see the operation of vehicles and drones or ships in an automated and autonomous transportation systems. As the legal landscape will continue to shift and change, all players involved will need to be cognizant of potential product-liability risks and the shift of liability, and with the sheer amount of data being exchanged, safeguards for data usage will need to be in place. Regulators, insurers and those involved in the operation of vehicles and drones or ships in an automated and autonomous transportation systems must work together to offer the best possible solutions for encompassing the risks and challenges embodied within the innovation.[4]

References

Chaney R (2020) Maarten Stassen and Evan Chuck, Autonomous vehicles: driving regulatory and liability challenges. Automotive World. https://www.automotiveworld.com/articles/autonomous-vehicles-driving-regulatory-and-liability-challenges/

de Vos J, Hekkenberg R, Valdez Banda OA (2021) The impact of autonomous ships on safety at sea – a statistical analysis. Reliab Eng Syst Saf 210:1

Kopelias P, Demiridia E, Vogiatzisa K, Skabardonis A, Zafiropouloua V (2020) Connected & autonomous vehicles – environmental impacts – a review. Sci Total Environ 712:1–6

Royal Academy of Engineering (2020) The journey to an autonomous transport system: identifying challenges across multiple modes. https://raeng.org.uk/media/kzujkic2/nepc-the-journey-to-an-autonomous-transport-system.pdf

Zwicker A, Farber HB, Hamm JA (2018) Comparing public concern and support for drone regulation to the current legal framework. Behav Sci Law 37:109–124

[4] Chaney (2020).

Drones: The New Delivery Men? A South African and UK Perspective

Samantha Huneberg

Abstract Delivery drones are the new frontier in the logistics area. Worldwide, online retail companies are realising the lucrative nature of utilising drones for delivery purposes. The benefits offered are monumental; however, the unique risks involved cannot be ignored. Amazon is just one of these retail giants recognising the potential delivery drones may offer their clientele. Regulators worldwide need to ensure that adequate regulations are in place to ensure the safe and effective use of these systems. From an insurance perspective, insurers need to be privy to the unique risks delivery drones pose to provide the delivery companies with adequate insurance coverage. Therefore, insurance policies need to be well-drafted and suitable for this innovation in logistics. Regulators and insurers will need to come together to harmonise the future of delivery drones.

Keywords Drones · Insurance · Delivery · Risks · South Africa · United Kingdom

1 Introduction

Delivery drones are going mainstream and are said to be the next big thing in delivery services. In June 2019, Amazon announced that it will be launching a new delivery drone that is capable of flying vertically, like a helicopter, and in a new aerodynamic airplane mode.[1] The use of drones to deliver packages by the online giant, Amazon, will become the next big thing in delivery and transportation.[2] Walmart and Dominos are also testing out this new delivery technique for their products.[3] It is no surprise that drone usage is on the rise. Everyone from film makers

[1] Vincent and Gartenberg (2019).

[2] Vincent and Gartenberg (2019).

[3] Burda (2021).

S. Huneberg (✉)
Department of Mercantile Law, University of Johannesburg, Johannesburg, South Africa
e-mail: shuneberg@uj.ac.za

K. Noussia, M. Channon (eds.), *The Regulation of Automated and Autonomous Transport*, https://doi.org/10.1007/978-3-031-32356-0_2

to photographers to insurers[4] are using drones for various reasons. This technology is also developing at a rapid pace and is becoming extremely sophisticated by the day.[5] The latest in drone technology is incorporating sonars, thermal cameras, and depth cameras to detect hazards, making it the most sophisticated to date.[6] The latest development from Amazon will also use machine learning models, which will enable the onboard computers to automatically identify obstacles and navigate around them.[7]

Delivery drones are the new frontier in drone usage, transportation and deliveries. Packages ordered online can now be delivered straight to the customer's door by a drone, eliminating a major human resource element and vehicle transportation component from the delivery process.[8] This can be extremely advantageous from a cost perspective, as well as from an environmental viewpoint. Fewer cars on the roads means less carbon emissions and that can only be a good thing for our environment. Furthermore, in the last year with the COVID-19 pandemic, delivery of goods has seen an exponential increase and with the need for less face-to-face contact, delivery drones are more needed than ever.

However, the risks involved cannot be discounted. There are novel risks which need to be identified and understood. Additionally, there is also the newly identified element that each of these deliveries will need to be insured, which means that there is a new possibility for insurance companies in respect of this realm. The current drone laws may need to be developed to account for the increase in drone usage for this specific industry as well as to assess the specific risks that delivery drones will bring to the marketplace.

This chapter will first consider the imminent introduction of delivery drones by Amazon to the market, the possible risks posed by these systems and then consider the advantages that delivery drones will bring to the market. Further, the insurance implications of these drones will be considered. Specifically, the current drone regulations in South Africa will be assessed to determine whether the existing regulations will suffice considering the potential growth that drone and delivery industries will see with this innovation. A comparative study of the United Kingdom (UK) drone regulations will also be undertaken to consider whether the UK drone laws are more effective in this regard. Finally, recommendations will be made for the future of this innovation.

---

[4] Insurers use drones to assess the state of impact after natural disasters in areas where access is difficult to acquire as well as in the assessment of damages and risks.

[5] Hodgkinson and Johnston (2018), p. 111.

[6] Vincent and Gartenberg (2019).

[7] Vincent and Gartenberg (2019).

[8] Wing is already undertaking this in countries such as Australia and Finland. See https://wing.com.

2 Conceptualising Delivery Drones

2.1 Drone Use

The use of drones is not a new phenomenon. These technologies were initially developed for military use back in the 1900s.[9] There is a general negative connotation attached to the word 'drone' because of its origin in the military sector; however, this is changing as it is increasing in the civilian sphere.[10] However, in recent years there has been a major increase in the commercial use of drones.[11] The increase in the use of drones is significant, mainly because they have become far more accessible. Drones are being used for commercial purposes in the building, agriculture and entertainment industries, among others.[12]

The thriving drone industry can be attributed to the general increase in availability of these systems.[13] They have become far more accessible to the layperson and the technology within these systems is becoming more sophisticated by the day. As mentioned above, various industries are utilising drones in their commercial undertakings, and this is also set to have an increase in the number of jobs which may be created using these systems.[14]

Delivery drones are just one of those areas set to soar in coming years. Amazon, as the biggest online retailer, has realised the lucrative nature of utilising drones to deliver packages to their customer base in a more efficient and effective manner.[15]

For context, it is worth noting that within the aviation industry, drones fall within their own category known as Unmanned Aerial Systems (UAS) and are therefore viewed as a type of 'aircraft'.[16] UAS have the ability to navigate autonomously, without human control and beyond the line of sight.[17] Some drones even use machine learning to inform their autonomy, allowing them to even recognise and learn information as the fly.[18] The navigation of drones remains a critical issue in the development of these systems. Uber have stated that autonomous operation technology together with electronic motor technologies are considered to be the most critical advanacements in overcoming challenges related to safety, emissions and the

---

[9]Keane and Carr (2013), pp. 559–561—showing how the early predecessor to the modem drone was the military aerial torpedo, designed in 1916.

[10]Hodgkinson and Johnston (2018), p. 1.

[11]North (2015), p. 334.

[12]Burzichelli (2016), p. 165.

[13]Burzichelli (2016), p. 164.

[14]Hodgkinson and Johnston (2018), p. 112.

[15]Hodgkinson and Johnston (2018), p. 11.

[16]Marsh (2015), pp. 1–5.

[17]Huneberg (2017), p. 596.

[18]Hodgkinson and Johnston (2018), p. 113.

overall vehicle performance.[19] Therefore, the development of an effective navigation system is crucial in mitigating the risks of possible collisions.[20]

It is suggested that the biggest growth is predicted to be in the commercial applications of drones, which a recent PWC report, *Clarity from Above*, it is estimated that the grow in value will be from $2 billion in 2016 to $127 billion by 2020.[21] Due to this significant increase in the use of these systems, it is vital to have adequate regulations in place governing the proper and adequate use of drones.[22]

2.2 *Amazon Prime Air: Delivery Drones*

Amazon Prime Air is part of Amazon's online retail platform to deliver packages of up to 5 pounds safely and conveniently to customers within 30 min or less through autonomous drones.[23] This delivery system is set to provide customers with 'nearly instant order fulfilment'.[24] The plan is to have accessible fulfilment centres where packages can be transported from.[25] These centres would have to be in relatively close proximity to the customer base to provide the service. Currently, it is estimated that 86% Amazon's orders fall within this weight restriction and would therefore be eligible for delivery through drones.[26] This means that Amazon will need enough fulfilment centres to cater for the masses.[27] This does not seem to be a major obstacle for a company like Amazon, they can build fulfilment centres within their most dense delivery zones.[28] The key obstacle of such a venture appears more to be within the confines of the use of the airspace and the regulations that come with the use of drones.[29]

Prime Air state that safety is their top priority.[30] Their drones will be built with sophisticated 'sense and avoid' technology.[31] This is one of the key technological

[19] Hodgkinson and Johnston (2018), p. 113.

[20] Hodgkinson and Johnston (2018), p. 113.

[21] Kennedys (2017), p. 3.

[22] Such as the new regulations in South Africa contained in Part 101 of the South African Civil Aviation Authority (CAA).

[23] See Amazon Prime Air website. https://www.amazon.com/Amazon-Prime-Air/b?ie=UTF8&node=8037720011.

[24] Burzichelli (2016), p. 166.

[25] https://www.amazon.com/Amazon-Prime-Air/b?ie=UTF8&node=8037720011.

[26] Burzichelli (2016), p. 166.

[27] Hodgkinson and Johnston (2018), p. 128.

[28] Burzichelli (2016).

[29] Burzichelli (2016).

[30] https://www.amazon.com/Amazon-Prime-Air/b?ie=UTF8&node=8037720011.

[31] See Amazon Prime Air website. https://www.amazon.com/Amazon-Prime-Air/b?ie=UTF8&node=8037720011.

developments within delivery drones. Worth noting, in August 2020, Amazon received a Part 135 Air Carrier Certificate from the United States Federal Aviation Administration (FAA).[32] This certification is an important step forward for Prime Air and indicates the FAA's confidence in Amazon's operating and safety procedures for an autonomous drone delivery service that will one day deliver packages to customers around the world.[33] It is therefore apparent that the widespread use of Amazon Prime Air is something we may see implemented in the very near future.[34] The imminent introduction of delivery drones requires a thorough understanding of how these systems will be introduced into the aerial space as well as the how the regulators will have to respond to the increased number of drones taking to the skies. Thus, the risks involved with the use of these systems cannot be ignored. It is imperative that the unique risks of delivery drones are understood as these will ultimately influence the insuring of drones for companies like Amazon, as well as the insuring of the packages being delivered. It is also worth mentioning that delivering packages by drones brings about a new product for insurers. Insurers can tap into this market and offer specific insurance products for delivery companies in the realm of insuring the products being delivered.

It therefore becomes crucial that with the introduction of delivery drones, regulators, retail companies like Amazon, as well as insurers all work together to create an environment that allows for flexibility in the industry but also accounts for the possible risks that delivery drones may pose. Insurers will need to work closely with delivery drone companies to understand how exactly the systems are being used, the value of the items being carried, the amount of control the company has over the drone being flown, how far the drone is being flown out of the direct line of sight as well as all obstacles in the way. This will assist insurers in understanding and underwriting the risks involved specifically with delivery drones.

2.3 *Risks Associated with Delivery Drones*

It is known that drones possess capabilities to cause destruction and damage either in the wrong hands of certain individuals or if flown in a reckless manner.[35] Some of the general concerns around drone usage are: drones can be used in such a way as to invade another person's privacy,[36] be used as surveillance for terrorist schemes,[37]

[32]This gives Amazon the ability to carry property on small drones 'beyond the visual line of sight' of the operator. Palmer (2020).

[33]See Amazon Prime Air website. https://www.amazon.com/Amazon-Prime-Air/b?ie=UTF8&node=8037720011.

[34]Miller (2020), pp. 140–141.

[35]In 2018, drones brought Gatwick Airport to a standstill.

[36]Mathews (2015), p. 586.

[37]Barrett (2016).

cause havoc to the public[38] and even cause severe physical damage to people or property.[39] Drones are also open to the possibility of being hacked which can then result in damage to the drone itself, third parties or even theft of the drone.[40]

From an insurance perspective, there are two main safety concerns: either drones involved in mid-air collisions with any other types of aerial systems, or a loss of control by the operator of the drone.[41] The first risk worth consideration is that of a mid-air collision. This may occur if the operator cannot see exactly what is happening around the drone mid-flight and cannot avoid a manned aircraft in time, especially those that normally fly below 500 feet, such as helicopters, agricultural aircraft and aircrafts landing or taking off.[42] According to Hodgkinson and Johnston, 'this is further compounded by the ability (or inability) for drones to cope with inclement weather and the limitations of the technology to facilitate drone operations'.[43]

Over recent years there have been numerous reports of UAS sightings from pilots, citizens, and law enforcement agents. Even more alarming is that there have also been several near-miss incidents around the world including in China, Dubai and the UK. [44] This type of incident can cause significant damage and loss. Damage to property as well as possible harm caused to third parties are all plausible scenarios that may occur. Further, the amount of damages caused to property (such as to airplanes or other manned aircraft) can be exorbitant, especially when considering the value of the property involved. Therefore, it is necessary for insurers to understand these risks as well as being able to assess and quantify such losses when it comes to the insuring of drones.

The second major safety concern or risk that insurers need to consider is the loss of control over the UAS. This generally results from a system failure or in instances where the UAS flies beyond signal range[45] or through frequency interferences.[46] If an operator loses control of a drone due to a system failure, then it is highly likely that the drone can fall from the sky causing damage to another's property or to oneself. Once again, the damages that may occur in such an instance require some

---

[38] Mathews (2015), p. 594.

[39] Fox (2017), pp. 687–700.

[40] Michaelides-Mateou (2015), p. 435; Yiannakis (2019), p. 16.

[41] Huneberg (2017), p. 597.

[42] Huneberg (2017), p. 597. In *Leading Prospects Trading 38 (Pty Ltd) v Centrique Insurance Company and Tromp* GPPHC 8 April 2015 (case 302 unreported), the pilot flew below 500 feet in a manoeuvre that did not form part of landing or take-off and crashed into power lines. As it is a well-known fact that power lines are below 500 feet above ground, the only logical conclusion that could be made was that the pilot was at fault.

[43] Hodgkinson and Johnston (2018), p. 34.

[44] Allianz Global Corporate (2016), pp. 3–7.

[45] Allianz Global Corporate (2016), pp. 4–9.

[46] This happens when in the radio frequency spectrum, a disturbance is generated by an external source that affects an electrical circuit by electromagnetic induction, electrostatic coupling or conduction.

form of insurance to protect oneself against such possibilities. These are factors insurers need to consider when underwriting the risks involved in using drones.

Now, in recent years, insurers have for the most part figured out how to deal with these types of risks and how to study the underwriting process for these scenarios.[47] Understandably, drone insurance offerings are constantly evolving with the ever-changing technological developments and uses thereof. In January 2018, Flock launched Europe's first app-based 'pay-as-you-go' drone insurance model in conjunction with Allianz.[48] This is just one example of how insurers are providing viable insurance offerings for policyholders.

However, with the introduction of delivery drones, there are additional and unique risks which need to be considered by insurers. These include operation beyond the line of sight;[49] packages being delivered may either be lost mid-flight or perhaps even being misdelivered and the incorrect person receiving the package.[50] With the increase in the number of drones taking to the skies for delivery purposes, there is an increase in risk of collisions with other drones.

To understand a few of these risks, let us say that a retail company is delivering a product to a consumer within a 5 km radius to deliver this package, the drone will have to fly beyond the operator's line of sight, meaning he may not be fully aware of all obstacles in the way as well as not being able to foresee what else may be flying in the direct line of the drone (whether it be another drone or a microlight or a bird).[51] This clearly brings in a new element of risk, here the risks are not known and the operator may encounter certain obstacles along the journey increasing the risk of a mid-air collision. This is where it becomes imperative that the drone technology itself is sophisticated enough to sense and avoid obstacles in its way as well as ensuring that the operators have the skills and ability to manoeuvre the drones swiftly and safely.

It is also crucial that the insurers and company operators understand the possible risks that may occur in such instances and therefore account for these risks in the underwriting process. Further, the fact that this drone will be carrying valuable goods means that there is another risk involved here. The package may fall mid-air and be lost or stolen, or a package can be mis-delivered and the costs of regaining the package will need to be considered. Now let us add that there may be numerous delivery drones operating at any given time and with this increase, there is now an even bigger possibility of mid-air collisions with other drones.

---

[47] This is evident from the current insurance offerings provided by big and small insurers.

[48] Thompson (2021). On this mobile app, pilots (commercial and recreational operators) can purchase liability insurance on demand. The cost of cover is 'exposure-based' which is assessed according to the risk of each flight considering real-time data and algorithms to assess the specific risk involved.

[49] On this risk, it will be important to consider whether the drone uses any technology for recognising air traffic which will then force certain actions to avoid collisions.

[50] This will also require proper identification of the receiver and authentication.

[51] Thompson (2019–2020), pp. 8–9.

While there are various dangers associated with delivery drones and implications resulting therefrom, the liability aspect is arguably the biggest factor. These risks mentioned above can be managed and mitigated through well drafted regulations and collaboration with the insurance industry.[52] Already, companies like Allianz recommend considering particular risks when considering the insuring of delivery drones. Collaboration throughout the various industries remains critical. The only way to really manage these risks is through cooperation and possible alliances between insurers, regulators and delivery companies using drones.

It is therefore crucial that insurers be prudent in their assessment of the risks involved with delivery drones. With this technology reaching new frontiers in growth, insurers and regulators should understand the unique risks involved to provide comprehensive cover for companies like Amazon.

2.4 Advantages of Drone Delivery

From the above, some might believe that drones can only cause mayhem, however, drone usage is not entirely dismal. These systems are a sophisticated piece of technology that have various capabilities. Some of the general advantages of drones include the following: using drones to deliver medical supplies to people in rural areas;[53] exploring dangerous underground mines and pipelines; assisting in search and rescue missions; and even in assisting countries with border patrol.[54]

The insurance industry is one example which is recognising the potential benefits of drone usage in their businesses. For example, insurers are using drones to better control risk through enhanced data collection as well as in reducing operating costs through improved efficiency and effectiveness in the claims handling stage.[55] Drones are being utilised in the pre-loss phase for natural disaster management, as well as aerial assessments.[56] In the post-loss phase, drones are utilised for risk assessment, inspection, as well as claims adjudication.[57]

The advantages that delivery drones may provide include less carbon emissions than motor vehicles and therefore offer an environmentally friendly solution to package deliveries by ordinary motor vehicles. There is also increased efficiency

[52] This will be further elaborated on in Sect. 6 below.

[53] Mathews (2015), p. 580. Just recently, India has utilised drones to deliver COVID-19 vaccinations to people in rural areas.

[54] Michaelides-Mateou (2015), p. 426; Yiannakis (2019), p. 17; Jung-Sup Um (2019), p. 27.

[55] Deloitte (2018), pp. 3–7; Yiannakis (2019), p. 17.

[56] Deloitte (2018), pp. 3–7. This can help reduce the premium for policyholders; Yiannakis (2019), p. 17.

[57] Business Tech (2019) where in South Africa, Old Mutual iWyze have applied for CAA approval to be allowed to use drones for car crash investigations and crop loss assessments.

in delivering packages to consumers in a more cost-effective and timely manner.[58] Delivering packages by drones means that consumers can receive their packages almost instantaneously.

Additionally, there is the element of less face-to-face contact when using delivery drones. This is something that has become increasingly necessary in the last two years due to the global COVID-19 pandemic. Consumers increasingly utilise online shopping and receive their goods via delivery. Therefore, if goods can now be delivered by a drone, that would provide consumers with more peace of mind as they do not have to deal with as much human contact. Delivery drones therefore appear to be beneficial in numerous respects.

With all these considered, it is thus clear that adequate regulations are required as well as well-drafted drone insurance policies—similar to those in traditional logistics but taking into account the unique risks associated with the delivery drones. Before considering the insurance implications, it is necessary to understand the current drone regulations in South Africa.

3 Current Drone Regulations in South Africa

The South African Civil Aviation Authority (CAA) issued a media statement on 2 April 2014, which indicated that under the then current aviation laws, operations of drones in the South African civil aviation airspace were prohibited and therefore, illegal.[59] This statement indicated that regulations were being developed to deal with the use of these systems. The important apprehensions of the CAA were surrounding the safety, security and privacy issues associated with drone use.[60]

The previous airspace regulations concentrated on the specific requirements for the operation of a manned aircraft in South African airspace, and drones were not included therein. The CAA therefore collaborated with the drone industry and formulated regulations which officially came into effect on 01 July 2015.[61]

3.1 Introduction to Part 101 of the Regulations

Part 101 of the Regulations as part of the Civil Aviation Act specifically deals with remotely piloted aircraft systems (RPAS). An RPAS is defined as 'an unmanned

[58] Miller (2020), p. 140.

[59] South African Civil Aviation Authority Media Statement (2014).

[60] South African Civil Aviation Authority Media Statement (2014).

[61] Part 101 (Remotely Piloted Aircraft Systems) of the Regulations to the Civil Aviation Act 13 of 2009.

aircraft which is piloted from a remote pilot station, excluding model aircraft and toy aircraft'.[62] A drone, therefore, falls within the definition of an RPAS.[63]

The Regulations state that an RPAS may be operated for: commercial operations; corporate operations; non-profit operations; and private operations.[64] The Regulations further state that certain sections of the Regulation will not apply to private operations.[65] Therefore, there is a clear distinction between two categories of operations in South Africa—specifically, private operations and commercial operations.[66] It is clear that the Regulations appear to be far more exhaustive towards commercial operations.

It is interesting to note that the Regulations place an obligation on the seller of a drone in South Africa to, via written notification or on the packaging, inform the buyer of the drone that there are specific regulations applicable to the use thereof.[67] The aim thereof is to ensure that the buyer is notified to acquaint themselves with the relevant regulations. This is problematic and the enforcement thereof cannot be managed. If the operator does not read the relevant regulations, there is a risk that the operator may be flying the drone in an unsafe and possibly illegal manner.

3.2 *Private Use of Drones*

On the provisions regulating the private use of drones, an operator may only use the RPA: for individual and personal gain;[68] on a property which the operator owns, or has permission to fly on; within restricted visual line of sight—which means within 500 metres of the operator, and never above the height of any obstacle within 300 metres of the operator; while observing all statutory requirements relating to liability, privacy and any other laws enforceable by any other authority.[69]

[62] Part 1.01.1 of the Regulations to the Civil Aviation Act 13 of 2009. The term 'toy aircraft' is defined as a product designed or intended for use in play by children. The term 'model aircraft' means a non-human carrying aircraft capable of sustained flight in the atmosphere and used exclusively for air display, recreational use, sports or competitions.

[63] SA-CATS 101 of the Regulations to the Civil Aviation Act 13 of 2009 provides for various classifications of drones according to their size and speeds.

[64] Part 101.01.1(2).

[65] Part 101.01.2.

[66] Commercial operations include—commercial, corporate and non-profit operations, which are all essentially operations that will carry some form of commercial outcome, interest or gain.

[67] Part 101.01.5 states that: 'No RPA shall be sold within the Republic unless the seller has, by way of a packaging label, or in the case of the resale thereof, by way of written notification, notified the buyer of the requirements as prescribed in Document SA-CATS 101'.

[68] See the definition of 'private operation' in Part 1.01.1 of the Regulations to the Civil Aviation Act 13 of 2009.

[69] Part 101.01.7 of the SA-CATS 101.

These restrictions are inconsequential when compared to those of an RPAS utilised for commercial operations. Private use RPAS do not require registration with the CAA and furthermore, their operators are not required to obtain an RPAS pilot's licence.[70]

3.3 Commercial Use

The use of an RPAS for commercial purposes carries more stringent requirements, for example: the drone must be registered and approved by the CAA;[71] the drone must be appropriately marked;[72] the operator must obtain a remote pilots licence (RPL)—which requires being over the age of 18, obtaining medical certification, obtaining a certificate of proficiency in aeronautical radiotelephony, providing proof of proficiency in the English language, completing flight training and also passing a theoretical knowledge examination and skills test.[73]

An RPL lasts for 24 months, afterwards an operator must undergo a revalidation check 90 days prior to the expiry of the RPL to renew it.[74] The Regulations also require commercial operators to maintain a pilot logbook detailing each flight.[75]

There are various other provisions surrounding the administration, documentation, safety management, security, power reserves, first aid kits and fire extinguishers which are required when the operator is a licensed RPAS operator.[76]

On insurance implications for drones, it is required that a licensed commercial operator must always be adequately insured for third-party liability.[77] Therefore, it is evident that the insurance requirement in the Regulations only applies to commercially licensed operators. On the possibility of delivery drones entering the South African market, it must be noted that these would fall under the commercial purpose use under the Regulations. Therefore, for this chapter, emphasis will be placed on commercial use and not on private use.

---

[70] While it may seem fair that there are less administrative requirements on private use operators, it cannot be ignored that these individuals may pose a higher risk than a commercial operator who has undertaken the required formal training.

[71] Part 101.02.4.

[72] Part 101.02.4(5).

[73] Part 101.03.2.

[74] Part 101.03.6.

[75] Part 101.03.7.

[76] Part 101 Subpart 4 and 5.

[77] Part 101.04.12.

3.4 General Restrictions Applicable to Both Private and Commercial Operators

On general restrictions, the Regulations provide thorough prohibitions applicable to both private and commercial operators. The general restrictions indicate that no drone shall be operated above 400 ft. (120 metres) of the surface; within 10 km of an aerodrome; within restricted airspace; or adjacent to or above a nuclear power plant, prison, police station, crime scene, court of law, national key point or strategic installation.[78] There is also a requirement for a pre-flight inspection prior to each and every flight,[79] as well as the fact that a drone must always give way to manned aircraft.[80]

Part 5 of the Regulations further indicate that no flights should be operated within a lateral 50-metre radius from any person,[81] public road,[82] building or structure.[83] Should any of the aforementioned operations be required, this may only be conducted by a valid holder of a remote operator certificate and with the prior permission of the Director of the CAA.[84]

It is evident that there are significant restrictions placed on commercial drone operators when operating these systems. These restrictions are necessary to protect others, as well as the drone operator.

3.5 General Evaluation of the Current Regulations

Part 101 provides a rather comprehensive set of rules to govern drones and their flight patterns. The regulations contain in-depth rules more for commercial operations but manage to regulate both types of operations effectively and comprehensively. However, enforcement of these regulations remains difficult and problematic for the CAA. The fact that drone flights can be carried out by untraceable drone operators and do not require the same infrastructure as a manned aircraft would[85] makes the monitoring, compliance and policing thereof very difficult.[86] There do not appear to be very stringent rules around the policing of these systems, and this is

[78] Part 101.05.10(3).

[79] Part 101.05.17.

[80] Part 101.05.20. However, if there is a danger of a head on collision, each aircraft must alter its course to the right.

[81] Part 101.05.13.

[82] Part 101.05.15.

[83] Part 101.05.14.

[84] Part 101.05.13–15.

[85] Drones, unlike manned aircraft, do not require an airport, runway or any formal record of a flight itinerary or path.

[86] Fox (2017), p. 709; Huneberg (2018), p. 279.

something that would be extremely valuable to ensure compliance with the Regulations. This is an aspect that must be addressed by the CAA, and proper enforcement of the regulations must be prescribed by regulating authorities.

From the perspective of delivery drones, these types of drones will evidently fall under commercial operations in South African law. This is because delivery companies would be undertaking the use of delivery drones for a commercial purpose and gain. The current commercial operation regulations are extensive and rather comprehensive; however, it may be necessary for South Africa to implement further regulations with regard to the risk of the flights being undertaken by commercial operators. This will require the CAA to perhaps consider various types of commercial operations and divide commercial operations into various classes depending on the specific risk that the flight possesses. For example, a delivery drone may face more risks than a drone being flown for aerial photography. Delivery drones will be flying very closely to buildings, people's homes, even possibly near traffic; this will increase the risk of the flight. There is also the additional factor that delivery drones will be carrying goods of value. Therefore, it is possible to lose these goods during flight or possibly delivering the goods to the incorrect consumer. These are all factors which increase the risk and the value of operation being undertaken. Therefore, although Part 101 includes numerous provisions on safe flying of commercial drone operations and requires strict adherence of the competence requirements for commercial operators, it may be necessary for the regulations to be developed to account for the increased risks which delivery drones bring. It may be necessary for the regulations to further divide commercial operations into various risk classes and have further requirements for riskier flights.

On insurance requirements, Part 101 currently prescribes compulsory third-party liability insurance for commercial operations. This is a laudable provision as it will be necessary and vital for drone delivery companies to procure the necessary insurance for these flights due to the risk associated with drones taking to the skies. The possible risks that drone flights possess are endless—from crashing into structures, to causing destruction to buildings and even the possibility of injuring people—therefore insurance against these risks is essential to protect the owner against liability that may be incurred. The fact that the regulations also compel commercial operators to be licensed, maintain the necessary logbooks of flights and that the drones must be adequately marked is once again pertinent for such operations. It will be necessary to know which company is responsible for a collision or damage caused. These requirements will assist insurers in ensuring that safe and accountable drone flights are being undertaken. Delivery drones would require the permission of the CAA since they would be flying closely to buildings and people. Therefore, although the current regulations are comprehensive, to an extent, on commercial operations, further development is needed especially considering the growing nature of these systems. South Africa's drone regulations were implemented in 2015. Since then, there has been significant growth around the nature of use of these systems. Drones are now being used for purposes which were not foreseen back in 2015 and therefore, it is imperative that the regulators keep up with the evolution of these systems and their uses, and amend the regulations as required. This is something that the SACAA does not appear to be doing, they are not updating the regulations on a continuous basis, although there is a growing demand for new uses in drone systems.

3.6 *The Possibility of Delivery Drones in South Africa*

From the above evaluation of the current drone regulations in South Africa, it is evident that the regulations are comprehensive but are also lacking in continual updates regarding the constant innovation within the industry. So, is there a viable opportunity for delivery drones in the country? In 2020, a new smart drone system was approved by the SACAA which sees South Africa being one step closer to the implementation of delivery drones. United Drone Holdings have been granted a waiver by SACAA for a smart computer system which can be used for long-range commercial drone flights with vertical take-off and landing capabilities.[87]

This technological innovation is a welcoming introduction for automated commercialised delivery drones in South Africa, which need to deliver payloads safely and which fly beyond the visual line of sight (BVLOS) of the pilot. This technology was developed by the San Francisco-based company, Iris Automation, which enables the 300-g computer to be attached to a drone. It includes an onboard camera equipped with a machine learning identification system that can identify, track and predict if another plane is on a collision course with it.[88] If there is a possibility of an obstacle in the drone's course, then the technology can make intelligent decisions to alter its course and avoid collisions, as well as alert the pilot on the ground. Effectively, it can assist the drone to see and react to its surroundings in the same way pilots do. Therefore, it ultimately helps overcome safety concerns raised about drones that may lose connection.

The device has received SACAA approval for operations of up to 15 kilometres and is able to demonstrate its capabilities of avoiding collisions with manned aircraft during a live flight operation.[89]

The approval of this innovative device by SACAA signifies the regulators advancement towards this technology. There is great opportunity for delivery drones, and this type of device is exactly what is required to ensure the safety of such flights undergoing delivery trips. This is a monumental step forward for South Africa and the possibility for the introduction of delivery drones; however, it is only the first step, and it also shows that the regulators collaboration is much needed for delivery drones to take to the skies. Therefore, delivery drones might be in South Africa's near future if the regulators continue to liaise with industry and novel innovations.

[87] Caboz (2020).

[88] Caboz (2020).

[89] Caboz (2020).

4 The Insurance Implications of Delivery Drones

A contract of insurance integrates various terms into an insurance policy, these include: firstly, a term that the insurer will indemnify or satisfy the insured for either a patrimonial or non-patrimonial loss,[90] secondly, a term which provides for the payment of a premium by the insured,[91] and lastly, a term that the insurer's performance is dependent upon the occurrence of an uncertain event.[92] These terms are generally known as the *essentialia* of an insurance contract and it is these speciufic terms that identify an insurance contract as that.

The purpose of non-life (or short-term) insurance contracts in South Africa is generally to protect the insured from any possible risks which may arise in terms of the insured items, and which ultimately cause damage or loss for the insured.[93] The insured party will therefore procure insurance to cover himself/herself against any risk or peril which may arise during the subsistence of the insurance policy.[94] If we turn to the insuring of aviation systems (that being manned as well as unmanned systems), these are regulated through an insurance policy, in which the owner of the manned/unmanned aerial systems enters into a contract with the relevant insurer. This policy will provide that the specific insurer undertakes to cover the insured against any risk which may arise to the manned/unmanned systems. Therefore, aviation policies are a frequent practice for commercial insurers and are of the utmost importance to the insured parties to protect themselves against any possible risks which may occur.

Due to licensing requirements imposed by legislation, insurers will only undertake to cover such a risk if the insured has the necessary licence required for the use of such systems.[95] Manned aerial systems (such as airplanes, helicopters and microlights)[96] have been regulated by legislation for decades[97] and, by now, insurance companies have very good knowledge of the risks involved with these types of systems. Knowledge of risk enables insurers to sell policies covering these systems with relative ease. However, since drones have not been in use for exceptionally long, insurers find it more difficult to calculate and predict the risks involved. It is not

[90]Reinecke et al. (2013), p. 82. *Ackerman v Loubser* 1918 OPD 31 34.

[91]Reinecke et al. (2013), p. 86. *Lake v Reinsurance Corporation Ltd* 1967 (3) SA 124 (W) 127–128.

[92]Reinecke et al. (2013), pp. 89–90. *Prudential Insurance Co v Inland Revenue Commissioners* [1904] 2 KB 258662.

[93]Reinecke et al. (2013), p. 9. Millard (2013), p. 116.

[94]Reinecke et al. (2013), p. 540.

[95]Marsh (2015), pp. 8–11.

[96]All of these require third-party liability insurance under the regulations.

[97]See the current Civil Aviation Act 13 of 2009. This statute was preceded by the Aviation Act 74 of 1962. Other aviation legislation includes the Air Services Licensing Act 115 of 1990, the Air Traffic and Navigation Services Company Act 45 of 1993, the South African Civil Aviation Authority Levies Act 41 of 1998 and the South African Maritime and Aeronautical Search and Rescue Act 44 of 2002.

sufficient to simply use existing manned aerial system policies and make them applicable to a drone insurance policy as this 'cut-and-paste' method does not involve a proper assessment of the specific risks associated with UAS. The unique nature of these systems requires prudent consideration. Despite this hindrance, insurers have stepped up in recent years and are offering comprehensive cover for drones as previously mentioned.

On the current South African insurance legislation, one can refer to the Insurance Act, which defines non-life insurance policies[98] as 'any arrangement under which a person, in return for provision being made for the rendering of a premium to that person, undertakes to meet insurance obligations that fully or partially indemnifies loss on the happening of an unplanned or uncertain event, other than—(a) a life event; or (b) a death event or disability event not resulting from an accident, and includes a renewal or variation of that arrangement' and further divides such policies into various classes of business.[99] Drone insurance will fall directly within the aviation class of business.[100] The aviation class of business is further split into two subclasses—namely personal lines and commercial lines.[101]

As mentioned previously, drones in South Africa are either used for private or commercial use. Drones used for private use will fall within the personal sub-class under the aviation class of business because an individual person will be using the drone for private and personal recreational purposes. A drone operator that intends to use a drone for commercial purposes, interest or gain will need to insure their drone/s under an appropriate commercial lines policy. Delivery drones will therefore fall within the latter sub-class. The liability factor will need to be examined next.

4.1 An Evaluation of Liability Insurance

Liability insurance is frequently referred to as third-party insurance.[102] It is necessary to differentiate that in third-party insurance, the insured is referred to as the 'first' party, the insurer is the 'second' party, and the 'third' party is the one towards whom the insured is liable.[103] As far as the object of liability insurance is concerned,

[98] 18 of 2017.

[99] See Schedule 2 of the Insurance Act 18 of 2017.

[100] Class 6 of Schedule 2 of the Insurance Act 18 of 2017.

[101] The Insurance Act 18 of 2017 defines personal lines insurance as 'non-life insurance business where the policyholder is a natural person, acting otherwise than solely for the purposes of the person's own business'. Commercial lines business is defined as 'non-life insurance business other than in respect of personal lines'. Yiannakis (2019), pp. 19–20.

[102] Reinecke et al. (2013), p. 538.

[103] Padayachee (2021), p. 13.

this refers to the interest of the insured in not incurring liability towards third parties.[104]

Liability insurance undertakes a considerable financial role in the distribution of risk and most notably in the burden of civil liability.[105] The importance of liability insurance can be attributed to the hasty increase of rights awareness amongst consumers and insured's alike, consumer-centric legislation and an expanding litigious society.[106] A liability policy typically provides cover to an insured against their legal liability for third-party damages. It is imperative to understand that mere loss or damage suffered by the third party is insufficient. The insured must be legally liable for the third party's loss or damage and the third party must be capable of recovering damages from the insured prior to the latter's liability insurer incurring any liability under the policy issued.[107]

The liability class is distinguished and detailed under table 2 of the Insurance Act.[108] Within the various liability sub-classes, public liability is listed therein.[109] The liability class within the Act describes the sub-class as covering liability to another person.[110] Therefore, when it comes to the insuring of drones and the requirement of third-party liability insurance, this is where one finds application within the legislation.

Reinecke et al. describe liability insurance that which is concerned with negative elements which come into being as part of the insured's patrimony.[111] Insured parties will procure such insurance to protect themselves against economically debilitating consequences of sustaining legal liability towards others.[112]

As mentioned in paragraph 3 above, Part 101 of the Regulations requires that a commercial drone operator must have third-party liability insurance.[113] Therefore, if delivery companies in South Africa were to utilise drones for delivery purposes, they would require adequate third-party liability insurance to protect themselves against the possible harm or losses they may incur towards others. This is where it is necessary that the insurance policies offered by insurers are well-drafted, comprehensive and clear on the specific perils and risks involved for delivery drones.

---

[104] Reinecke et al. (2013), p. 539. *Boshoff v South British Insurance Co Ltd* 1951 (3) SA 481 (T).

[105] Reinecke et al. (2013), p. 538. *Cape Town Municipality and Another v Allianz Insurance Co Ltd* 1990 (1) SA 311 (C).

[106] Jacobs (2011), p. 464.

[107] Millard (2013), p. 45. *Watson NO v Shaw NO* 2008 (1) SA 350 (C).

[108] 18 of 2017. It details the classes and sub-classes of insurance business for non-life insurance.

[109] Table 2 of Schedule 2 of the Insurance Act 18 of 2017. Yiannakis (2019), pp. 19–20.

[110] Table 2 of Schedule 2 of the Insurance Act 18 of 2017.

[111] Reinecke et al. (2013), p. 10.

[112] Reinecke et al. (2013), p. 540.

[113] Part 101.04.12 of the Regulations to the Civil Aviation Act 13 of 2009.

4.2 Drone Insurance Cover

Current drone insurance policies provided by South African insurers generally relate to the physical theft, loss or damage of the drone or its control system;[114] the theft, loss or damage of any specified equipment that is added to the drone;[115] and any legal liability that may occur on the drone owner or pilot due to the flying of the drone.[116]

Insurers need to consider various factors which may influence the policy and its limitations. Yiannakis states that some of these factors include: the cost of the drone;[117] where it will be largely flown;[118] the purpose for which it will be flown;[119] how often it will be flown;[120] and the capabilities of the drone will all play an important role in the selection of an appropriate insurance policy.[121]

Insurers are able to adjust their risk calculations and provide better coverage limits at better premiums for commercial operators because they are required to abide by more restrictive regulations, such as registering each drone and obtaining the necessary operator and pilot licenses required.[122] One example in the South African insurance realm is Hollard Insurance Company which currently only provides drone insurance to commercial operators that are appropriately registered and licensed with the Civil Aviation Authority.[123]

4.3 Insuring of Delivery Drones

The insuring of delivery drones will look slightly different from the insuring of ordinary commercial drones. The risks associated with delivery drones have been

---

[114] Either while in flight or on the ground.

[115] Such as improved camera systems, gimbals or sensors.

[116] Such as third-party liability cover for damage to other people's body or property. Yiannakis (2019), pp. 20–21.

[117] Owners of cheaper drones might not be worried about the risk of theft or loss.

[118] The risks associated with commercial flights in populated areas are much higher than a private operator flying in an open field for recreational purposes.

[119] Commercial or private use.

[120] The more often the drone is flown might weigh upon the insured and their desired cover limit. For example, an operator that plans to fly their drone daily carries more risk than one who will only fly their drone a few times a year.

[121] Drones that can go further (flying many kilometres outside line of sight) will be susceptible to more risk than a drone that can only operate 50 metres away from the pilot. Yiannakis (2019), p. 21.

[122] An example is Hollard's current offerings for 'Specialist Sector Insurance'.

[123] Yiannakis (2019), p. 21. An example is Hollard's Specialist Sector Insurance available at https://www.hollard.co.za/business-insurance/specialist-sector-insurance/drone. This is an example of how insurers are relying on the drone regulations to provide adequate cover to operators considering the risks posed by such operators.

highlighted above and it is due to these unique risks that the insuring thereof will differ from ordinary commercial drones.

Some of the factors include the fact that with delivery drones one will likely be flying out of the line-of-sight zone which increases the risk of collisions. This is where it is vital that the technology of the drone is able to navigate and avoid; something that Amazon have mentioned their delivery drones will possess.

Insurers will therefore need to consider these unique characteristics and risks when providing delivery drone insurance offerings.[124]

5 Comparative Study: The United Kingdom

5.1 Introduction

The United Kingdom (UK) have taken a very proactive approach to drone regulation. This is evident through their continual update of the domestic drone laws. In November 2019, there was an update on the Air Navigation (Amendment) Order which became effective on 30 November 2019.[125] Another Air Navigation (Amendment) Order was undertaken in 2021 again. Further, there were changes to the legislation in terms of the EU Regulation 2019/437 which became operative on 31 December 2020.

Similarly to South Africa, the UK also have their own Civil Aviation Authority (UK CAA) whose primary purpose in relation to drones is to 'enable the full and safe integration of all UAS operations into the UK's total aviation system'.[126] It is interesting to note that the UK CAA, the Department for Transport, the Home Office and the Police all signed a memorandum of understanding which sets out their duties and responsibilities on drone related issues.[127] The memorandum ensures that the Police are responsible for enforcement of contraventions of the legislation such as drone misuse and will set aside resources to do so.[128] They have also taken an active

[124]Flock offers drone insurance in Europe at monthly and annual subscriptions as well as usage-based offerings (pay-as-you-use). The monthly and annual subscriptions provide worldwide cover for unlimited flights and policies are priced according to the safety and size of the operations being undertaken. In July 2019, Flock also introduced a product called 'Enterprise' which is set to provide scalable exposure-based insurance for connected drone fleets.

[125]The Air Navigation Amendment Order 2018 No. 623.

[126]United Kingdom Civil Aviation Authority 'An introduction into unmanned aircraft systems' (2022).

[127]Civil Aviation Authority 'CAP 1421 - Annual Report & Accounts 2015/16' 11.

[128]Haylen (2019), p. 10. They are also to receive new powers, including to land, inspect and seize drones, under the Traffic Management and Unmanned Aircraft Bill. Drone pilots could face on the spot fines of up to £1000 for offences such as not having or displaying a flyer ID, not being able to provide proof of permissions and exemptions and, of course, for flying dangerously and/or in restricted locations.

stance in analysing the potential use of drones for criminal and terrorist purposes.[129] This is something that South Africa have not undertaken and may well benefit from following suit.

5.2 Drone Regulations

Aircraft regulation in the UK is dealt with under their Civil Aviation Act.[130] Drone specific legislation was previously found in the Air Navigation Order (ANO),[131] as well as the amendment to this published as Statutory Instrument 2019,[132] and titled 'The Air Navigation (Amendment) Order 2019'.[133] However, many of the detailed drone rules were moved to the Commission Implementing EU Regulation 2019/437 of 24 May 2019. This regulation sets out the rules and procedures for the operation of unmanned aircraft. The regulation has been retained in the UK despite the UK leaving the EU. This is under the European Union Withdrawal Act 2018.[134]

The most significant change in the new regulations is an emphasis on the risk of the flight. Therefore, the new regulations are moving away from the purpose of the flight and therefore the distinction between commercial and leisure flying has been removed.[135] Now, the emphasis is on the potential risk of the flight, the size, weight and speed of the drone and those factors become the driving force in understanding the potential to cause harm. Therefore, the rules that a drone operator should be following will be determined by the drone itself.

The new regulations state that from 30 December 2020, any operator of a drone which weighs less than 250 g but has a camera must be registered. Therefore, the person responsible for the UAV must register as an operator to get an operator ID.[136]

---

[129] Haylen (2019), p. 10.

[130] 1982.

[131] 2016 (SI 2016/765).

[132] No. 261.

[133] The amendment came into force on 13 March 2019. This most recently amended order details new specific regulations that come into force on 30 July 2019 and 30 November 2019, respectively.

[134] One can refer to the UK Consolidation Regulation 2019/947. Refer further to CAP1789A.

[135] The new rules remove the requirements of PfCO (Permission for Commercial Operations).

[136] There is a £9 fee for this, and one must be 18 or over to register for an operator ID. Drone owners in the UK will need two IDs before flying outdoors, the Flyer ID and an Operator ID. The person or organisation that is responsible for a drone or model aircraft that requires an operator ID must register to get an operator ID. The flyer ID is for the person flying the drone. The operator and flyer need not be the same person. The operator ID is valid for 12 months, but flyer ID lasts for five years. The two exceptions are for toy drones that weigh under 250 g and do not have a camera and need neither ID; and drones that weigh below 250 g and have a camera which only need an operator ID.

Additionally, any person intending to fly a drone must pass an online theory test and get a flyer ID.[137] All drones must be labelled with the Operator ID Number and the label must be: visible from the outside, or within a compartment that can easily be accessed without using a tool; clear and in block capitals taller than 3 mm; secure and safe from damage and on the main body of the aircraft.

There are now three categories of operations providing a framework related to the level of risk involved in the flight. These categories are open, specific and certified. UAS operations in the open category is further subdivided into A1, A2 and A3, and present a low or no risk to third parties and does not require authorisation.[138] Operations under the specific category present greater risk than those under the open category. This means that the drone operator must ensure safety by obtaining an operational authorisation from the national competent authority before starting the operation. To obtain the authorisation, the drone operator is required to conduct a safety risk assessment, which will determine the requirements necessary for safe operation of the drone.[139] Operations under the certified category present the same risk as manned operations and so will be subject to the same regulatory regime, including certification of aircraft and the operator, licensing of the pilot.[140] It would therefore appear as if delivery drones would fall under either the specific or certified category.

The previous Permission for Commercial Operations has been replaced by an operational authorisation for the specific category. This operational authorisation requires a CAA application and submission on an Operations Manual to demonstrate how the operators drone actions will be carried out safely and with due regard to legislation.

On airspace around all aerodromes and airports, there is an exclusion zone (restricted airspace) of 2–2.5 nautical miles, with a 5 km long and 1 km wide exclusion at both ends of each runway.[141]

---

[137] Anyone under the age of 13 is still required to get a Flyer ID but must have someone over 18 with them to take the test.

[138] It is worth noting that the Delegated Regulation lays down the requirements for the design and manufacture of UAS and rules on making UAS intended for the use under the open category, including additional components. The regulation also sets out requirements for importers and distributors.

[139] Brown (2020).

[140] The UK regulations relating to the Certified Category are still being developed and are not yet published. Until unique UAS regulations are available, the principles set out in the relevant manned aviation regulations for airworthiness, operations and licensing will be used as the basis for regulating the Certified Category. From 1 January 2023 all new drones will be required to meet a set of product standards and will be classified from C0 to C4. These classes will be based on the weight and capability of the drone and will determine how and where you can fly.

[141] Restricted airspace means that one cannot fly in these areas unless one has permission from the relevant authority. These areas include around airports, critical infrastructure and military installations.

Although the UK is no longer a Member State of the EU, EU legislation as it applied to the UK on 31 December 2020 is now a part of UK domestic legislation.[142] The EU Regulation 785/2004 places insurance obligations for third-party liability on commercial drone operators. It must be noted that this regulation was adopted in the UK following Brexit and therefore, the provisions in this Regulation are still applicable in the UK. The Regulation requires that all commercial drone operators purchase third-party liability insurance.[143] The regulation defines limits for the minimum amount of third-party liability insurance required based on the mass of the aircraft at take-off. For drones weighing less than 500 kg, the minimum cover required is approximately €660,000.[144] It must be noted that model aircraft, weighing less than 20 kg are not required to have third-party liability insurance.[145] The reason for this exception is because these model aircraft systems are being flown for recreational purposes and not for a commercial benefit. This exception implies that the authorities do not believe these 'model' aircrafts pose as big a risk as commercial aircraft do.

This implies that commercial users therefore need to take cognisance hereof and ensure their drones are suitably insured. Interestingly, the insurance regulation has not changed; therefore, it is still a legal requirement to obtain drone insurance if one is undertaking commercial operations. It is important to note that the term 'commercial operation' means any [operation of an aircraft other than] for public transport—(a) which is available to the public; or [(b) which, when not made available to the public, is performed under a contract, between an operator and a customer, where the latter has no control over the property] in return for remuneration or other valuable consideration.[146]

5.3 Delivery Drones in the UK

The recent amendments to the drone regulations in the UK indicates the proactive approach that is being taken by the UK government and aviation authority to ensure that the law stays up to date with the risks involved with drone flights.

Amazon set up a testing centre for delivery drones in Cambridge in 2016 due to the UK's innovative regulations around drones. It is worth mentioning that drone

[142] See EU legislation and the UK law. Section 3 of European Union Withdrawal Act 2018.

[143] Regulation (EC) No. 785/2004 of the European Parliament and of the Council of 21 April 2004 on insurance requirements for air carriers and aircraft operators.

[144] Article 7 Regulation (EC) 785/2004.

[145] Article 2 Regulation (EC) 785/2004. Take note that the word 'model' has been interpreted by the aviation authorities to mean 'for sport or recreational use only'. Therefore, any drone that is being used for commercial purposes needs to comply where 'commercial' has been defined as 'an operation for remuneration and/or hire'.

[146] See Air Navigation (Amendment) Order 2021 section 7.

deliveries in the UK will also fall under the commercial operations definition and therefore third-party liability insurance will be compulsory.

Thus, it is evident that the UK drone regulations are stringent yet flexible to cater for various types of drone flights. This means that operators will have to abide by the lengthy and stringent requirements all based on the risk of their flights. The risks associated with delivery drones does appear to be higher than most since flights will be out of line of sight and the drone will be carrying packages of value. This will likely move drone deliveries into the specific or certified category based on the drone and the risk involved.

It is clear that the UK regulations are far more nuanced and comprehensive than the South African drone regulations. Both recognise the need for compulsory third-party liability insurance in the case of commercial operations, but the fact that the UK includes operator IDs, flyer IDs, police involvement and focus on the risk of each flight allows for more accountability as well as flexibility for operators.

6 Recommendations and Conclusion

6.1 Challenges in Insuring Delivery Drones

The most significant challenges that insurers must consider with delivery drones relate to the distinctive risks that they possess. The fact that delivery drones will be flown out of line of sight is one of the biggest concerns. This is also where the technology of the system needs to be of such a nature that it can give way to other obstacles and fly in a safe manner.[147] The technology must be able to recognise other air traffic and consequently force certain actions to avoid a collision. This technology is known as 'sense-and-avoid' technology and this technology becomes essential for the successful large-scale introduction of delivery drones entering the airspace.[148] This technology is now available for drones and therefore when it comes to delivery drones, this will be an essential element of the drone to ensure safe and efficient flights.

Insurers can intervene by ensuring that the policies state that if drones are to be flown out of line of sight, then they must be equipped with the technology that can still navigate in a safe and effective manner. The insurers can also impose higher premiums for delivery drones since the risks are higher with such systems.

Some of the other additional challenges insurers will need to consider are the possibilities of dropping the packages, which then includes the risk of damaging the package as well as the requirement for the receiver of the package to be authenticated upon delivery. This is where it is essential that insurers draft drone policies in a way that they cater for these unique risks and adjust premiums accordingly.

---

[147] Burda (2021).

[148] Hodgkinson and Johnston (2018), p. 114.

Insurers like Allianz are already considering these risks and recommend drafting policies that include provisions on maximum take-off mass, using high or low altitude, experience of the pilot, operating within or beyond line of sight.[149]

In South Africa, Lawrenson and De Oliveira suggest that insurers could assist in ensuring drone operators abide by the regulatory frameworks by incorporating the regulations into their policy conditions thereby ensuring that the operators are aware of all legal requirements.[150] This is a simple way of ensuring that drone operators are familiar with the regulations. A further possibility is even to include a concise, plain language inclusion of the regulations for the operators to understand. This can simplify the regulations for all users as well as point out the most important regulations for the operators. It therefore becomes pertinent for insurers and regulators to work together in this regard. Simplifying the regulations for policyholders and providing ease in understanding them can only be beneficial for the operators.

It is laudable to see that insurance companies like Allianz are already considering the unique risks of delivery drones. This once again shows the industry's commitment to the innovative developments occurring within the drone sphere.

6.2 Development of Current Drone Regulations Necessary?

On the question of whether the development of the current drone regulations is necessary, it is important to differentiate between the regulations in South Africa and the UK. The fact that the UK regulations are constantly evolving is proof that the regulators are continually considering the development of the technology itself, the growing uses of drones and the future impact that they are to have is commendable. The UK regulators are clearly recognising the potential uses of these systems and the current regulations appear to be in line with the future development of drones. The fact that the UK have moved away from commercial vs private use of drones and now focus more on the risk of the flight means that there is far more flexibility in the use of these systems.

South Africa's current drone regulations will need to be developed to allow for more accountability and perhaps introduce flyer and operator IDs. South African regulators should follow the UK's approach of updating the regulations continuously considering the rapid development and uses of these systems. Enforceability of the regulations definitely remains an issue in South Africa and accountability must be ensured by the CAA to safeguard the interests of the public, as these are the people that are susceptible to possible harm and damage caused by the unsafe operations of drones. It is interesting to note that SACAA is open to the development of drone technology as referred to in paragraph 3.6. above. This innovation and openness of

[149] Burda (2021).

[150] Lawrenson and De Oliveira (2018).

the SACAA to accept such developments is a step in the right direction for the future implementation and growth of delivery drones in the South African airspace.

As far as the current UK drone regulations are concerned, they generally appear to be adequate in terms of being stringent but also flexible enough to allow for innovation in the industry. The regulations call for flyer as well as operator IDs and this is vital. As far as the insurance requirements are concerned, both jurisdictions recognise the need for third-party liability insurance for commercial drone operators.[151]

The fact that compulsory third-party liability insurance is required in both the UK and South Africa is proof of the fact that both jurisdictions recognise the safety concerns around the use of commercial drones and that damage and harm is a reality. Further, commercial operations are generally riskier, and delivery drones particularly bring further risks to the table. These risks will require further investigation, understanding and analysing for insurers to underwrite the risks properly.

Therefore, it may be preliminarily concluded that the current UK drone laws appear to be adequate for the implementation of delivery drones but perhaps it may be even more worthwhile for the UK CAA to provide further clarification on bringing the insurance requirements in line with the new categories of flights; that is, open, specific and certified. This will enhance the understanding and implications of compulsory insurance for different types of flights, emphasising the element of risk for each flight undertaken. As far as the South African regulations are concerned, these should undergo more frequent updates to account for the developing nature of these types of systems and allow for more flexibility in the market. The fact that the South African regulations have not been updated since they were introduced is a cause for concern. Due to the ever-changing nature of drones and their uses, it goes without saying that continuous updates of the regulations are required.

6.3 The Way Forward?

It is known that the development of drone technology (and technology in general) outpaces the regulators. It is a challenge for regulators to stay up to date with the technological innovations occurring. Therefore, regulations are always one step behind. However, this should not be a deterrent for the advantageous use of drones. Delivery drones bring new frontiers to the industry, but they also bring in new risks for insurers to understand and ultimately, underwrite.

Current drone policies focus on the risk of personal injury and property damage, related to the drone itself or the damage caused by it. However, with the growing demand for delivery drones, it goes without saying that the insurance policies for these types of drones need to be wider and cover a larger range of risks. Insurers have

[151] Lawrenson and De Oliveira (2018).

a difficult task in front of them in trying to understand the exact risk assessment associated with delivery drones. It is recommended that each industry using delivery drones should be analysed on a case-by-case basis taking into account all relevant factors and circumstances. Therefore, insurers need to be astute and prudent in drafting delivery drone policies. This is a daunting task for insurers but a valuable one eventually.

Some of the imminent factors that insurers can utilise is ensuring delivery drone policies state that operators must have flight logbooks that provide the insurer with data around flights undertaken as this may assist the insurers in understanding the causes of any incidents involving the system as well as having a compulsory registration scheme (something the regulations also recognise).[152]

Insurers can provide annual policies to bigger delivery companies that undertake daily deliveries via drones, or they can offer on-demand cover for smaller companies entering the market. That way insurance can be provided on a pay-as-you-fly basis. This type of model will play a fundamental role in the insurance industry from a general perspective. As far as insuring delivery drones is concerned, this may allow smaller and new entrants into the market.

Furthermore, to properly control liability exposure of delivery drones, it is vital that operators are adhering to the regulations and flying drones in a safe and legal manner. Therefore, insurers can through proper policy wordings ensure that operators adhere to the strict regulations and judge each company undertaking drone deliveries on a case-by-case basis.

Since the operation of delivery drones is automated, it will be essential that the companies using delivery drones ensure that the operators/pilot's function almost as air traffic controllers would. It is known that air traffic controllers play a pivotal role in the safe functioning of manned aircraft and therefore, if the same role was to be established for drone operators, then that can ensure the safety of such flights. Evidently, this can bring in a further element of safety and ensure that routes are always determined beforehand considering the distance to be travelled, weather conditions at the time as well as the airspace being utilised. It is vital that drone delivery companies adhere to the regulations and guidelines as put forward in every jurisdiction and although the technology is already in use, the regulatory framework will ultimately determine how pervasive and how quickly delivery drones will take to the sky.

6.4 Conclusion

Delivery drones as an emerging technology do create additional challenges for regulators worldwide. Regulators are required to understand the novel risks posed by delivery drones and create a regulatory framework which does not stifle

[152] Kennedys (2017), pp. 12–14.

innovation. The advantages that delivery drones may create is monumental. Delivery drones have the potential to reduce inequalities in access to quality products, particularly for vulnerable people such as the elderly, people with disabilities or health issues and even for those individuals in more rural and remote areas. The potential of delivery drones is exciting, but regulators and insurers must now seek to focus on the various risks and liability concerns and create a regulatory framework which can accommodate the growing sector. The drone regulations in the UK appear to be more favourable for delivery drones than the South African regulations since they are constantly evolving and rather concentrate on the risks of the flight as opposed to the reason for the flight.

With these considerations, it appears as if compulsory public liability insurance for commercial operations is a non-negotiable and the fact that the UK as well as South Africa do recognise the need for such compulsory insurance is laudable.

For delivery drones to take flight, it will be necessary for the insurers and the regulators to come together and work hand-in-glove to ensure that policies are drafted in such a way to cover the distinctive risks and that the regulations are strict enough to ensure accountability yet also flexible enough to allow for growth and innovation.

Legislation

Air Navigation (Amendment) Order 2021.
Air Navigation Amendment Order 2018 No. 623.
Civil Aviation Act 13 of 2009.
Civil Aviation Authority CAP 1421 - Annual Report & Accounts 2015/16.
EU Regulation 2019/437.
Insurance Act 18 of 2017.
Part 101 of the Civil Aviation Authority Regulations.
Regulation (EC) No 785/2004 of the European Parliament and of the Council of 21 April 2004.
UK Consolidation Regulation 2019/947.

Case Law

Ackerman v Loubser 1918 OPD 31 34.
Boshoff v South British Insurance Co Ltd 1951 (3) SA 481 (T).
Cape Town Municipality and Another v Allianz Insurance Co Ltd 1990 (1) SA 311 (C).
Lake v Reinsurance Corporation Ltd 1967 (3) SA 124 (W).
Leading Prospects Trading 38 (Pty Ltd) v Centrique Insurance Company and Tromp GPPHC 8 April 2015.

Prudential Insurance Co v Inland Revenue Commissioners [1904] 2 KB 258.
Watson NO v Shaw NO 2008 (1) SA 350 (C).

References

Articles

Burzichelli CD (2016) Delivery drones: will amazon air see the national airspace. Rutgers Comput Technol Law J 42(1):162
Fox SJ (2017) The rise of the drones: framework and governance - why risk it. J Air Law Commer 82:683
Haylen A (2019) Civilian drones. House of Commons Library Briefing Paper CBP 7734
Huneberg S (2017) On drones, new risks and insurance. THRHR 80(4):592
Huneberg S (2018) The rise of the drone: privacy concerns. THRHR 81:263
Jacobs W (2011) Legal protection insurance in the context of liability insurance: possible solutions in English law? SA Merc Law J 23:464
Keane JF, Carr SS (2013) A brief history of early unmanned aircraft. Johns Hopkins Apl Technical Dig 32(558):559–561
Mathews BD (2015) Potential tort liability for personal use of drone aircraft. St Mary's Law J 46: 573
Michaelides-Mateou S (2015) Ignorantia Juris non excusat: remotely piloted aircraft – safety concerns, violations, and the need for awareness. J Air Law Commer 80:435
Miller BM (2020) Drone delivery and the takings clause. Tex A&M J Prop Law 6:139
North D (2015) Private drones: regulations and insurance. Loy Consum Law Rev 27:334
Thompson D (2019–2020) Rethinking the highway: integrating delivery drones into airspace above highways. Ind Law J Supp 95:8

Books

Hodgkinson D, Johnston R (2018) Aviation laws and drones: unmanned aircraft and the future of aviation. Routledge, London
Millard D (2013) Modern insurance law. Juta, Cape Town
Reinecke MFB, Van Niekerk JP, Nienaber P (2013) South African insurance law. LexisNexis, Cape Town
Um JS (2019) Drones as cyber-physical systems: concepts and applications for the fourth industrial revolution. Springer, Singapore

Online Sources

Allianz Global Corporate (2016) Rise of the drones: managing the unique risks associated with unmanned aircraft systems. Available at Rise of the drones (allianz.com). Accessed on 24 July 2022

Barrett B (2016) When good drones go bad. Wired. Available at https://www.wired.com/2016/01/when-good-drones-go-bad/. Accessed on 24 July 2022
Brown J (2020) New UK & EU unmanned aircraft regulations and the implications for coverdrone insurance. Available at https://www.coverdrone.com/de/new-uk-eu-unmanned-aircraft-regulations-and-the-implications-for-coverdrone-insurance/. Accessed on 24 July 2022
Burda J (2021) Insurance for drones: why insuring drone deliveries is the next big thing. Available at https://www.comarch.com/finance/articles/why-insuring-drone-deliveries-is-the-next-big-thing/. Accessed on 24 July 2022
Business Tech (2019) Old Mutual to use 'insurance drones' to investigate car accidents in South Africa. Available at https://businesstech.co.za/news/technology/336009/old-mutual-to-use-insurance-drones-to-investigate-car-accidents-in-south-africa/. Accessed on 24 July 2022
Caboz (2020) Smart delivery drones are one step closer in SA, thanks to the approval of a new autopilot system. Business Insider South Africa. Available at https://www.businessinsider.co.za/automatic-delivery-drones-are-one-step-closer-in-sa-thanks-to-approval-of-a-new-autopilot-system-2020-2. Accessed on 24 July 2022
Deloitte (2018) Insurance industry drone use is flying higher and farther. Available at https://www2.deloitte.com/us/en/pages/financial-services/articles/infocus-drone-use-by-insurance-industry-flying-higher-farther.html. Accessed on 4 November 2021
Kennedys (2017) Taking Flight: The Rising importance of drone insurance. Available at kennedys_droneswhitepaper.pdf (kennedyslaw.com). Accessed on 24 July 2022
Lawrenson & De Oliveira (2018) Without drone-ing on: a legal overview of drones in South Africa *Clyde & Co.* Available at http://www.mondaq.com/southafrica/x/746350/Aviation/Without+DRONEing+On+A+Legal+Overview+Of+Drones+In+South+Africa. Accessed on 24 July 2022
Marsh (2015) Dawning of the drones: the evolving risk of unmanned aerial systems. Available at http://bit.ly/2nXQQoO. Accessed on 24 July 2022
Palmer (2020) Amazon wins FAA approval for Prime Air drone delivery fleet. Available at https://www.cnbc.com/2020/08/31/amazon-prime-now-drone-delivery-fleet-gets-faa-approval.html. Accessed on 24 July 2022
South African Civil Aviation Authority Media Statement (April 2014) Civil Aviation Authority to crackdown on illegal drone flying. Available at http://www.caa.co.za/Media%20Statements/2014/Civil%20Aviation%20Authority%20to%20crackdown%20on%20illegal%20drone%20flying.pdf. Accessed on 24 July 2022
Thompson (2021) Drones and Emerging Insurance Risks Clyde & Co. Available at https://www.clydeco.com/en/insights/2021/04/drones-and-insurance. Accessed on 24 July 2022
United Kingdom Civil Aviation Authority (2022) An introduction into unmanned aircraft systems. Available at An introduction to remotely piloted aircraft systems | Civil Aviation Authority (caa.co.uk). Accessed on 24 July 2022
Vincent J, Gartenberg C (2019) Here's Amazon's new transforming prime air delivery drone. The Verge. Available at https://www.theverge.com/2019/6/5/18654044/amazon-prime-air-delivery-drone-new-design-safety-transforming-flight-video. Accessed on 24 July 2022

Other

Padayachee, D (2021) Are liability insurers obliged to prove prejudice when repudiating a policyholder's claim based on late notification? A comparative analysis of the legal position in the United Kingdom and South Africa. Dissertation. University of Johannesburg
Yiannakis YA (2019) Does the current drone legislation in South Africa and the United Kingdom assist insurers and their underwriters to assess and address the liability risks associated therewith? A comparative study. Dissertation, University of Johannesburg

Unmanned Air Transports: The Use of Drones and Legal Issues Arising Thereof

Viktoria Chatzara

Abstract Unmanned Aerial Vehicles (UAVs)—most commonly known as drones—are increasingly being used in daily activities, with new, innovative drone-based solutions disrupting a number of core industries, such as agriculture, urban development, traffic and disaster management, mining and oil research, etc. Drone solutions are now expected to be adopted by the transport industry in various applications, including in healthcare transports, with small drones already being used for the delivery of goods and small packages. The increased integration, however, of unmanned aircrafts into transports will create or enhance concerns for aviation safety, security of people, property, environment and wildlife, and privacy, which will need to be duly addressed in applicable laws and regulations. This chapter aims to examine any prior authorization and/or notification (Sect. 2), and operational requirements (Sect. 3) for drone activities under characteristic examples of existing legislations (namely, EU, USA, and ICAO), and to highlight some of the key issues that may need to be reviewed by legislators and competent authorities in view of the forthcoming developments in drone cargo and passenger transports, also taking into account the global nature of the transports sector.

Keywords Unmanned Aerial Vehicles · Unmanned Aircraft Systems · Remote pilots · Authorization requirements · Operational rules

Thanks to Sotiria Bouranta, for her assistance in the academic research for this chapter.

V. Chatzara (✉)
Rokas Law Firm, Athens, Greece
e-mail: v.chatzara@rokas.com

K. Noussia, M. Channon (eds.), *The Regulation of Automated and Autonomous Transport*, https://doi.org/10.1007/978-3-031-32356-0_3

1 Introduction

Unmanned Aerial Vehicles (UAVs) or—as most commonly known—drones, are being increasingly used in everyday life, for multiple functions, uses, and purposes ranging from purely personal to commercial, government and military. A UAV is an aircraft without any human pilot, crew, or passengers on board,[1,2] and is commonly defined as a powered, aerial vehicle that does not carry a human operator, uses aerodynamic forces to provide vehicle lift, can fly autonomously or be piloted remotely, can be expendable or recoverable, and can carry a lethal or nonlethal payload, presenting two basic functions: flight and navigation.

Drones do not operate at a standalone basis, but they rather constitute a component of a larger system, which additionally includes a ground control station, a ground-based controller and a system of communications-data links (i.e., a transmission center allowing the drone to communicate with the ground control station and/or operator while in flight). The operation of drones as a system and the regulation thereof, has been recognized by legislators, although under different terms/names and including different components: the Commission Delegated Regulation (EU) 2019/945[3] refers to "Unmanned Aircraft Systems" (UAS) as *"an unmanned aircraft and the equipment to control it remotely,"* Part 107 of the USA Federal Aviation Regulations[4] regulates the operation of *"small unmanned aircrafts"*[5] and *"small unmanned aircraft systems"* defined as a small unmanned aircraft and its associated elements[6] that are required for the safe and efficient operation thereof in the national airspace system, while the Canadian Aviation Regulations[7] refer to a *"remotely piloted aircraft system or RPAS"* as a set of configurable elements consisting of a remotely piloted aircraft, its control station, the command and control links and any other system elements required during flight operation.

---

[1] At least until the time this chapter is being drafted.

[2] Similar definition is also used in Henri Eisenbeiss (2004), *A mini unmanned aerial vehicle (UAV): System overview and image acquisition*, at International Workshop on "Processing and Visualization using High-Resolution Imagery," 18–20 November 2004, Pitsanulok, Thailand, available at: https://citeseerx.ist.psu.edu/viewdoc/download;jsessionid=115A5B4916F3005B45E7EEEB50C5D3C6?doi=10.1.1.145.4342&rep=rep1&type=pdf, last accessed on 05.08.2022.

[3] Commission Delegated Regulation (EU) 2019/945 of 12 March 2019 on unmanned aircraft systems and on third-country operators of unmanned aircraft systems.

[4] Available at: https://www.faa.gov/regulations_policies/faa_regulations/, last accessed on 05.08.2022.

[5] Meaning unmanned aircrafts weighing less than 55 pounds on takeoff, including everything that is on board or otherwise attached to the aircraft.

[6] Including communications links and the components that control the small unmanned aircraft.

[7] Available at: https://lois-laws.justice.gc.ca/eng/regulations/SOR-96-433/FullText.html#s-900.01, last accessed on 05.08.2022, see Part I—General Provisions, Subpart 1—Interpretation, Article 101.01(1) thereof.

At the same time, both the drones per se and generally the UAS depend on the use and interoperability of numerous technological components, which may include electronic speed controllers, GPS modules, battery, antenna, receiver, altimeter, accelerometer, sensors,[8] and visual positioning systems being used for landing purposes.

1.1 Categories of Drones

Drones can be classified in different categories, depending on the basic criterion for the categorization being used.[9] The two main types of drone platforms are rotor[10] or fixed-wing, including the hybrid Vertical Take-Off and Landing drones.[11]

Drones may be also classified, depending on their weight (e.g., nano weighing up to 250 g, miniature UAV between 2 and 25 kg, medium between 25 and 150 kg, etc.), the altitude and range they may reach (hand-held: 600 m altitude and about 2 km range, close: 1500 m altitude and up to 10 km range, Medium Altitude-Long Endurance: up to 9000 m altitude and over 200 km range, High-Altitude-Long Endurance: over 9100 m altitude and indefinite range, etc.), or their function (target and decoy, reconnaissance, logistics, civil and commercial, et. Al).

One of the most important classifications, also from a legal point of view, is based on the degree of autonomy in the operation of the drones. Depending on the use and on the technology employed, drones may range from remotely piloted to fully autonomous,[12] with intermediate degrees of autonomy, or combinations of remote piloting and autonomy, according to the International Civil Aviation Organization (ICAO).[13] In any case, drones often have built-in autonomous operations, such as stabilization, altitude hold, hover/position hold, take-off and landing, return-to-home, etc.

[8]Depending also on the use of each drone, such as obstacle detection and collision avoidance sensors.

[9]See, for different classifications of UAVs, Watts et al. (2012).

[10]Including single-rotor or multi-rotor.

[11]Earls Alan, Ben Lutkevich (last updated in December 2021), Definition of "drone," Internet of Things Agenda, and available at: https://internetofthingsagenda.techtarget.com/definition/drone, and at: https://www.techtarget.com/iotagenda/definition/drone, last accessed on 05.08.2022.

[12]See relevant reference Peter Neenan (2021), *UK drone regulations summarised by Peter Neenan in Getting the Deal through*, available at: https://www.stewartslaw.com/news/uk-drone-regulation-2022-in-lexologys-getting-the-deal-through/, last accessed on 05.08.2022.

[13]See relevant reference in the context and for the purposes of developing UAS Regulation: https://www.icao.int/safety/UA/UASToolkit/Pages/Narrative-Regulation.aspx, last accessed on 05.08.2022.

1.2 The Global Drones Market

Although drone technology was originally developed in the context and for the purpose of military activities, over the last decade it has increasingly penetrated private and public life and is nowadays being employed also for a series of commercial applications. Recently, the COVID-19 pandemic and the related imposed restrictions in movements have demonstrated—or accelerated—the need for the further expansion of drone use in a number of situations, including in the transport and delivery of goods.

According to recent figures, in the United States (USA) alone, more than 100,000 remote pilot certificates had been issued by 2019 by the Federal Aviation Administration (FAA), while in Denmark, according to information from the Danish Transport, Construction and Housing Authority, such remote pilot certificates doubled from approximately 14,000 in 2017 to more than 28,000 in 2018. From an intellectual property point of view, in 2019 a research in global patent databases identified almost 35,000 patent families associated with drone and UAS technologies; this figure skyrocketed from 307 published patents in 2010 (and only 40 in 2000) to 11,393 in 2018.[14]

The rapidly increasing use of drones in a number of industry sectors generates, in its turn, an increase in research and development activities aiming to provide more innovative, cost-efficient and time-saving solutions. According to available data, the global commercial drone market is expected to grow at a Compound Annual Growth Rate (CAGR) of 16–17% to 2023, while the global drone market has been estimated to reach an amount of US $21.47 billion within 2021. The value of drone-powered solutions specifically in the transport industry was estimated in 2018 to amount to US $13.0 billion.[15] According to more recent reports with respect to the drone service market, its size is expected to grow to US $63.6 billion by 2025, with consumer drone shipments estimated to reach an amount of US $29 million by 2021.[16]

---

[14] Association for Unmanned Vehicle Systems International in cooperation with the Danish Technological Institute, *Global Trends of Unmanned Aerial Systems*, 2019, available at: https://www.auvsi.org/global-trends-unmanned-aerial-systems, last accessed on 05.08.2022.

[15] Pricewaterhouse Coopers (2018), *Flying high: Drones to drive jobs in the construction sector*, available at: https://www.pwc.in/assets/pdfs/publications/2018/flying-high.pdf, last accessed on 05.08.2022.

[16] Business Insider – Insider Intelligence, *Drone market outlook in 2021: industry growth trends, market stats and forecast*, updated 15.04.2022, and available at: https://www.businessinsider.com/drone-industry-analysis-market-trends-growth-forecasts, last accessed on 05.08.2022.

1.3 Drone Uses

Apart from the oldest, most well-known (and probably most controversial) military use of drones, and their recreational use, drones have effectively disrupted and impacted on business and operating models in various sectors of the economy, including urban development, traffic and disaster management,[17] mining and oil research, healthcare, etc. One of the main economy sectors employing drones is the agricultural, with drones being used to spray pesticides on fields,[18] carry out field surveys, seed over fields, track livestock and estimate crop yield.

Drones are also being used in emergency situations, where human intervention cannot (timely) take place, for instance, to search for trapped individuals in avalanches, or in fighting wildfires. During natural disasters, the insurance sector is also expected to avail of the possibilities provided by drone solutions to provide faster and more accurate property assessments, claims handling, and settlement.

Drones further constitute a cost-efficient solution to wildlife conservation, rendering possible the monitoring of wildlife populations and the tracking of roaming animal populations, while at the same time assisting in preventing and addressing the poaching issue. In the same context, drones are being employed in reforestation efforts, dropping seed vessels, fertilizers and nutrients in decimated by fires areas, in a much more efficient way.[19]

1.4 Drones in Transports

In the transport sector, the use of drones has also provided new, innovative solutions, and is expected to rapidly develop within the next time. Currently, delivery drones are usually autonomous vehicles used to transport food, packages or other goods to consumers/ buyers from nearby stores or warehouses (known as “last mile” flying vehicles). In August 2020 Amazon was granted approval from the FAA to operate its Prime Air drone delivery fleet of drones to deliver packages to its customers in areas with low population density (and for packages weighing up to 5 pounds), after it had begun testing delivery drones in 2013.[20] Amazon aims to widely use the Prime Air delivery option to deliver packages to its customers around the world in 30 min or

[17] See Outay et al. (2020), pp. 116–129.

[18] As early as the 1980s in Japan—see PwC, as above. See also del Cerro et al. (2021), p. 203.

[19] Built In, *Drones. What is a drone? What Are Uses For Drones?*, as updated on 28.07.2022 at: https://builtin.com/drones, last accessed on 05.08.2022.

[20] See in this relevance: https://www.cnbc.com/2020/08/31/amazon-prime-now-drone-delivery-fleet-gets-faa-approval.html, last accessed on 05.08.2022.

less using small drones;[21] however, reports state that the further development of the project is currently (at the time this chapter is being drafted) uncertain.[22]

On the contrary, Alphabet's (Google's parent company) drone delivery company, Wing, which operates in Australia, the US and Finland, reached in August 2021 the milestone of 100,000 deliveries since its establishment in 2012.[23] Following the submission of an order via the Wing mobile application, a Wing drone picks up the package and delivers it to the customer within 10 min (with a record of 2 min and 47 s from order to arrival). Wing's drones are designed to deliver small packages weighing up to 1.2 kg, including fresh foods, medicines, household items and tools.[24]

Apart from consumer good delivery services, the next major development in transports is expected to be the use of large drones and UAS. According to a relevant report of the Aerospace Industries Association (AIA),[25] air cargo transports for short-haul flights at a relatively low altitude are expected to be adopted between 2025–2031, at the same time as intra-city taxi/ shuttle service flights. Following 2032, long-haul passenger and cargo drones will be rolled out, still having, however, pilots on board even if the aircrafts are equipped with full autonomy characteristics.

Furthermore, the use of drones in the healthcare sector is widely discussed, to address inherent problems, such as delivering medical suppliers to hard-to-reach areas, or to improve the speed and quality of the healthcare services, for instance, by ensuring prompt delivery of necessary blood products (packed red blood cells, plasma, etc.) in case of accidents, hazardous materials, or organs for transplant patients,[26] and to enhance the efficiency and accuracy of diagnostic tests, particularly the ones that are time and temperature sensitive, by effecting rapid transfer of the patient samples. Drone solutions have already been used to deliver vaccines in remote areas, including for the delivery of COVID-19 vaccines in the USA with

---

[21] According to the publicly available information at: https://www.amazon.jobs/en/teams/prime-air, last accessed on 05.08.2022.

[22] Andrew Kersley (2021), *The slow collapse of Amazon's drone delivery dream*, available at: https://www.wired.co.uk/article/amazon-drone-delivery-prime-air, last accessed on 05.08.2022.

[23] James Vincent (2021), *Alphabet's drone delivery service Wing hits 100,000 deliveries milestone*, available at: https://www.theverge.com/2021/8/25/22640833/drone-delivery-google-alphabet-wing-milestone, last accessed on 05.08.2022.

[24] Wing, Frequently Asked Questions about Delivery, as available on 09.11.2021 at https://wing.com/about-delivery/, last accessed on 05.08.2022.

[25] Aerospace Industries Association, *Think Bigger – Large Unmanned Systems and the Next Major Shift in Aviation*, March 2018, available at: https://www.aia-aerospace.org/uasreport2018/, last accessed on 05.08.2022.

[26] Thiels et al. (2015).

the use of UPS Flight Forward,[27] and is under wraps in Africa and Southeastern Asia at the time this chapter is being drafted.[28]

1.5 Legal Issues

The increasing use of drones has created a number of concerns and the enactment of new regulations concerning their operation. Even the use of civil, private drones with cameras being able to take photographs (particularly over private properties) either for recreational purposes or (even more so) by professional photographers, has raised privacy concerns.

Safety concerns on air traffic could also arise if drones are being used to transport larger cargo and passengers, considering that even the private use of drones has caused significant disruptions in air traffic circulation, with the 2019 Heathrow airport incident, which halted all departures for approximately an hour due to a mere drone sighting close to the airport, being an indicative example.[29] Drones may also pose an increased risk to aviation safety, by flying improperly or getting too close to other aircrafts, given the multiple sources from which an error can occur, be it the drone itself, a malfunction of the algorithm regulating its course (in case of full autonomous vehicles) or in the systems transmitting data to and from the ground control station, human error (in case there is a pilot or a controller), or other.

Security concerns also arise from the risk for drones and their supporting systems to be hacked, resulting in privacy breaches or threats to public security. The development of drones able to carry larger cargo, could also create the risk for such solutions to be used to carry hazardous materials close to security-sensitive locations and/or infrastructure aiming to terrorist actions.[30]

Considering the increased and different risks that (are likely to) derive from the continuous integration of drone solutions into the transport chain, a number of legal and regulatory issues also need to be addressed to prevent or mitigate any adverse effects. This chapter aims to examine, at a high level, some of the major issues arising, in connection with any prior authorization/ notification, and the main operating requirements. The above issues will be reviewed in the light of the existing

---

[27] See in this relevance: UPS (2021) operates first ever U.S. drone COVID-19 vaccine delivery, available at https://about.ups.com/us/en/our-stories/innovation-driven/drone-covid-vaccine-deliveries.html, last accessed on 05.08.2022.

[28] Bill Wimberley (2021), *Drones in healthcare: Blood transportation and beyond*, available at: https://healthcaretransformers.com/healthcare-business/drones-healthcare/, last accessed on 05.08.2022.

[29] BBC, *Heathrow airport: Drone sighting halts departures*, 08.01.2019, available at: https://www.bbc.com/news/uk-46803713, last accessed on 05.08.2022.

[30] Deloitte, *Unmanned Aircraft Systems (UAS) Risk Management: Thriving Amid Emerging Threats and Opportunities*, November 2018, available at: https://www2.deloitte.com/us/en/pages/public-sector/articles/drone-risk-assessment.html, last accessed on 05.08.2022.

laws and regulations in the EU, the Federal Aviation Regulations in USA, and the International Civil Aviation Organization (ICAO) model UAS Regulations.[31]

Civil liability and compulsory insurance aspects, as well as issues arising from telecommunications, network and system security, regulatory/ supervisory laws, will not be treated herein, as this chapter does not aim to proceed with an exhaustive outline of all legal issues arising from the use of drones, but rather with an indicative review that may act as trigger for further thought. The below analysis further refers to civil/ commercial drone operations, and does not cover state-related operations, such as military, customs, police, firefighting, etc.

2 Prior Authorization and Notification Requirements

One of the main legal issues relating to the operation of drones is the issue of any prior authorization and/or notification requirements that need to be fulfilled for such operation to be in compliance with applicable laws and regulations. Considering the risks arising from the operation of drones, and the main objectives of aviation regulations to establish and maintain a high level of civil aviation safety, and enhance public confidence to aviation, most jurisdictions have acknowledged the need to provide for such requirements, depending on the characteristics of each drone category.

2.1 *The European Union Legislation*

Following an agreement between the EU Member States endorsed by the European Parliament, the European Union (EU) was granted competence to regulate the operation of all UAS,[32] as provided in the Regulation (EU) 2018/1139[33] of the European Parliament and of the Council of 4 July 2018 *on common rules in the field of civil aviation and establishing a European Union Aviation Safety Agency* (the "Basic Regulation").

[31] Which currently constitute a compilation drawn from UAS Regulation in effect in Vanuatu, New Zealand, Australia, Canada and the USA, thus providing an insight into these jurisdictions as well. Available at: https://www.icao.int/safety/UA/Pages/ICAO-Model-UAS-Regulations.aspx, last accessed on 05.08.2022.

[32] Except for the ones being used for "state" operations, such as military, customs, police, firefighting, etc.

[33] See Studio Pierallini - Francesco Grassetti (2021), *Drone Regulation: European Union*, available at: https://www.lexology.com/library/detail.aspx?g=10ac2457-86ad-44d5-8eb1-bfc674951fc6, last accessed on 05.08.2022.

Under Articles 55–58 of the Basic Regulation, and following consultation with the European Aviation Safety Agency,[34] the Commission Delegated Regulation (EU) 2019/945 of 12 March 2019 *on unmanned aircraft systems and on third-country operators of unmanned aircraft systems* (the "Delegated Regulation"), and the Commission Implementing Regulation (EU) 2019/947 of 24 May 2019 *on the rules and procedures for the operation of unmanned aircraft* (the "Implementing Regulation") were issued, providing, among others, for certification and licensing requirements for drones. Under the Implementing Regulation,[35] the following categories[36] of UAS operations are distinguished:

- "Open" category, not subject to any prior operational authorization, or operational declaration;
- "Specific" category, requiring an operational authorization from the competent authority or an authorization received following a request of a model aircraft club or association, or a prior declaration; and
- "Certified" category, requiring the certification of the UAS pursuant to the Delegated Regulation, the certification of the operator and, where applicable, the licensing of the remote pilot.

The Implementing Regulation further provides that it falls with the designated national competent authorities in each EU Member State to ensure the enforcement of its provisions, including to develop a risk-based oversight system for UAS operators, establish audit planning, carry out any necessary inspections, and implement a system to detect and examine non-compliance incidents.

2.1.1 The "Open" Category of UAS Operations

For a UAS operation to be characterized as "open" and, thus, not be subject to prior authorization/ declaration, the following requirements shall be met:[37]

- the UAS must be privately built or pertain to one of the classes C0-C4 provided in the Annex of the Delegated Regulation;[38]

[34] See Opinion 01/2018, *Introduction of a regulatory framework for the operation of unmanned aircraft systems, in the "open" and "specific" categories*, available at: https://www.easa.europa.eu/sites/default/files/dfu/Opinion%20No%2001-2018.pdf, last accessed on 05.08.2022.

[35] See Articles 3 et. seq. thereof.

[36] See Hogan Lovells, *Drones in German skies: new EU regulations take flight*, available at: https://www.hoganlovells.com/-/media/hogan-lovells/pdf/2020-pdfs/2020_06_01_iplr77_drones-in-the-german-skies-new-eu-regulations-take-flight.pdf, last accessed on 05.08.2022.

[37] See Article 4 of the Implementing Regulation.

[38] Parts 1–5 of said Annex provide for the C0–C4 UAS categories, based on the criteria including their maximum take-off mass (MTOM), maximum speed in level flight, maximum attainable height above the take-off point, and additional characteristics in each of the categories. Alternatively, the Implementing Regulation provides that if the requirements of the Delegated Regulation are not met, a UAS may still fall into the scope of the "open" category, if it has been placed on the market before

- the UAV must have a maximum take-off mass (MTOM) of up to 25 kg;
- the remote pilot ensures that the UAV is kept at a safe distance from people and it is not flown over assemblies of people;
- the remote pilot keeps the UAV in Visual Line of Sight Operation (VLOS) at all times, except when flying in follow-me mode or when using a UAV observer;[39]
- during flight, the UAV is maintained within 120 m from the closest point of the surface of the earth, except when overflying an obstacle;
- during flight, the UAV does not carry dangerous goods and does not drop any materials.

UAS operations falling into this category include flights by the general public performed for recreational purposes, model flying activities, drones being used by photographers,[40] private flights with drones lighter than 900 g (e.g., a DJI Mavic Air) within residential areas, drone operations for mapping a construction site, or for performing coastal surveillance flights,[41] etc.

Although the UAS operations classified into the "open" category are not subject to prior authorization/ declaration obligations, they are not completely non-regulated. Under the provisions of the Implementing Regulation, depending on the sub-category each "open" operation falls into (A1, A2, or A3), they are subject to different operational limitations, which will be outlined further below.

2.1.2 The "Specific" Category of UAS Operations

In any case where a UAS operation cannot be characterized as "open" or "certified," it is considered to fall into the scope of the "specific" category, and the UAS operator is required to obtain an operational authorization from the national competent authority of the Member State where it is registered,[42] or submit a declaration.

01.07.2022, and provided it is in sub-category A1 and has a maximum take-off mass (MTOM) up to 250 g (including payload), or in sub-category A3 and has a MTOM up to 25 kg (including fuel and payload). Sub-categories A1, A2, and A3 of the "open" UAS operations are classified under the Implementing Regulation, based on operational limitations, requirements for the remote pilot and technical requirements for UAS.

[39] According to the relevant operational-piloting requirements set out in Part A of the Annex to the Implementing Regulation.

[40] See in this relevance Dawn M.K. Zoldi (2021), Drone Life – Drones are Flying High in Europe, in dronelife.com, available at: https://dronelife.com/2021/05/12/european-drone-regulations-easa-basic-regulation-and-whats-next/, last accessed on 05.08.2022.

[41] See Airhub (2021), UAS operations in the Open Category, available at: https://dronelife.com/2021/05/12/european-drone-regulations-easa-basic-regulation-and-whats-next/, last accessed on 05.08.2022.

[42] Each Member State is required to appoint one or more entities as the national competent authority, having, among others, the competence to issue, suspend or revoke UAS operators certificates and licenses of remote pilots, issue, amend suspend, limit or revoke operational authorizations, maintain documents, records and reports concerning UAS operational authorizations, declarations, certificates, etc.

Such a prior authorization or declaration is not required for UAS operators holding a light UAS operator certificate (LUC)[43] with appropriate privileges, and for operations conducted in the framework of model aircraft clubs and associations that have received an authorization.

"Specific" UAS operations can take place, for example, for line inspections, aerial works, or even transport of goods (depending on the size of the packages being transferred) purposes.

Operational Authorization A UAS operator shall obtain an operational authorization before the launch of "specific" UAS operations.[44] Along with the relevant application, the UAS operator shall perform an operational risk assessment, describing the characteristics of the UAS operation,[45] proposing adequate safety objectives, identifying the risks of the operation on the ground and in the air[46] and a range of possible risk mitigating measures,[47] and determining the necessary level of robustness of the selected mitigating measures, in such a way that the operation can be conducted safely, and that it can be determined whether the proposed mitigating measures are commensurate with the safety objectives and the risks of the intended operation. Such risk assessment must also propose a target level of safety, equivalent to the corresponding level in manned aviation. The application must also contain information on the UAS operator registration number, the name of the accountable manager or the name of the UAS operator (in case of an individual), an operation manual (if necessary), and a confirmation that an appropriate insurance cover will be in place at the launch of the operation.

After the filing of the application, the competent authority shall evaluate the risk assessment and the proposed mitigating measures and will grant an operational

[43] A certificate issued to the UAS operator by a national competent authority, under the conditions of Part C of the Annex to the Implementing Regulation.

[44] Or an updated authorization in case of any significant changes to the UAS operation or to the mitigation measures listed in the initial operational authorization.

[45] Article 11 par. 2 of the Implementing Regulation details the minimum information the description of the UAS operation shall contain, including indicatively the nature of the activities performed, the operational environment and geographical area for the intended operation, the technical features of the UAS, the competence of the personnel involved, etc.

[46] Such identification must consider the extent to which third parties or property on the ground could be endangered by the activity, the complexity, performance and operational characteristics of the UAV involved, the purpose of the flight, the type of UAS, the probability of collision with other aircrafts, the type, scale and complexity of the UAS operation or activity (including, if applicable, the size and type of traffic handled by the responsible organization or person), and the extent to which the persons affected by the above risks are able to assess and exercise control over those risks. Article 11 par. 4 of the Implementing Regulation further details the information to be provided along with the identification of the risks, including any unmitigated ground risk of the operation, the unmitigated air risk of the operation, etc.

[47] Such as containment measures, strategic operational limitations to the UAS operation, capability to cope with possible adverse operating conditions, organization factors, etc. (see Art. 11 par. 5 of the Implementing Regulation).

authorization if it concludes that the operational safety objectives considering the operational risks, the proposed mitigation measures in combination with the competence of the personnel involved and the technical features are adequate and sufficiently robust to ensure safety, and the UAS operator has stated that the intended operation complies with any applicable EU and national laws and regulations. The national competent authority will grant the authorization, provided it also concludes that a procedure is in place for coordination with the relevant service provider for the airspace, if the intended operation will take place, wholly or partially, in controlled airspace.

The operational authorization shall detail its scope, any specific conditions and operational limitations that may apply, information on the UAS operator registration number and the technical features of the UAS, a reference to the operational risk assessment, the mitigation measures that need to be implemented, the location (s) where the operation is authorized to take place, and all documents and records relevant for the type of operation and events that should be reported. Furthermore, the competent authority shall specify whether the operational authorization concerns the approval of a single operation or a group of operations, or the approval of a LUC.

If the competent authority considers that the risk of the operation cannot be adequately mitigated without the certification of the UAS and of the UAS operator and (if applicable) the licensing of the pilot, then the UAS operation will be characterized as "certified" and the relevant provisions will apply.

Prior Declaration Under certain conditions, the UAS operator may only submit a prior declaration of compliance with a standard scenario[48] to the national competent authority of the Member State of operation, and not be subject to prior operational authorization. Such option applies in operations of a UAV with maximum characteristic dimension up to 3 m in VLOS over controller ground area (except over assemblies of people[49]), maximum characteristic dimension up to 1 m in VLOS (except of assemblies of people), maximum characteristic dimension up to 1 m in Beyond VLOS (BVLOS, over sparsely populated areas), and maximum characteristic dimension up to 3 m in BVLOS (over controlled ground area). It is also possible to submit a prior declaration in relation to operations performed in an altitude of below 120 m, and in uncontrolled aerospace, or in controlled airspace after coordination and individual flight authorization, in accordance with published procedures for the area of operation.

The prior declaration must provide information on the UAS operator, and a statement that the intended operation meets the operational requirements under the

[48] As referred in Appendix I to the Annex of the Implementing Regulation.

[49] An assembly of people relates to a group of individuals which is so dense, that the ability of the assembled individuals to freely leave or distance themselves, is limited. Assemblies of people are considered to be, for example, sports, cultural, religious or political gatherings, gatherings in beaches or parks in sunny days, commercial streets when stores are open, ski centers, etc. See in this relevance Questions and Answers issued by the Hellenic Civil Aviation Authority, available at: https://uas.hcaa.gr/Faq, last accessed on 05.08.2022.

Implementing Regulation. The UAS operator must also commit to comply with the mitigation measures required for the safety of the operation and confirm that an appropriate insurance cover will be in place at the launch of the operation. The UAS operator is further required to notify, without delay, the competent authority of any changes to the above information.

After receiving a prior declaration, the competent authority needs to verify that it contains all the information required and provide the UAS operator with a confirmation of receipt and completeness, without undue delay, after which the intended operation may begin.

2.1.3 The "Certified" Category of UAS Operations

For a UAS operation to classify as "certified," the UAS involved must be certified by having a characteristic dimension of 3 m or more and be designed to operate over assemblies of people, or for transporting people, or for transporting dangerous goods and requiring a high level of robustness to mitigate the risks. Furthermore, the intended operation must be conducted over assemblies of people, or involve the transport of people, or the transport of dangerous goods that may result in high risk for third parties in case of an accident.

"Certified" UAS operations would include air taxi activities, international cargo and passenger transport operations, package deliveries over people, etc.

Under the relevant provisions of the Delegated Regulation,[50] a UAS subject to certification must comply with the requirements under Commission Regulation (EU) No. 748/2012,[51] the Commission Regulation (EU) 2015/640,[52] and Commission Regulation (EU) No. 1321/2014.[53] Regulation (EU) No. 748/2012 provides for the technical requirements and the administrative procedures to be followed for the certification of aircrafts and related products, including the issue of temporary and restrictive permits to fly if the certification requirements are not met,[54] while Regulation (EU) No. 1321/2014 establishes common technical requirements and administrative procedures for ensuring the continuing airworthiness of aircrafts,[55] and Regulation (EU) 2015/640 establishes additional airworthiness specifications.[56]

[50] See Article 40 thereof.

[51] Commission Regulation (EU) No. 748/2012 of 3 August 2012 laying down implementing rules for the airworthiness and environmental certification of aircraft and related products, parts and appliances, as well as for the certification of design and production organizations.

[52] Commission Regulation (EU) No. 2015/640 of 23 April 2015 on additional airworthiness specifications for a given type of operations.

[53] Commission Regulation (EU) No. 1321/2014 of 26 November 2014 on the continuing airworthiness of aircraft and aeronautical products, parts and appliances, and on the approval of organizations and personnel involved in these tasks.

[54] See Article 1 and Annex I thereto.

[55] See Article 1 thereto.

[56] See Article 1 thereto.

However, even if the certification of the UAS falling into this category is completed, their lawful operation will still require amendments to almost all the aviation regulations.

As described by the European Union Aviation Safety Agency,[57] this activity is designed to take place in multiple phases, in connection with the different types of "certified" UAS operations. At a first stage, the Agency intends to address the Operations type #1 concerning international flights of certified cargo drones conducted in instrumental flight rule (IFR) in airspace classes A-C[58] and taking off and landing at aerodromes under the Agency's scope, the Operations type #2 covering drone operations in urban or rural environments, using pre-defined routes in airspaces where U-space services are provided and including operations of drones carrying passengers or cargo, and Operations type #3 which shall be the same as type #2 but conducted with an aircraft with a pilot on board. In this context, a public consultation is currently (at the time this chapter is being drafted) open (until 31.12.2021) with the aim of a future comprehensive EU policy package regulating the rapid developments in drone technology, drone applications and transport services.[59]

The "certified" UAS are also registered[60] in the relevant registries maintained by the national competent authority,[61] along with information on their manufacturer's name, the manufacturer's designation of the unmanned aircraft, the unmanned aircraft's serial number, and full details of the individual or legal entity under whose name the unmanned aircraft is registered. EU Member States are also obliged to ensure that such registries are digital and interoperable and allow for mutual access and exchange of information.

---

[57] See in this relevance: https://www.easa.europa.eu/domains/civil-drones-rpas/certified-category-civil-drones, last accessed on 05.08.2022.

[58] Class A being the one where only IFR flights are permitted, all flights are provided with air traffic control service and are separated from each other, continuous air-ground voice communications (manned) are required, and flight shall be subject to air traffic control (ATC) clearance, Class B being the one where both IFR and VFR (Visual Flight Rules) flights are permitted, all flights are provided with air traffic control service and are separated from each other, continuous air-ground voice communications are required and all flights shall be subject to ATC clearance, and Class C being the one where IFR and VFR flights are permitted, all flights are provided with air traffic control service, IFR flights are separated from other IFR flights and VFR flights, VFR flights are separated from IFR flights and receive traffic information for other VFR flights and traffic avoidance advice upon request, continuous air-ground voice communications are required, VFR flights are subject to a speed limitation (with exceptions), and all flights shall be subject to ATC clearance. See in this regard Commission Implementing Regulation (EU) 2016/1185 of 20 July 2016 as regards the update and completion of the common rules of the air and operational provisions regarding services and procedures in air navigation (SERA Part C).

[59] European Commission, *Roadmap – A Drone Strategy 2.0 for Europe*, available at: https://ec.europa.eu/info/law/better-regulation/have-your-say/initiatives/13046-A-Drone-strategy-20-for-Europe-to-foster-sustainable-and-smart-mobility_en, last accessed on 05.08.2022.

[60] By the owner of the unmanned aircraft, whose design is subject to certification.

[61] While an unmanned aircraft cannot be registered in more than one Member States at a time.

2.1.4 Cross-Border Operations

The Implementing Regulation further regulates UAS operations taking place on a cross-border basis between EU Member States. That is, a duly authorized "specific" UAS operation which is intended to, wholly or partially, take place across Member States borders, must take place following an application submitted by the UAS operator to the competent authority of the receiving Member State, including the operational authorization already granted under the Implementing Regulation, and information on the location(s) of the intended operation, and updated mitigation measures specific to the local airspace, terrain, population, and climate characteristics. The receiving competent authority shall assess such application and confirm to the competent authority of the Member State of registration and to the UAS operator that the proposed mitigation measures are considered adequate. Upon receipt of such confirmation, the UAS operator may launch the intended activity. If only prior declaration requirements are applicable, the UAS operator shall submit to the competent authority of the Member State where the operation is intended to take place a copy of such duly filed declaration, and a copy of the relevant confirmation of receipt and completeness.

It is further provided in the Delegated Regulation that UAS operators having their principal place of business, being established or residing in a third country, must also comply with the requirements of the Implementing Regulation with respect to UAS operations taking place within the single European sky airspace. For competence purposes, it is explicitly provided that the national competent authority for such activities shall be considered to be the designated by the Member State where the third-country UAS operator first intends to operate.

2.2 *The USA Federal Aviation Regulations*

The Federal Aviation Administration (FAA) has issued a set of rules, incorporated into the Federal Aviation Regulations[62] concerning the rules and regulations applicable to the operation of UAS. Generally, it is noted that certified remote pilots, including commercial operators may fly a small drone (less than 55 pounds) for work or business purposes, in accordance with the Part 107 guidelines,[63] applicable to the registration, airman certification, and operation of civil small unmanned aircraft systems within the USA, as well as to the eligibility of civil small unmanned aircraft systems to operate over individuals in the USA.

[62] Available at: https://www.faa.gov/regulations_policies/faa_regulations/, last accessed on 05.08.2022.

[63] Drone Laws (2021), Drone Laws in the USA, available at: https://drone-laws.com/usa-drone-laws-in-usa/, last accessed on 05.08.2022.

Registration and Remote Identification For drones being used for non-recreational purposes, and falling into the scope of the *"small unmanned aircraft"* notion, to be lawfully used, the FAA provides that the UAS must be registered with the FAA, regardless of their weight, with such registration being valid for three (3) years. The relevant registration number needs to be clearly marked on the outside of each UAS, in an easily readable manner, so that the operator may be traced and located, for instance, if involved in an accident or unlawful activity.[64]

In December 2020, the FAA further finalized its Remote ID rule, according to which UAS will be required to broadcast information in real-time concerning their operations, including information on their location and the location of their operator, and a unique identification number that law enforcement officials may be able to cross-reference. The Remote ID rule is not yet in force; drone manufacturers must comply with the relevant requirements as of 16.09.2022, while all drone pilots must meet the relevant operating requirements as of 16.09.2023.[65]

Prior Authorization On prior authorization, par. 107.41 of Part 107 of the Federal Aviation Regulation stipulates that any operations of small unmanned aircrafts are subject to prior authorization from the Air Traffic Control (ATC), if intended to take place in Class B,[66] Class C,[67] or Class D[68] airspace, or within the lateral boundaries of the surface area of Class E[69] airspace designated for an airport. However, apart from these cases, commercial drone operators may proceed with the intended UAS operations without prior authorization, provided that they meet the strict operational requirements under Part 107. In the same relevance, it is provided that UAS operations in prohibited or restricted areas are not permitted, unless with the permission from the using or controlling agency, as appropriate.[70]

FAA has also developed the Low Altitude Authorization and Notification Capability (LAANC), a form of collaboration with the market directly supporting UAS

---

[64] See *Drone Laws in the United States of America*, available at: https://uavcoach.com/drone-laws-in-united-states-of-america/, last accessed on 05.08.2022, as well as Federal Aviation Administration, How to register your drone, available at: https://www.faa.gov/uas/getting_started/register_drone, last accessed on 05.08.2022. See also, for historical data, Herrmann (2017).

[65] Federal Aviation Administration, UAS Remote Identification Overview, available at: https://www.faa.gov/uas/getting_started/remote_id/, last accessed on 05.08.2022.

[66] The airspace utilizing the space surrounding the busiest airports and beginning from the surface to 10,000 feet Mean Sea Level (MSL)—see ATP Flight School, Types of Airspaces and How Are They Defined, available at: https://atpflightschool.com/become-a-pilot/flight-training/airspace-classes.html, last accessed on 05.08.2022.

[67] The airspace surrounding busy airports that are not quite as busy as Class B airports—see ATP Flight School as above.

[68] The airspace existing around airports that still have an operating control tower but are not as busy as the Class C airports—see ATP Flight School as above.

[69] Any controlled airspace not classified as the airspaces surrounding airports—see ATP Flight School as above.

[70] Federal Aviation Regulations, Part 107, par. 107.45.

integration into the airspace.[71] LAANC authorizations are provided to drone pilots through companies approved by the FAA (FAA-Approved UAS Service Suppliers), which have built desktop and mobile applications, with the use of which drone pilots submit their authorization request to operate below 400 ft. within controlled airspace around airports. The relevant requests may be submitted up to 90 days in advance of the planned flight and are checked against multiple airspace data sources to ensure better airspace coordination.

Cross-Border Issues As provided in par. 107.1 of Part 107, the Federal Aviation Regulations on small unmanned aircraft systems apply to the registration, certification, and operation thereof within the United States.

2.3 *The ICAO Model UAS Regulations*

The model UAS Regulations of the International Civil Aviation Organization (ICAO) were drafted following a request from member states to the ICAO, to develop a regulatory framework for UAS operating outside the Instrument Flight Rules (IFR) International area and they offer a template for member states to implement or supplement their existing UAS laws. Under the ICAO model UAS Regulations a UAS operation taking place according to the *standard unmanned aircraft operating conditions* (Section 101.7), under which the UAV is operated within the visual line-of-sight of the operator, at or below 120 m by day, and not within 30 m of a person not directly associated with the operation, not within a prohibited, restricted, over a populated area or within 4 km of the movement area of a controller airport, not operated over an area where a fire, police or other public safety or emergency operation is being conducted, and where the UAV operator does not operate another UAV, is not subject to any prior authorization.

2.3.1 Prior Authorization Requirements

General Rules Any UAS operations not falling into the scope of Section 101 of the model UAS Regulations (i.e., for UAVs exceeding a gross mass of 25 kg on takeoff) are subject to prior authorization or a prior UAS operator certificate (UOC) issued by the competent aviation authority, provided the requirements it sets out are met and any necessary fees are paid. The application for authorization must detail the name address for service in the country of the applicant, the details of the operation for UAS authorization or UOC, the application as required by the competent authority, and any other information the competent authority may require. Furthermore, considering the nature, degree and risk of the intended operation, the application

[71] Federal Aviation Administration, UAS Data Exchange (LAANC), available at: https://www.faa.gov/uas/programs_partnerships/data_exchange/, last accessed on 05.08.2022.

shall also address issues such as the identities of the persons that are primarily responsible for any part of the UAS operation, or who have or are likely to have control over the exercise of the privileges under the certificate, the physical locations to be used in the operations, an operational risk assessment,[72] reporting procedures to the competent authority, operating requirements, details of the UAV and the control system to be used, among others.[73]

Following receipt of an application according to the above, the competent authority may issue the requested UAS authorization or UOC, or opt to impose requirements or additional conditions, specify procedures to be followed by the UAS operator, or conclude that the drone to be used shall display identification markings to ensure aviation safety. The UAS authorization or the UOC issued shall include details of the physical location(s) of the holder's principal base of operations and address for service, a list of any business names under which the holder operates, the privileges and operations the holder is entitled to perform,[74] along with an indication of the date the authorization becomes effective and the exact duration of the authorization. The Model UAS Regulations also provide for procedures concerning amendments to the application for UAS authorization or UOC, and the filing of requests for renewal thereof.[75]

UAS Operations in Controlled Airspace Any UAS operations in controlled airspace are subject to prior authorization from the ATC that is responsible for that airspace. The interested UAS operator shall hold an appropriate qualification for the use of an aeronautical radio,[76] maintain a listening watch on a specified frequency (-ies), and broadcast on a specified frequency(-ies) and/ or maintain other ways of communication requested by the competent ATC.[77]

The Regulations also provide that a UAS operator may apply to the competent authority for the approval of a certain area of operation in general, or for a specific UAS category.[78]

---

[72] Identifying the known and likely consequences to hazards to people, property and other aircraft of the proposed operation, and including a description of mitigating measures to be implemented.

[73] See ICAO Model UAS Regulations, Section 102.24.

[74] Such as the number, type and description (including serial number and registration, if applicable) of every authorized drone, the geographical areas of approved operations, any exemption from any operational requirement(s), and any additional condition the competent authority deems necessary for aviation safety purposes—see Section 102.27 of the Model UAS Regulations.

[75] See Sections 102.34 and 102.35 thereof.

[76] In the sense of an aeronautical radio operator certificate, a remote pilot license (or flight crew license), an air traffic control license, a military qualification equivalent to any of the previously mentioned licenses, or a flight service license.

[77] ICAO Model UAS Regulations, Section 101.13.

[78] If such approval is granted, it shall have effect as of the date it was issued or a different date stated by the competent authority, and it shall remain valid for an either indefinite or specific period. The competent authority may also impose conditions on the approval, and it reserves the right to amend

2.3.2 Prior Notification Requirements

If an intended UAS operation is to take place more than 4 km from an aerodrome boundary and above 120 m Above Ground Level (AGL), such operation shall remain within uncontrolled airspace, in a segregated airspace designated for that purpose. Alternatively, at least 24 h before the operation an appropriate notification must be provided to the air navigation service provider (ANSP) for the issuance of a NOTAM, containing information on the name, address and telephone number of the UAS operator, the location of the proposed operation, the date, time and duration of the proposed operation and the proposed maximum height AGL for the operation.[79]

2.4 *Concluding Remarks*

It derives from the above analysis that the existing regulatory framework in both the EU and the USA does not support the operation of large drones, suitable for long-distance transports of goods or passenger transports. Particularly in the EU, although the certification requirements for such drones have been established by reference to the respective requirements for aircrafts, the lawful operation thereof is subject to further amendments of the applicable aviation regulations or the enactment of further rules. The European legislative procedure for such regulatory developments has already been launched, it should not, however, be expected to be completed soon as it is currently (as of Nov. 2021) at an initial, public consultation stage.

The same situation is encountered in the USA,[80] with the existing Federal Aviation Regulations regulating *small unmanned aircrafts*, thus excluding large drones appropriate for cargo and passenger transports. Further, although the ICAO Model UAS Regulations do not explicitly exclude large drones, it is doubtful that the prior authorization and notification procedures would suffice to adequately and safely regulate such operations. Thus, these sets of rules also need to be duly amended or supplemented to support any forthcoming developments in the cargo and passenger drone transport sector, while at the same time safeguarding aviation safety.

Furthermore, and considering the global character drone transports are expected to have, the divergence between the two main sets of rules, i.e., the EU regulation on the one hand being more detailed and exhaustive and, on the other, the US and ICAO rules providing more for operational requirements and general principles rather than

---

such conditions or even revoke the approval it has granted—see ICAO model UAS Regulations, Section 101.9.

[79] ICAO Model UAS Regulations, Section 101.27(c).

[80] U.S. Government Accountability Office, *Unmanned Aircraft Systems: Current Jurisdictional, Property, and Privacy Legal Issues Regarding the Commercial and Recreational Use of Drones*, available at: https://www.gao.gov/assets/b-330570.pdf, last accessed on 05.08.2022.

strict authorization procedures, cannot be overlooked. Compliance with both sets in the context and for the purposes of international cross-border activities by the interested UAS operators could result in a burdensome task, increasing the research and development costs, and possibly hindering the development of such a global market, in view also to any possible administrative and/ or criminal sanctions in case of non (full)-compliance.

In this relevance, and to ensure the efficient development and the safe operation of international drone transports, the competent regulatory and supervisory authorities will need to develop appropriate cooperation and coordination mechanisms, particularly in the field of inspections and sanctions, and provide adequate and efficient guidance for compliance to interested UAS operators.

3 Operational Requirements

Apart from the prior authorization and notification requirements, another major legal issue concerning UAS operations relates to the operational limitations, the obligations of the UAS operator, and the competence requirements and other obligations of remote pilots, as set out in the applicable regulations.

3.1 The European Union Legislation

3.1.1 Operational Limitations

"Open" Category UAS operations in the "open" category are divided into three sub-categories A1–A3, each of which is subject to different limitations and requirements.[81] With respect to altitude levels, drones may fly up to 120 m from the closest point of the surface of the earth, which may increase if the drone flies within a horizontal distance of 50 m from an artificial obstacle taller than 105 m, up to 15 m above the height of the obstacle.[82]

Depending on the sub-category, different limitations apply with respect to the distance the drone must keep from uninvolved individuals[83] or assemblies of people:

[81] According to Annex-Part A to the Implementing Regulation, points UAS.OPEN.020, UAS. OPEN.030, and UAS.OPEN.040 of Annex-Part A to the Implementing Regulation.

[82] Except for unmanned sailplanes (i.e., UAVs supported in flight by the dynamic reaction of the air against its fixed lifting surfaces, the free flight of which does not depend on an engine) with a MTOM of less than 10 kg, which may be flown at a distance above 120 m from the closest point in the earth's surface, provided they are not flown at a height greater than 120 m above the remote pilot.

[83] An uninvolved individual is considered to be a person not participating in drone operations, or not acquainted with the safety guidelines and precautions provided by the UAS operator, while an

in A1 operations the drone must not overfly assemblies of people or uninvolved persons, or solely assemblies of people,[84] while A2 operations must take place at a safe horizontal distance of 25–30 m from uninvolved persons (depending on the drone function), and A3 operations must be conducted in areas where it is reasonably expected that no uninvolved persons will be endangered, and at a safe horizontal distance of at least 150 m from residential, commercial, industrial or recreational areas.

"Specific" Category The "specific" UAS operations are subject to the operational limitations set out in the operational authorization thereof, and the mitigation measures proposed by the UAS operators. Where the "specific" UAS operation is subject to a prior declaration of compliance with a standard scenario, it shall be performed below 120 m from the surface of the earth, and either in uncontrolled airspace, or in controlled airspace following coordination and individual flight authorization.

Such UAS operations are also subject to the operational obligations outlined in Commission Implementing Regulation (EU) No. 923/2012,[85] according to which the operation of aircraft shall comply either with the VFR or the IFR. The applicable provisions also prohibit negligent or reckless operation of aircrafts in a way that may endanger life or property of others, as well as flights in prohibited or restricted areas, provide for specific conditions under which an aircraft may proceed with dropping or spraying, oblige the pilots to engage in collision avoidance maneuvers (as necessary), detail right-of-way rules, rules concerning lights to be displayed by an aircraft (depending on its type), rules on flight plans, among others.

3.1.2 UAS Operator Obligations

Registration Obligations UAS operators are subject to registration obligations in registers maintained by the national competent authorities of Member States, if they perform "open" category operations with the use of drones with a MTOM exceeding 250 g, or which can transfer to an individual kinetic energy exceeding 80 J in case of

"involved" individual is considered to be the one participating in the activity, being aware of the risk, and being able to check the drone's location during flight. Therefore, for an individual to be considered as "involved," they must have granted his/her consent to participate in the activity (i.e., to have the drone flying over them), receive safety guidance and precaution measures to be applied in emergency situations from the drone operator, and not be engaged in other activities, so that they may monitor the drone's location and take measures to avoid impact with the drone in case of any incident. See also the relevant Q&As from the Hellenic Civil Aviation Authority, available at: https://uas.hcaa.gr/Faq, last accessed on 05.08.2022.

[84] If the drone has a maximum take-off mass of less than 25 kg, including its payload.

[85] Commission Implementing Regulation (EU) No. 923/2012 of 26 September 2012 laying down the common rules of the air and operational provisions regarding services and procedures in air navigation

impact, or equipped with a sensor to capture personal data, as well as "specific" UAS operations. The registers include information on identity and contact details, insurance policies UAS operators keep (if required so under applicable law), a confirmation that all personnel involved are competent to perform their tasks, as well as information on any operational authorizations, certifications, etc. the operator holds. UAS operators are provided with a unique digital registration number, which they must display on every drone falling into one of the categories that trigger the operator's registration obligation.

"Open" Category Part A of the Annex to the Implementing Regulation provides additional obligations of the UAS operators in relation to "open" UAS operations. Particularly, they must develop operational procedures proportionate to the risks deriving from the operations, ensure that all operations effectively use and support the efficient use of radio spectrum, designate a remote pilot for each UAS, ensure that remote pilots and all other personnel performing supporting tasks are familiar with the user's manual of the UAS manufacturer and the UAS operator's procedures, and are adequately trained, update the information into the geo-awareness system (where applicable), and ensure that the UAS being used are in compliance with any technical requirements under the Delegated Regulation.[86] Furthermore, and particularly in case of A2 or A3 operations, the UAS operator must ensure that all persons involved that are present in the area of the operation, have been duly informed of the risk and have explicitly agreed to participate.

"Specific" Category Annex-Part B to the Implementing Regulation respectively provides responsibilities of the UAS operators that are active in "specific" UAS operations. The UAS operator shall establish appropriate procedures and limitations[87] aiming to ensure safety of the operations, designate a remote pilot for each operation or ensure that the pilot's obligations before and during flight are properly allocated in case of autonomous operations, ensure that all operations effectively use and support the efficient use of radio spectrum, as well as compliance of the remote pilots and other personnel[88] with the necessary requirements, and compliance of the operations with the limitations and conditions provided in the operational authorization. They shall further keep a record of information on the UAS operations, as required by the operational authorization, use appropriately designed UAS and maintain them in a suitable for the operations condition.

[86] Namely, the technical requirements concerning the classes defined in Parts 1–5 of the Delegated Regulation.

[87] Including operational procedures to ensure the safety and security of the operations, measures to protect against unlawful interference and unauthorized access, procedures to ensure compliance with the General Data Protection Regulation (GPDR) (EU) No. 2016/679, and guidelines for its remote pilots to plan UAS operations in a manner that minimizes nuisances to people and animals.

[88] Such as competency and training requirements, understanding of the UAS operator's operational manual, and information on any designated for safety, security, privacy or environmental reasons geographical zones concerning the intended UAS operation.

3.1.3 Remote Pilot Requirements and Obligations

The Implementing Regulation provides for general requirements and obligations for remote pilots regardless of whether they operate in the "open" or in the "specific" category of UAS operations, namely, remote pilots shall be at least 16 years old, except if operating in the A1 "open" subcategory, or with privately-built UAS with a MTOM of up to 250 g, or under the direct supervision of a remote pilot complying with the Implementing Regulation.[89]

Competency Requirements Remote pilots must also comply with the competency requirements under Annex-Part A of the Implementing Regulation with respect to "open" UAS operations, and Part B thereof as far as "specific" UAS operations are concerned.

For "open" UAS operations, the remote pilot shall be familiarized with the user's manual provided by the UAS manufacturer, and:

- In A1 operations performed by a class C1 drone and A3 operations, have completed an online training course, followed by successful completion of an online theoretical knowledge examination provided by the national competent authority or a recognized entity on subjects including air safety, airspace restrictions, aviation regulation, privacy, operational procedures, UAS general knowledge, insurance;
- In A2 operations hold a remote pilot competency certificate issued by the national competent authority or a duly recognized entity,[90] and proceed with a self-practical training in the A3 subcategory operating conditions, and an additional theoretical knowledge examination by the competent authority or a recognized entity aimed at assessing the pilot's knowledge of technical and operational mitigations of risk;

Operational Obligations Remote pilots are further subject to operational obligations both before and during the flight; namely, before the UAS operation, the remote pilot shall obtain from the UAS operator (or the competent authority) updated information concerning the intended UAS operation and any prohibited or restricted geographical zones, observe the operating environment (including any obstacles) and ensure that it is compatible with any authorized or declared limitations and conditions (in case of "specific" UAS operations), ensure that the UAS's condition is

[89]The Implementing Regulation also grants Member States the possibility to lower the minimum age, based on a risk-based approach, by up to four years for pilots operating in the "open" category, and by up to two years for pilots operating in the "specific" category. However, in such case, said pilots will only be able to operate within the specific Member State. For instance, in Greece, it is stated that no minimum age limited applies in A1 "open" UAS operations in which drones with in the C0 class are being used—see Hellenic Civil Aviation Authority at https://uas.hcaa.gr/Faq, last accessed on 05.08.2022.

[90]Following an online training course and an online theoretical knowledge examination as for A1 and A3 sub-category operations.

appropriate to safely complete the intended operation, and check if the remote identification properly functions (if applicable). In "open" UAS operations, the remote pilot must also ensure that the drone's mass does not exceed the MTOM limitations set out by its manufacturer or its class if an additional payload is fitted, while in "specific" UAS operations the pilot must ensure that the information on the operation has been made available to the relevant air traffic service (ATS) unit, other airspace users and relevant stakeholders, as required by the operational authorization or the standard scenario.

During the flight, remote pilots shall not perform their duties under the influence of alcohol or psychoactive substances and shall refrain from performing them if unfit due to injury, fatigue, medication, sickness or other causes. They shall also comply with any limitations concerning prohibited or restricted geographical zones, the UAS operator's procedures, and refrain from areas where an emergency response effort is ongoing.[91] In "open" UAS operations, remote pilots shall further keep the drone in VLOS, maintain a thorough visual scan of the airspace surrounding the drone, discontinue the flight if it poses risk to another aircraft, people, animals, environment or property, be able to maintain control of the drone, and operate the UAS in accordance with the manual provided by the manufacturer. Respectively, remote pilots in "specific" UAS operations shall comply with any limitations specified in the operational authorization or the declared standard scenario, avoid any risk of collision with any manned aircraft, and discontinue a flight that may pose any risk.

3.2 *The USA Federal Aviation Regulations*

3.2.1 Operational Limitations

Part 107 of the Federal Aviation Regulations provides for a series of operational limitations to UAS operations falling into scope, both similar and different to the ones established under the applicable EU Regulations. For instance, it is provided that small UAS shall not be operated in a reckless or careless manner that may endanger the life or property of another, or allow an object to be dropped from a small UAV in a way creating undue hazard to another,[92] while small drones must not carry hazardous materials.[93] It is further provided that drones shall not be operated in a way that interferes with air traffic operations and patterns at any airport, heliport or seaplane base, or in a prohibited or restricted areas or in certain airspace classes, unless explicitly authorized, or in a way not compliant with flight restrictions in the proximity of certain areas, designated by notice to airmen.[94]

---

[91] Unless permitted to do so by the responsible emergency service.

[92] Par. 107.23.

[93] Par. 107.36.

[94] Par. 107.41–107.47.

Furthermore, drones may not be flown over individuals, unless such individuals are directly participating in the operation or located under a covered structure or a stationary vehicle,[95] or so close to other aircrafts that may create a collision hazard. Drones must also yield the right of way to any other aircrafts.[96] Operational limitations are also established in terms of the time during which drones may be flown, as daylight operation is the rule and flight during twilight hours are permitted under strict conditions.[97] It is also provided that drones must not be flown from a moving aircraft or a moving land or waterborne vehicle.[98]

Apart from the above, operating limitations also include maximum speed (100 miles/h), maximum altitude (400 ft), minimum flight visibility (no less than 3 statute miles), and minimum distance from clouds.[99] The remote pilot must be also able to see the drone (without visual aids) throughout the entire flight to know its location, and observe the airspace for any hazards, either himself or with the use of a visual observer.[100]

3.2.2 UAS Operator and/or Remote Pilot Obligations

UAS operators and remote pilots must ensure and be able to demonstrate compliance of the UAS operations with the applicable provisions, and to provide the competent authorities with any documentation, information, inspection right, etc. that is necessary to prove such compliance.[101] Furthermore, they shall ensure that the UAS is in a condition allowing for safe operation and cease the flight upon any indication that this is not the case.[102]

UAS operators shall designate a remote pilot before or during the flight, with a pilot not being permitted to act as such in the operation of more than one drones at the same time.[103] Before the flight, the remote pilot must assess the operating environment,[104] ensure that all participants are informed about the operating conditions, check the functionality of the control links, ensure the drone has enough available power for the operation, and ensure that any object attached to the drone

[95] Par. 107.39.

[96] Par. 107.37.

[97] Particular provisions apply for Alaska—see par. 107.29.

[98] Par. 107.25.

[99] Par. 107.51.

[100] Meaning a person designated by the remote pilot in command to assist him/ her and the person manipulating the flight controls of the drone, to see and avoid other air traffic or object aloft or on the ground.

[101] Par. 107.7.

[102] Par. 107.15.

[103] Par. 107.35.

[104] Including weather conditions, local airspace and flight restrictions, location of persons and properties in the ground, other ground hazards—see par. 107.49.

is secure. The remote pilot shall also refrain from his/her tasks in case of a physical or mental condition that may interfere with the safety of the operation, and not be subject to the influence of drugs or alcohol.[105]

During the flight, the remote pilot is directly responsible for the operation, and shall ensure that no risk is caused to any person, property, or other aircraft, and that the operation takes place in accordance with the applicable regulations.[106] In case of an accident, the remote pilot shall file a relevant report to the competent authority, within 10 calendar days.[107]

Registration and Certification Requirements UAS operators must hold an appropriate and current airworthiness certificate, having the registration number assigned to the relevant aircraft.[108] At the same time, remote pilots are obliged to hold a remote pilot certificate, issued in accordance with the provisions of Subpart C of Part 107, and provided they comply with all the other certification requirements of said Subpart C.[109]

3.3 The ICAO Model UAS Regulations

The ICAO model UAS Regulations provide for similar operational limitations and flight obligations to the UAS operators and remote pilots as the USA Federal Aviation Regulations. The following divergences are indicatively noted:

- Accident reporting obligations provide for a shorter deadline of 48 h after the incident;
- Night operations are permitted, provided they take place indoors or in a shielded operation;
- UAS operators are obliged to implement a safety management system.

---

[105] Par. 107.17 and 107.27.

[106] Par. 107.19—Except in case of an in-flight emergency, in which case the remote pilot may deviate from any applicable rule, and report such deviation to the competent authority, upon relevant request (par. 107.21).

[107] In case the UAS operation involved serious injury to any person or loss of consciousness, or damage to any property, unless the cost of repairs does not exceed $500 or the fair market value of the property does not exceed $500 in the event of total loss (par. 107.9).

[108] See par. 107.13 and 91.2013.

[109] Including not having any conviction for offences involving alcohol or drugs, be at least 16 years old, be able to speak, read, write and understand the English language, not have (to their knowledge) a mental or physical condition that would interfere with the safe operation of the drone, and demonstrate aeronautical knowledge, by passing the necessary initial and recurrent tests.

3.4 Concluding Remarks

Although the existing rules do not regulate the operation of large cargo and passenger drones, they can still provide guidance as to the main operational requirements that will need to be met for such future operations. The cross-border nature of drone transport operations, in conjunction with the segregation of the regulation that is currently being observed, could further hinder the use of drones in the transport sector, particularly if the administrative costs to issue and maintain the necessary national/supranational (EU)/federal registrations and/or remote pilot licenses would become significantly high.

Furthermore, issues arising from other applicable legislation would also need to be considered. Such would be the applicable data protection laws, obliging UAS operators, among others, to adopt and implement appropriate policies and procedures regulating their data collection and processing activities, to identify and implement appropriate legal bases for each activity, to appoint a Data Protection Officer, or to conduct data protection impact assessments (when required under the applicable provisions). UAS operations should be designed and implemented taking account of the privacy by design and by default principles, adopting appropriate technical and organizational measures to implement data protection principles, both when determining the means for processing and during the processing itself (*by design*),[110] and to ensure that, by default, only personal data necessary for each specific processing purpose are being processed (*by default*).[111] UAS operators (or other data controllers) would also need to think out of the box with regard to their obligation to duly inform affected individuals of the processing of their personal data,[112] considering that traditional notice mechanisms (privacy notices, written forms, etc.) would be inadequate to provide proper transparency and control over the personal data processing activities, particularly in cases where a drone would be flying over residential areas, event spaces, etc.,[113] so as to ensure that the

[110] The privacy design notion, explicitly referred to in Art. 25 of the GDPR, was originally used by Ann Cavoukian, Information and Privacy Commissioner in Ontario, Canada, and was broken down in the so-called "seven foundation principles": (1) Proactive not Reactive, Preventative not Remedial; (2) Privacy as the Default Setting; (3) Privacy Embedded into Design; (4) Full functionality—positive sum, not zero sum; (5) End-to-end security—full lifecycle protection; (6) Visibility and transparency—keep it open; (7) Respect for user privacy—keep it user centric. See in this respect: https://www.ipc.on.ca/wp-content/uploads/2018/01/pbd.pdf, last accessed on 05.08.2022.

[111] See European Data Protection Board (2020), Guidelines 4/2019 on Article 25 Data Protection by Design and by Default, adopted on 20.10.2020, and available at: https://edpb.europa.eu/sites/default/files/files/file1/edpb_guidelines_201904_dataprotection_by_design_and_by_default_v2.0_en.pdf, last accessed on 05.08.2022.

[112] In line with the general principle of transparency which, although not explicitly defined in GDPR provisions, constitutes an overarching principle, applying irrespectively of the applicable legal basis and throughout the processing activities.

[113] For instance, if an event takes place on a beach, and is being videotaped with the use of a drone, both data minimization and information issues would arise towards the collection of images of other individuals on the same beach, but not participating in the event.

information solutions being adopted provide concise, transparent, intelligible and easily accessible information to the data subjects.[114]

Any applicable provisions regulating system and network safety would also be of relevance, while if radio frequencies are used, UAS operators would also need to comply with the applicable provisions concerning proper allocation and use of such limited resources under applicable national laws. Considering that, as of now, such allocation takes place at a national level, this could pose additional obstacles to the wider use of cross-border, international drone transport activities.

4 Summary and Conclusions

Drones have largely escaped their initial military use and have increasingly been introduced into daily activities, including in the transport and delivery of goods. With the COVID-19 era acting as additional accelerator for technological advances, the notion of Advanced Air Mobility (AAM) is expected to rapidly develop and lead into innovative, transformative airborne solutions in the transportation of people and goods. At the same time developments in Artificial Intelligence technology will lead to increased automation levels in future UAS solutions for transportations.

Along with new solutions, come new legal issues and concerns relating to aviation safety, security of people, properties and the environment, privacy, e-communications, liability, insurance, risk sharing, oversight, regulatory/ supervisory capacity and many more, which will need to be duly addressed before the integration of drones in large cargo and passenger transports, thereby avoiding the drowning of a newly emerging and indeed boosting activity.

On prior authorization requirements, the existing regulations both in the EU and the US do not support the operation of drones suitable for large or long-distance cargo and passenger transports, creating the need for amendments or enactment of new legislation. Apart from any new regulation that may be introduced, the global nature of particularly cargo transports will require interested UAS operators to ensure that any devices, systems and operations are designed and conducted in line with different and diverging sets of rules, thus increasing both the time needed for research & development and roll-out, and the related costs.

At the operational side of things, existing laws can still provide guidance for the main operational requirements that would have to be complied with, including registration and identification requirements, remote pilot certifications and licensing, pre- and in-flight obligations, etc. As with authorization requirements, the cross-border, international nature of the transports sector would render the exercise of

[114] See in this regard Working Party 29, Guidelines on transparency under Regulation 2016/679, 17/EN, WP260 rev.01, adopted on 29.11.2017 and last revised and adopted on 11.04.2018, as endorsed by the EDPB, available at: https://ec.europa.eu/newsroom/article29/items/622227, and last accessed on 05.08.2022.

ensuring compliance of drone solutions with all applicable regulations a quite burdensome and even costly one.

Apart from drone-related specific regulation, other laws would also need to be considered during the design and implementation of any new solutions. Drone devices, control links, systems, operational policies and procedures need to be designed with a view to the data protection principles, aiming to ensure that any and all data processing activities take place in a lawful and transparent towards any affected individuals way and that no personal data are being collected and processed by UAS operators, other than the necessary ones. UAS operations will also need to be conducted in a way that ensures compliance with network and system safety provisions, as well as in line with regulations on the use of radio frequencies, numbering and any other relevant limited operational resources regulations.

With the development of new unmanned, and increasingly automated transportation systems and solutions being increasingly discussed, the question on the existence of laws and regulations permitting and even supporting their use still remains to be answered. In the absence of a concise regulatory framework setting out the exact certification, authorization and operational requirements that will need to be fulfilled by UAS operators, the transport sector cannot begin to design the adoption of such new solutions and its adaptation to the new reality that will come. Further to the simple enactment of new provisions, the international nature of drone transports, along with the significant transformation in airborne transports they are expected to cause, also call for further cooperation and coordination between the competent authorities and legislators to ensure high level of aviation safety and the efficient integration of new technologies in air transports.

Regulatory Guidance and Decisions

Canadian Aviation Regulations. Available at: https://lois-laws.justice.gc.ca/eng/regulations/SOR-96-433/FullText.html#s-900.01

Commission Delegated Regulation (EU) 2019/945 of 12 March 2019 on unmanned aircraft systems and on third-country operators of unmanned aircraft systems.

Commission Implementing Regulation (EU) 2019/947 of 24 May 2019 on the rules and procedures for the operation of unmanned aircraft.

Commission Implementing Regulation (EU) 2016/1185 of 20 July 2016 as regards the update and completion of the common rules of the air and operational provisions regarding services and procedures in air navigation (SERA Part C).

Commission Regulation (EU) 2015/640 of 23 April 2015 on additional airworthiness specifications for a given type of operations.

Commission Regulation (EU) No. 1321/2014 of 26 November 2014 on the continuing airworthiness of aircraft and aeronautical products, parts and appliances, and on the approval of organizations and personnel involved in these tasks.

Commission Implementing Regulation (EU) No. 923/2012 of 26 September 2012 laying down the common rules of the air and operational provisions regarding services and procedures in air navigation.

Commission Regulation (EU) No. 748/2012 of 3 August 2012 laying down implementing rules for the airworthiness and environmental certification of aircraft and related products, parts and appliances, as well as for the certification of design and production organizations.

European Data Protection Board (2020) Guidelines 4/2019 on Article 25 Data Protection by Design and by Default, adopted on 20.10.2020. Available at: https://edpb.europa.eu/sites/default/files/files/file1/edpb_guidelines_201904_dataprotection_by_design_and_by_default_v2.0_en.pdf

Federal Aviation Regulations. Available at: https://www.faa.gov/regulations_policies/faa_regulations/

Hellenic Civil Aviation Authority, Questions and Answers on the use of drones. Available at: https://uas.hcaa.gr/Faq

Information and Privacy Commissioner of Ontario (2018) Privacy by design – seven foundational principles, 01.2018. Available at: https://www.ipc.on.ca/wp-content/uploads/2018/01/pbd.pdf

International Civil Aviation Organization, Model UAS Regulations and Advisory Circulars. Available at: https://www.icao.int/safety/UA/Pages/ICAO-Model-UAS-Regulations.aspx

Regulation (EU) 2018/1139 of the European Parliament and of the Council of 4 July 2018 on common rules in the field of civil aviation and establishing a European Union Aviation Safety Agency.

Working Party 29, Guidelines on transparency under Regulation 2016/679, 17/EN, WP260 rev.01, adopted on 29.11.2017 and last revised and adopted on 11.04.2018, as endorsed by the EDPB. Available at: https://ec.europa.eu/newsroom/article29/items/622227

References

Aerospace Industries Association (2018) Think bigger – large unmanned systems and the next major shift in aviation. Available at: https://www.aia-aerospace.org/uasreport2018/

Airhub (2021) UAS operations in the open category. Available at: https://dronelife.com/2021/05/12/european-drone-regulations-easa-basic-regulation-and-whats-next/

Amazon, Prime Air delivery option. Available at: https://www.amazon.com/Amazon-Prime-Air/b?ie=UTF8&node=8037720011

Association for Unmanned Vehicle Systems International, Danish Technological Institute (2019) Global trends of unmanned aerial systems. Available at: https://www.auvsi.org/global-trends-unmanned-aerial-systems

ATP Flight School, Types of airspaces and how are they defined. Available at: https://atpflightschool.com/become-a-pilot/flight-training/airspace-classes.html

Built In (2021) Drones. What is a drone? What are uses for drones? As available on 9 Nov 2021 at: https://builtin.com/drones

Business Insider – Insider Intelligence (2021) Drone market outlook in 2021: industry growth trends, market stats and forecast. Available at: https://www.businessinsider.com/drone-industry-analysis-market-trends-growth-forecasts

Del Cerro J, Cruz Ulloa C, Barrientos A, de León Rivas J (2021) Unmanned Aerial Vehicles in agriculture: a survey. Agronomy 11:203. https://doi.org/10.3390/agronomy11020203

Deloitte (2018) Unmanned Aircraft Systems (UAS) risk management: thriving amid emerging threats and opportunities. Available at: https://www2.deloitte.com/us/en/pages/public-sector/articles/drone-risk-assessment.html

Drone Laws (2021) Drone laws in the USA. Available at: https://drone-laws.com/usa-drone-laws-in-usa/

Earls A, Lutkevich B (2019) Definition of "drone." Available at: https://internetofthingsagenda.techtarget.com/definition/drone

Eisenbeiss H (2004) A mini Unmanned Aerial Vehicle (UAV): system overview and image acquisition. Available at: https://citeseerx.ist.psu.edu/viewdoc/download;jsessionid=115A5B4916F3005B45E7EEEB50C5D3C6?doi=10.1.1.145.4342&rep=rep1&type=pdf

European Commission (2021) Roadmap – A Drone Strategy 2.0 for Europe. Available at: https://ec.europa.eu/info/law/better-regulation/have-your-say/initiatives/13046-A-Drone-strategy-20-for-Europe-to-foster-sustainable-and-smart-mobility_en

European Union Aviation Safety Agency (2018) Opinion No. 01/2018, Introduction of a regulatory framework for the operation of unmanned aircraft systems, in the "open" and "specific" categories. Available at: https://www.easa.europa.eu/sites/default/files/dfu/Opinion%20No%2001-2018.pdf

European Union Aviation Safety Agency, Certified category – civil drones. Available at: https://www.easa.europa.eu/domains/civil-drones-rpas/certified-category-civil-drones

Federal Aviation Administration, How to register your drone. Available at: https://www.faa.gov/uas/getting_started/register_drone

Federal Aviation Administration, UAS remote identification overview. Available at: https://www.faa.gov/uas/getting_started/remote_id/

Federal Aviation Administration, UAS Data Exchange (LAANC). Available at: https://www.faa.gov/uas/programs_partnerships/data_exchange/

Herrmann M (2017) A comparison of Unmanned Aerial Vehicle regulations in the United States and Europe. In: 53rd ASC annual international conference proceedings

Hogan Lovells, Drones in German skies: new EU regulations take flight. Available at: https://www.hoganlovells.com/-/media/hogan-lovells/pdf/2020/pdfs/2020_06_01_iplr77_drones-in the-german-skies-new-eu-regulations-take-flight.pdf

Kersley A (2021) The slow collapse of Amazon's drone delivery dream. Available at: https://www.wired.co.uk/article/amazon-drone-delivery-prime-air

Neenan P (2021) UK drone regulations summarised by Peter Neenan in Getting the Deal through. Available at: https://www.stewartslaw.com/news/uk-drone-regulation-2022-in-lexologys-getting-the-deal-through/

Outay F, Mengash HA, Adnan M (2020) Applications of Unmanned Aerial Vehicle (UAV) in road safety, traffic and highway infrastructure management: Recent advantages and challenges. Transp Res A 141:116–129

Pricewaterhouse Coopers (2018) Flying high: drones to drive jobs in the construction sector. Available at: https://www.pwc.in/assets/pdfs/publications/2018/flying-high.pdf

Studio Pierallini - Francesco Grassetti, Drone regulation: European Union. Available at: https://www.lexology.com/library/detail.aspx?g=10ac2457-86ad-44d5-8eb1-bfc674951fc6

Thiels C, Aho J, Zietlow S, Jenkins D (2015) Use of unmanned vehicles for medical product transport. J Air Med Transport 34(2):104–108. https://doi.org/10.1016/J.amj.2014.10.011

U.S. Government Accountability Office, Unmanned Aircraft Systems: current jurisdictional, property, and privacy legal issues regarding the commercial and recreational use of drones. Available at: https://www.gao.gov/assets/b-330570.pdf

UAV Coach, Drone laws in the United States of America. Available at: https://uavcoach.com/drone-laws-in-united-states-of-america/

UPS (2021) UPS operates first ever U.S. drone COVID-19 vaccine delivery. Available at: https://about.ups.com/us/en/our-stories/innovation-driven/drone-covid-vaccine-deliveries.html

Vincent J (2021) Alphabet's drone delivery service Wing hits 100,000 deliveries milestone. Available at: https://www.theverge.com/2021/8/25/22640833/drone-delivery-google-alphabet-wing-milestone

Watts AC, Ambrosia VG, Hinkley EA (2012) Unmanned aircraft systems in remote sensing and scientific research: classification and considerations of use. Remote Sens 4:1671–1692. https://doi.org/10.3390/rs4061671

Wimberley B (2021) Drones in healthcare: blood transportation and beyond. Available at: https://healthcaretransformers.com/healthcare-business/drones-healthcare/

Wing (2021) Frequently asked questions about delivery. Available at: https://wing.com/about-delivery/

Zoldi D (2021) Drone life – drones are flying high in Europe. Available at: https://dronelife.com/2021/05/12/european-drone-regulations-easa-basic-regulation-and-whats-next/

The Civil Unmanned Aerial Vehicle (UAV) Law of China: A Comparative Study of the Mainland, Hong Kong, and Macao

Can Luo

Abstract Civil unmanned aerial vehicle (UAV) is a national strategic emerging industry in China. The *Made in China 2025* formulated by China State Council requires to promote the rapid development of UAV industrialization. Although it started late, China's civil UAV industry has made rapid progress. However, it also brings a series of legal problems as well. How to promote the long-term and healthy development of the industry and effectively supervise it has always been the theme and foothold of China's legislation.

China has highlighted the legislative supervision of UAVs and made a series of legislative achievements in three different regions. Both Hong Kong and Macao of China have made relative systematic norms to regulate the operation of small UAVs and remote-control drivers, while China's mainland is still committed to drafting further regulations. The laws and regulations in the above three regions of China mainly focus on the effective supervision of UAV safe flight, and each has its own characteristics. With the drafting of the legislative supervision of UAVs in China's mainland, legislators have been attempting to promote industrial development while ensuring safety of supervision. How to find the best balance between the two may become the focus of UAV legal reform in the future.

Keywords UAV · Supervision · Safe flight · Legal liability

1 Introduction

China's civil drone development has great potential and the drone market is growing rapidly. With the maturity of industrial chain and the development of UAV technology, China's civil UAV industry has shifted from simple production, processing, and manufacturing to independent research and development, and is in the forefront of civil UAV manufacturing in the world. With the mutual integration of information

C. Luo (✉)
School of Artificial Intelligence and Law, Southwest University of Political Science and Law, Chongqing, China

K. Noussia, M. Channon (eds.), *The Regulation of Automated and Autonomous Transport*, https://doi.org/10.1007/978-3-031-32356-0_4

technologies such as big data, Cloud Computing and mobile Internet and UAV technology, the civil UAV industry chain has turned from "vertical progress" to "horizontal progress." In recent years, the global drone market has maintained a high growth rate, especially the commercial drone market. "The Drone Market Report 2020–2025" published by Drone Industry Insights, a German drone research company, shows that the global drone market size was approximately US$18 billion in 2019 and is expected to double to US$42.8 billion by 2025.[1] According to the annual data of Economic Census from the National Bureau of Statistics (NBS) of China, China has 3,088,300 civil drones. In 2018, the Ministry of Industry and Information Technology of China stated in the "Guidance on Promoting and Regulating the Development of Civil Drone Manufacturing"[2] that "By 2020, the civil drone industry will continue to develop rapidly, with the output value reaching 60 billion RMB with an average annual growth rate of more than 40%... By 2025, the civil drone output value is expected to reach 180 billion RMB, with an average annual growth rate of more than 25%." China's drone industry maintained a fast growth rate in 2019, and the transaction size of its industry has exceeded 50 billion RMB. According to the NBS, the added value of the high-tech manufacturing industry grew by 7.1% in 2020 compared to the previous year, 4.3% points higher than the average level of all regulated industries. Among them, civil drones have achieved rapid growth, with the growth rate of more than one time.[3]

Civil drones are widely used since drones have powerful functions and currently the supply does not meet the demand. Generally, civil drones are divided into industrial-grade drones and consumer-grade drones. The functions of industrial-grade drones, such as aerial photography, wildlife detection,[4] power intelligent inspection,[5] drone patrols,[6] logistics, aerial remote sensing images, are unimaginable before. In addition, drone technology in traditional fields is also developing. For instance, the development of the agricultural industry is inseparable from the application of drone technology in China. The use of small unmanned helicopters for agricultural plant protection has many advantages, such as the low operating height, low drift, the ability to hover in the air, no need for special landing and take-off airports, the downward airflow generated by the rotor that helps to increase the penetration of the fog stream on the crop for high control effect, remote-control operation that helps spraying operators avoid the danger of exposure to pesticides and improve the safety of spraying operations.[7] Further analysis shows that with the mutual integration of drone technology and information technology such as big data, Cloud Computing, mobile Internet, the civil drone industry started to promote the

[1] Schroth (2020).

[2] The Ministry of Industry and Information Technology (2017).

[3] Lushan (2021).

[4] Wang (2021).

[5] Xu (2021).

[6] Lin (2021).

[7] Askci Consulting Co., Ltd (2019).

development of the overall national economy comprehensively and promote the development of emerging industries to meet the needs of social and economic activities.[8]

Chinese consumer drone technology is leading the way. In the consumer drone sector, Chinese drone technology is at the leading level and has a dominant market share. In 2016, a report on the North American drone market by brand sales in *Skylogic* showed that two of the top three drone companies in the North American drone market were from China. Chinese drone company DJI has secured 50% of the share, while the third ranked drone manufacturer is also from China—Yuneec Drones, with 4% of the market share. China has the largest market share of consumer drones in the world, accounting for seven of the top ten civil drone companies in the world. Among them, DJI, the leading consumer drone company, has an absolute advantage in both technology and market with a 69% share of the global drone market in 2019. Statistics of the Fifth Drone Market Sector Report data of *Droneanalyst* shows that,[9] even in 2020, although suffering from the U.-S. government, DJI still occupies 54% of the global consumer drone market share, ranking first in the global civil drone companies.

It cannot be ignored that the potential problems of Chinese civil drones are also highlighted. First, there are insufficient legal provisions on the supply side, and the law in the regulation of drones is lagging (Yang and Hao 2017). In the face of the emerging drone market, the current legal provisions can hardly meet the market demand for civil drones. As civil drones are easy to transform, it gives rise to chaotic phenomenon and potential hazards, which have intensified the demand for relevant legal regulations. In addition, the lack of safety requirements for civil drone products and the lack of regulation to guide the manufacturing and modification of civil drones are also obvious problems. Particularly, "in recent years, the number of drones in China has grown explosively, which has brought about a series of contradictions and problems while promoting economic and social development and national defense and military construction. On the whole, the management of drones in China is in a state of 'two lacks' (lack of primary responsibility, lack of laws and regulations), 'three chaos' (chaos in manufacturing, chaos in sales, chaos in using) and 'five difficulties' (difficulty of ground control, difficulty of air discovery, difficulty of active disposal, difficulty of traceability, difficulty of crackdown on punishment) (Li 2019). In particular, drones used for entertainment is not under supervision, which disturb military and civil aviation flight frequently, threatening the flight safety of military and civil aviation. There is an urgent need to formulate top-level laws and regulations to standardize and guide them, so as to better safeguard national security, public safety, and flight safety, and promote the rapid and healthy development of the drone's industry and related fields."[10]

[8] Askci Consulting Co., Ltd (2019).

[9] Benowitz (2021).

[10] Notes on the Drafting of the *Interim Regulations on the Flight Management of Unmanned Aerial Vehicles (Draft Submitted for Review).*

2 Legal Provisions on Civil Drones in Mainland China

With the development and maturity of the drone industry, the mainland of China has followed the standards and recommended measures of the International Civil Aviation Organization in recent years, and has made its first foray into domestic legislation. The Civil Aviation Administration of China has successively issued four major departmental regulatory documents aimed at civil drones. The "Regulations on the Operation of Small and Light UAVs (for Trial Implementation) (AC-91-FS-2015-31)" governs the operation and management of slow, light, and small drones flying at low-altitude. The "Regulation on Civil UAV Pilot Management (AC-61-FS-2018-20R2)," govern the qualification of civil drones' pilots. The "Regulations on the Administration of Real-Name Registration of Civilian UAVs (AP-45-AA-2017-03)" governs the registration of civil drones. The "Measures on the Administration of Air Traffic of Civilian UASs (MD-TM-2016-004)" governs the management of civil drones' aerial flights. Generally, the basic issues and terminology that are urgently needed for the large-scale application of civil drones are initially defined and international standards are introduced, but there are obvious problems such as the localization of legislation and its lagging behind social development; there are also shortcomings in the convergence of the specific contents of different norms; and the legal norms is of low effectiveness. On January 26th, 2018, the office of the air traffic control committee of China State Council and the Central Military Commission organized the drafting of the Interim Regulations on Flight Management of Unmanned Aircraft (Draft for Solicitation of Comments). At present, the Ministry of Industry and Information Technology are soliciting comments on the official website. These regulations are regarded as a breakthrough since there are no laws and regulations on UAV industry at the national level. And the content not only covers the human, aircraft, environment (airspace), and management (operation) of UAV, but also puts forward reasonable standards for supervision. Generally, the management of civil drones is implemented with reference to the management of manned aircraft, and can be broadly divided into four main categories of institutions: the Department of Flight Standards, the Department of Aircraft Airworthiness Certification, the Air Traffic Management Authority, and the Market Operations Management Agency (Wang 2021b).

2.1 Legal Provisions for the Manufacturing of Civil Drones

According to the WeChat team of the Airworthiness Technology Research and Management Center of the Civil Aviation Administration of the Ministry of Industry and Information Technology (MIIT), "since 2012, MIIT has initiated a study on the Basic Conditions for Civil Drone Development Units and Evaluation Methods regarding the access of drone enterprises. This study, led by the China Institute of Comprehensive Aviation Technology, aims to regulate the order of competition in the civil drone manufacturing market by evaluating the basic conditions of civil drone development units, guiding the basic resources and capacity needs of the industry as well as the development direction of resource allocation, technical research and management level, and promoting the healthy and rapid development of the domestic civil drone industry."[11] This regulation aims to regulate the low access threshold for civil drones, insufficient technical standards, and problems made by many companies that do not have the conditions that they should, also producing drones, resulting in uneven drone products in the market. However, this specification, which entered the technical discussion phase, was not completed as expected. In retrospect, the approach of restricting manufacturers with access conditions was less than ideal and created a strong technical barrier that killed the prospects of some small start-up drone developers.

In 2017, the Department of Equipment Industry, affiliated to the Ministry of Industry and Information Technology, issued the "Guidance on Promoting and Regulating the Development of Civilian Drone Manufacturing Industry" (hereinafter referred to as the "Guidance"), which emphasized the need to accelerate the development of advantageous enterprises, "study the formulation of industry standard conditions for civil drone production enterprises, publish the list of enterprises meeting the standard conditions, adjust dynamically at an appropriate time, and guide social resources to concentrate on enterprises meeting the standard conditions."[12] The guidance also specifically proposes the establishment of a perfect standard system, suggesting "accelerate the formulation of urgent standards such as classification of civil drones, product safety requirements, identity coding rules, basic conditions for development units and evaluation methods, and the construction of control platforms, etc. All in all, it is important to establish a sound system of standards for civil drones."[13] At present, for drone manufacturing enterprises in the mainland of China, it is not simply to emphasize entering and access or not, but to dynamically formulate a "list of enterprises that meet the standard conditions" to guide the concentration of resources and cultivate advantageous enterprises. And small and medium-sized enterprises that do not meet the basic conditions are not completely negated. At the same time, China still fosters the development of drones manufacturing industry, emphasizing the establishment and improvement of the standard system, while giving space to some enterprises to develop and reminding them to work hard to meet the standards.

In 2018, to further strengthen the management of the drone industry and implement the Guidance, the Ministry of Industry and Information Technology (MIIT) compiled the "Normative Conditions for Drone Manufacturing Enterprises (Draft for Comments)." It specifies the basic requirements for drone manufacturing enterprises,

[11] Airworthiness and Safety WeChat Team of Airworthiness Technology Research and Management Center of Civil Aviation Administration and Ministry of Industry and Information Technology (2014).

[12] The Ministry of Industry and Information Technology (2017).

[13] The Ministry of Industry and Information Technology (2017).

innovation capability, manufacturing capability, product requirements, quality control, personnel requirements, social responsibility and management norms, which are the eight basic conditions. Article 3 of the General Provisions states that "The government shall implement announcement management for drone manufacturing enterprises that meet the specification conditions, and shall carry out dynamic management of the announced specification enterprises. Enterprises apply on a voluntary basis."[14] The fact that this draft has not yet been implemented shows that there is a long way to go in developing legal regulations for drone manufacturing.

In addition, to develop a standard for safety requirements for civil drone products, the National Standard Plan "Safety Requirements for Civil Drone Products" was submitted and implemented by the National Technical Committee for Aircraft Standardization on 30 December 2020, with a project cycle of 24 months. According to the Interim Regulations on the Administration of Unmanned Aerial Vehicle Flights (Draft for Examination), "micro, light and small drones shall be subject to compulsory product certification management by the market supervision and management department under the State Council through designated compulsory product certification bodies, and only after obtaining compulsory product certification shall they be shipped, sold, imported or used in other business activities."[15] Therefore, a mandatory national standard must be developed as a basis for mandatory certification of civil drones. This technical standard will help to improve the regulations related to the manufacture of drones.

In March 2021, to regulate the manufacturing of civil drones, the Ministry of Industry and Information Technology issued a document on soliciting comments on the "Certain Provisions on the Manufacturing of Civilian Drones (Draft for Comments)," which can be expected to head out to make up for the shortcomings of the existing legal provisions within a few years.

2.2 Legal Requirements for Civil Drones' Pilot Qualification

Back in 2013, the Flight Standards Department of the Civil Aviation Administration of China (CAAC) issued the "Interim Regulation on Civil UAV Pilot Management (AC-61-FS-2013-20)" to implement guidance on the management of pilots of UAVs and their systems that are currently in existence. The first amendment to these regulations was made in 2016 in conjunction with the "Regulations on the Operation of Light and Small Unmanned Aircraft (for Trial Implementation) (AC-91-FS-2015-

[14] Article 3 of the *Conditions for the Normative Conditions for Drone Manufacturing Enterprises (Draft for Comments).*

[15] *Interim Regulations on the Administration of Unmanned Aerial Vehicle Flights (Draft for Examination).*

31)" issued by the Department of Flight Standards.[16] In year 2018, based on the experience of the previous authorized industry associations that have authorized qualified industry associations to implement management of some drone pilot licenses, the original "Regulation on Civil UAV Pilot Management (AC-61-FS-2016-20R1)" has been revised for the second time to further standardize the management of drone pilot licenses. The new Regulation on Civil UAV Pilot Management (AC-61-FS-2018-20R2) adjusts the supervision model, improving the related supporting systems and standards that the administration is directly responsible for the issuance of licenses, refining the requirements and procedures for issuance of licenses and grades, and clarifying that the original certificate issued by the industry association will be converted into the certificate issued by the administration.

First, the Article 1 of the regulation states that "regardless of whether the pilot is located inside or outside the aircraft, the UAS and the pilot must comply with the requirements of the corresponding section of the civil aviation regulations."[17] It means that the corresponding provisions of the Civil Aviation Regulations are applicable to the UAS and the pilot, and the situation they are not applicable is clarified in this document.

Second, Article 4 of its regulations clearly imposes classification management on UAS pilots. Primarily, no license management is required for three types of situations: drones operating indoors; Category I and II drones, i.e., civil drones with an empty mass of less than or equal to 4 kg and a take-off full weight of less than or equal to 7 kg; and drones conducting tests in sparsely populated, open, non-densely populated areas. These three types of drones either do not have the need for regulation, lack the interest of public interest, or belong to micro and small drones, if necessary to carry out UAS for record, there is no need to overly restrict the application of drones excessively. In addition, bureau supervision should be implemented for drones operating in isolated airspace and convergent airspace, except for Category I and II rather than a ternary management model among the administration, industry associations and no license management to build hierarchical management for pilots operating different UAS. It can be seen that the categorization is based on the principle of proportionality, with minimal mismanagement of the qualifications of civil drone pilots, and the degree of regulation varies according to the influence of social interests of civil drones. The principle of proportionality regulates the proportional relationship between administrative power and its exercise from the three dimensions of "purpose orientation," "legal consequences," and "value orientation."[18] For I and II UAVs that operate indoors and conduct experiments in sparsely populated, empty, non-populated areas, the risk is pretty low, and unlicensed supervision is sufficient to achieve the goal; while for other drones operating in isolated airspace and convergent airspace, except for Category I and II, the risk is high and the consequences of losses are serious, and stricter measures

[16] See *Regulation on Civil UAV Pilot Management (AC-61-FS-2016-20R1).*

[17] Article 1 of *Regulation on Civil UAV Pilot Management (AC-61-FS-2016-20R1).*

[18] Huang (2001), pp. 72–78.

must be taken to achieve the goal. Using administrator's supervision model. The management agency for the qualification of drone pilots is determined by the principle of proportionality and managed hierarchically.

Third, compared with previous regulations, CAAC's management of UAS pilots is pretty completed: the relevant supporting systems and standards that the bureau is directly responsible for the issuance of licenses have been improved, such as the license requirements of operating Visual Line of Sight and beyond Visual Line of Sight and corresponding requirements for training driver instructors (Song 2019). It refines license and grade issuance requirements and procedures, such as clarifying the information endorsed on the driver's license, and the bureau shall issue drone pilot licenses and grades to applicants who meet the requirements for corresponding qualifications, aviation knowledge, flight skills, and flight experience. The specific requirements are "Conditions for Issuing UAV Pilot License and Grade." In addition, the rules elaborate on the validity period of the license and its renewal, the renewal of the instructor level, the re-application after the instructor level expires, the proficiency check, the increase of the level, the application and approval of the license and level, the practical examination certificate, the flight experience record, the specific requirements of the general examination procedure, and seven attachments are attached to explain.

Generally, the management regulations have formulated complete and feasible related supporting systems and standards. The requirements and procedures for issuing licenses and grades are quite detailed, and they are operable and practical.

2.3 Legal Provisions on the Airworthiness Management of Civil Drones

2.3.1 Airworthiness Management Procedures

First, the legal provisions on the registration and management of civil drones. On 16 May 2017, the Department of Aircraft Airworthiness Certification of the Civil Aviation Administration of China (CAAC) has implemented a real-name registration system for owners of civil drones to strengthen the management of such aircraft. It applies to civil drones with a maximum take-off weight of 250 g or more (including 250 g). The provisions bind the Department of Aircraft Airworthiness and Certification to formulate specific policies and manage them; impose obligations on manufacturers of civil drones to fill in information, warn consumers of the obligation to register in real names and attach non-sticky adhesive; and require owners of civil drones to register, affix registration marks and update information.

Second, the newly introduced "Procedures for the Administration of Airworthiness Certification of Civil Drone Products (for Trial Implementation)" provides for an attempt to fill the gap in this area by providing for airworthiness certification. At the level of review principles, it stipulates that it adopts the principle of risk-based review. In other words, the airworthiness audit is organized in accordance with the project management approach and adopts a graded review approach based on the risk level of the project. The risk of a civil UAV product airworthiness certification project comes from both the complexity and operational impact of the civil UAV product for which approval is sought, as well as the maturity of the applicant's design and production capabilities and airworthiness management system. For the sake of simplicity and convenience, the risks of civil UAV product airworthiness certification projects are classified into three levels: low, medium, and high. In addition, the management procedure specifies the process and conditions for design production approval. The level of involvement of the bureau is specifically regulated, i.e., how the initial review and post-certification changes are carried out. Third, the airworthiness audit body and the airworthiness approval process are specified.

This is accompanied by the implementation of provisions to guide and regulate the risk assessment activities related to the "Guidance on Risk Assessment for Civil UAV Airworthiness Certification Projects (for Trial Implementation)" and to provide guidance to applicants in the preparation of project risk assessment reports. The airworthiness certification of UAS adopts risk-based principle, and its risk assessment includes two aspects: risk assessment of the applicant's management system and product risk assessment.[19] The former is mainly based on the applicant's management system risk element score sheet to evaluate the risk level, and the latter provides a product risk level through two dimensions: the energy level of the UAV and the possible level of collision in the expected operating environment. In the final specific review of the project's risk assessment, the applicant's management system risk assessment and product risk assessment should be considered comprehensively. It should be noted that the airworthiness certification project of the manned drone system does not require risk assessment and is considered high risk. The assessment guideline provides detailed guidelines and procedures for risk assessment, and provides corresponding templates, which play a very important role during the trial period and provides valuable experience for further airworthiness management in the future.

2.3.2 Part of Airworthiness Management Documents and Important Correspondence

As early as 2009, the Department of Airworthiness Certification of the Civil Aviation Administration issued the "Interim Provisions on Issues Relating to the Management of Civil Drones" as a transitional document dealing with airworthiness management. Notes of Civil Drone Airworthiness Management Meeting stipulates that:[20] "1. Before flying, any UAV should apply to the Civil Aviation Administration for provisional registration in accordance with the 'Procedures for

[19] Article 4 of *Guidance on Risk Assessment for Civil UAS Airworthiness Certification Projects (for Trial Implementation).*

[20] *Interim Provisions on Issues Relating to the Management of Civil Drones.*

Registration of Civil Aircraft Nationality' and should be properly painted in accordance with the 'Regulations for Registration of Civil Aircraft Nationality'. 2. Before flying, it should apply for a special flight permit from the Civil Aviation Regional Administration in accordance with the 'Procedures for the Issuance and Management of Airworthiness Certificates for Civil Aviation Products and Components'. The CAAC inspectors of airworthiness inspection may refer to the applicable part of the 'Procedures for the Issuance and Management of Airworthiness Certificates for Civil Aviation Products and Parts' to check whether the restrictions of use proposed by the applicant are accurate and to confirm that it can ensure flight safety when completing the intended mission. Upon satisfactory completion of the inspection, a Class I Certificate of Authorisation will be issued and the restrictions of use approved by CAAC will form part of the Certificate of Authorisation."

2.4 Legal Provisions Governing the Operation of Civil Drones

The legal provisions for the management of civil drone operations in the mainland of China are imperfect, with the "Regulations on the Operation of Light and Small Drones (for Trial Implementation)" issued by the Department of Flight Standards in 2015 mainly targeting low-altitude, slow-speed, micro-light, and small drones. Because this type of drone accounted for the majority of civil drones at the time and even today, it is impossible to legislate in its entirety at once at the initial stage of legislative involvement, and phased in the objects of regulation that needed to be addressed urgently. It is the country's first administrative regulation specifically for the management of drones.

First, the scope of application of Article 2 is mainly[21] defined as follows: (1) "UAVs with an empty weight of less than or equal to 116 kg and a take-off gross weight of not more than 150 kg that can be operated within or beyond visual range, with a corrected airspeed of not more than 100 km/h"; (2) "UAVs with a take-off gross weight of not more than 5700 kg and a height of not more than 15 m above the receiving surface"; and (3) "UAVs with an inflatable volume of not more than 4600 m^3." It also specifies the applicable UAV operational management classification, which is the first time that China has refined the classification of UAVs according to weight criteria. Particularly, three cases as follow are excluded: Category I UAVs do not need to be managed in accordance with the subsequent provisions of this advisory circular; are not applicable to radio-operated aerial models, unless they use an autopilot, command and control data link or autonomous flight equipment; and not applicable to UAVs operating in isolated spaces such as indoors or inside interception nets.

[21] Article 2 of the *Regulations on the Operation of Light and Small Unmanned Aircraft (for Trial Implementation).*

Second, graded operation management should be carried out by setting up: (1) an electronic fence system, for Category III, IV, VI, and VII drones, and for Category II and V drones operating in key areas and below airport clearing zones, electronic fences should be installed and used; (2) access to the drone cloud management, for Category II and V civil drones used in key areas and below airport clearing zones, and for Category III, IV, VI and VII civil drones shall be connected to the UAV cloud. Particularly, a passive feedback system should be added for Category IV UAVs; and (3) for civil drones that are not connected to the UAV cloud, an application should be submitted to the control department in advance before operation and effective monitoring means should be provided.

The Regulations on the Operation of Light and Small Drones (for Trial Implementation) also specify the duties and authority of the drone captain, the duties of drone pilots in operation, the basic requirements for safe operation (qualification requirements for pilots, instructions for drone use, prohibited acts, flight technology requirements), instrumentation, equipment and marking requirements for the operation of civil drones, and requirements for drone cloud providers.[22]

2.5 *Legal Provisions on Air Traffic Management of Civil Drones*

To strengthen the management of civil UAV flight activities and regulate its air traffic management, in 2016, the CAAC Air Traffic Management Office issued the Measures on Air Traffic Management of Civil Unmanned Aerial Vehicle Systems, replacing the Measures on Air Traffic Management of Civil UAVs, which were formulated in 2009. Its general provisions clarify the legislative purpose of the law, the scope of application, the subject of authority and responsibility, and the subject of liability. The scope of application of Article 2 of the law is "applicable to the air traffic management of civil unmanned aerial vehicle system activities within or affecting the operation of civil aviation airspace such as air routes, approaches (terminals) and airport control zones in accordance with the law."[23] Its Article 4 confirms that civil UAVs are only allowed to fly in segregated airspace, and the responsible entity is the unit and individual that organizes them. If more than one entity carries out civil drones' flight activities within the same airspace at the same time, one activity organizer shall be specified and be responsible for the safety of civil UAV flight activities within segregated airspace.

Chapter II provides the assessment and management. Article 5 stipulates that flight activities within the airspace for civil aviation as stipulated in Article 2 should be assessed by the Regional Administration for UAVs unless all the following

[22] Gao (2017), pp. 28–36.

[23] Article 2 of the *Regulations on Air Traffic Management of Civil Unmanned Aerial Vehicle Systems.*

conditions are met, namely: "1. Outside the airport clearing area; 2. The maximum take-off weight of the civil unmanned aerial vehicle is less than or equal to 7 kg; 3. Flying within visual range and weather conditions do not affect the continuous visibility of the unmanned aerial vehicle; 4. Flying during daytime; 5. Flying at a speed not greater than 120 km/h; 6. The civil drones meet the relevant requirements for airworthiness management; 7. The pilot meets the relevant qualification requirements; 8. The pilot completes an inspection of the civil drone system before conducting the flight; 9. No impact is caused to other aspects other than the flight activity, including ground personnel, facilities, environmental safety and social security, etc. 10. The operator shall ensure that its flight activity continues to meet the above conditions."[24]

Article 6 provides the procedure for civil drones that require assessment. It is worth noting that Article 8 states that the flight activities that require assessment should be remotely piloted aerial vehicle systems, not autonomous UAS, and an electronic fence as required should be set up. So, there is a gap in the corresponding flight assessment for UAVs.

Article 7 makes bottom-line requirements for the content of the assessment, requiring the assessment of: "1. the situation of civil unmanned aerial vehicle systems; 2. basic information and licence status of pilots and observers; 3. basic information of civil unmanned aerial vehicle system operators; 4. flight performance of civil unmanned aerial vehicles; 5. activity plans for civil unmanned aerial vehicle systems; 6. air traffic control safeguards; 7. communication, navigation and surveillance equipment and capabilities of civil UAS; 8. sensing and avoidance capabilities of civil UAS; 9. emergency procedures in case of civil UAS failure; 9. number and location of remotely operated stations and handover procedures between remotely operated stations; 11. other information on missions, noise, security, operational loads, insurance, etc. 12. other risk control measures."[25]

The last article requires that the Regional Authority should organize a review of the assessment report by the relevant authorities. The analysis of the assessment management is essentially a flight permit, and judging by the content of the above-mentioned article, our flight permit is still quite strict, with a provision pattern of not being allowed in principle unless all the specific ten conditions are met. This provision is for segregated airspace and does not leave room for permission to fly in non-segregated airspace, but there may be special reasons for such a need, so there may be some loopholes. Also, from the conditions of daytime flight in Article 4, it seems that assessment and management does not exclude night flights, and relevant units subjectively assess and manage according to objective conditions, which makes it possible for drones to fly in some specific conditions. In addition, the requirement of electronic fencing ("is a software and hardware system that draws a specific area in the corresponding electronic geographic range to block aircraft that are about to intrude into a specific area, and to cooperate with the flight control

[24] Article 5 of the *Measures on Air Traffic Management of Civil Unmanned Aerial Vehicle Systems.*

[25] Article 7 of the *Measures on Air Traffic Management of Civil Unmanned Aerial Vehicle Systems.*

system and secure the area"[26]) to locate drones is quite a novel and effective technology to be incorporated into this provision, using new technology to solve new problems.

The third chapter specifies the general principles and requirements for the establishment of segregated airspace on air traffic services. In addition, the operation of segregated airspace is required to comply with the prescribed procedures and safety requirements, to ensure that flights are conducted within the allocated segregated airspace and at a distance of more than 5 km from the horizontal boundary, and to prevent unintentional separation from segregated airspace. Article 13 provides for equivalent measures to prevent civil drones and other aircraft from crossing the boundaries of segregated airspace with each other. Also, in the case of non-compliant flights of civil drones, ATC (Air Traffic Controller) units are required to establish notification and coordination relations with airports and military aviation control units, and to formulate plans for information notification, assessment and disposal, and operational recovery.

Chapter 4 identifies provisions on issues related to radio management.

2.6 *Legal Provisions Governing the Operation of Civil Drones*

Prior to 2018, in China, commercial operational drones faced the embarrassing situation of being forced to "black flight," which means private flight without permission. For example, when enterprises need commercial aerial photography by drones, such as drone logistics,[27] they ought to get a business license before obtaining approval from the airspace management department. However, there was no access to obtain such a license due to relevant legal regulation. Even some national aerial photography projects such as "Aerial Photography of China" cannot declare from formal access. As a result, companies have been avoiding punishment in the name of "flight demonstration" but in fact conducted "black flight."[28]

There was a blank in the Regulation of the operation of unmanned aerial vehicles (UAVs), because the previous Regulation on the Administration of General Aviation Operation Licensing stipulates that the management of the use of civil drones for operational general aviation activities shall be separately regulated by the CAAC (Luan 2019). But there were no relevant regulations issued. In 2018, the Civil Aviation Administration of China's Measures for the Administration of Civil Unmanned Aerial Vehicle Operational Flight Activities (hereinafter referred to as the Measures), finally fills the relevant loopholes. Thus, the operational civil drone flights of enterprises are regulated in compliance with laws, and there are rules to abide by operational flight activities of UAVs. Later in 2021, the Ministry of

[26] *Measures on Air Traffic Management for Civil Unmanned Aerial Vehicle Systems.*

[27] Deng (2018), pp. 136–142.

[28] Civil Aviation Administration of China (2018).

Transport revised the Regulations on the Administration of General Aviation Operation Licensing and included the operation of civilian drones into its scope of regulation. The comparison has the following new features:

First, the scope of application has expanded, and it integrates management of aircraft. While the Measures's scope of application is the use of drones with a maximum empty weight of 250 g and above in China, whose business activities are operational activities that require obtaining a business license and drone pilot training type of business activities, the recently Regulation on the Administration of General Aviation Operation Licensing applies to all types of civil drones, and no longer distinguishes between business category and training category, all of which belong to other types of operating general aviation activities. Under this situation, civil aircraft referred to in the regulations include civil manned aircraft and civil unmanned aircraft,[29] leading to no more distinction for integrated management.

Second, operating permit conditions is more scientific. At first, the provisions simplify access conditions, specifying access conditions and simplifying the access conditions for drones' business permits, with ten items reduced to three items in the operations category and four items in the training category, retaining only the basic permit conditions for corporate entities, UAV registered under real names, certified training capabilities (training category) and ground-based third-party liability insurance. The requirement for business-owned aircraft has also been reduced from two to one, and an exemption from licensing access has been made for unmanned aerial vehicles under 250 g.[30] Nowadays, according to recently Regulations on the Administration of General Aviation Operation Licensing, it also requires companies to have pilots who are compatible with civil aircrafts, have undergone professional training and obtained corresponding licenses. To consider possible future situations, it requires meeting other conditions stipulated by laws and administrative regulations. On the one hand, the operating license conditions are improved, and on the other hand, the circumstances of non-acceptance which shows chaotic logic have been deleted (making the conditions for granting operating licenses clearer and more standardized).

Third, the management of Regulation on the Administration of General Aviation Operation Licensing is stricter and more standardized, but still keeping integrity supervision. The new regulation cancels the electronic application which was online and convenient, for such management is too loose, which is not conducive to protecting the public interest of society. Its convenience and the characteristics of the listed activities are, that according to the Regulation on the Administration of General Aviation Operation Licensing, the management department will implement integrity management for enterprises, specifically, the regulatory approach will establish and improve the evaluation system of general aviation integrity management, and organized the evaluation of integrity management. General aviation companies with good integrity evaluation records may be reduced in inspection

[29] Article 9 of *Regulations on the Administration of General Aviation Operation Licensing.*

[30] Same as 26.

frequency or exempted from inspections in accordance with the law; general aviation companies with poor integrity evaluation records may be increased in accordance with the law.[31]

3 Legal Provisions on Civil Drones in Hong Kong and Macao

3.1 Legal Provisions for Civil Drones in Hong Kong

L.N. 116 of 2021 in Hong Kong makes the Small Drones Order under the Civil Aviation Ordinance (Cap. 448) to regulate the operation of small drones and to provide for the operation of small drones and remotely piloted aircraft. It is divided into 72 articles in 5 parts, with Part 1 defining terms in the introduction, Part 2 defining the operation of small drones, Part 3 defining the requirements for registration, classification and licensing, Part 4 specifying the rules for enforcement and Part 5 containing other miscellaneous provisions. Under the new regulations, a risk-based model will be used to regulate small drone operations. The different risks of small drone operations will be regulated according to the weight of the small drone and its level of operational risk, including registration and labelling of small drones, registration of remotely piloted pilots, training and assessment, equipment, operational requirements, insurance, etc.[32]

3.1.1 Definition of Terms

The definition of the drone differs from that of the mainland of China. It is defined[33] as a powered machine that can be supported in the atmosphere by the reacting force of air (except the reacting force of air to the ground) and that is operated without a pilot on board. A small drone is defined as a drone that does not exceed 25 kg in weight at all times during a flight. The focus of the weight is more on the drone that is flying, i.e., the traditional concept of "full take-off weight."

It is supervised according to weight and is divided into Category A-1 drones (e.g., a small drone weighing no more than 250 g at all times during a flight), Category A-2 drones (e.g., a small drone weighing no more than 7 kg at all times during a flight but greater than 250 g), and Category B drones (e.g., a small drone weighing no more than 7 kg at all times during a flight). In addition, Part 1 provides the specific meaning of "operation," "flight," and "enclosure."

---

[31] Article 33 of *Regulations on the Administration of General Aviation Operation Licensing.*

[32] Transport and Housing Bureau, The Government of the Hong Kong Special Administrative Region of the People's Republic of China (2021).

[33] Article 2 of the *Small Unmanned Aircraft Order.*

3.1.2 Management of Registration, Grading, and Licensing Procedures

Registration of Drones

Its drone registration is restricted to natural persons who are at least 18 years of age. It is with specific age restrictions, and the Director will review whether it is allowed. And its UAV registration is valid for no more than five years from the issuance of the registration certificate. In addition, provisions are made for the renewal of drone registration and application for re-issuance of the label. The Director has the power to vary, temporarily revoke or cancel the registration of a drone in certain circumstances.

Registration of Remote-Control Drivers

First, Article 29 provides the requirements for the registration of a remotely piloted driver and for any person who has attained the age of 14 years to apply to the Director for registration, in accordance with the requirements under Article 39. Second, Article 29(3) gives the Director the right to review whether to register the person as a remotely piloted driver. If the requirements are met, a remote pilot certificate shall be issued and shall be valid for a period not exceeding three years from the date of issue of the certificate. Third, Article 30, which provides for the renewal of the registration of a remotely piloted driver, similarly gives the Director the power to vary, suspend or revoke the registration under certain conditions.

Classes of Remote Pilot

First, to satisfy the categories of flight operations that an RC pilot may be permitted to perform under each rating, a person may apply to the Director for a rating. Second, the Director must determine whether to award the corresponding requested rating in accordance with the scheme in Article 32(4), taking the resultant knowledge, experience, and ability of the person into account. The so-called "person" must be, or will be, a registered remote-control driver on the effective date of the rating.[34] Such a provision in fact maximizes the convenience of the application, as applicants do not necessarily have to apply for registration and rating in a sequential manner. In addition, the Director must specify in an appropriate manner the types of flight operations that a remote pilot may be permitted to perform under each class. Third, Article 33 clarifies the renewal provisions for classes, again giving the Director the power to vary, temporarily revoke or withdraw a class in specified circumstances.

[34] The *Small Unmanned Aircraft Order*.

Permission to Flight Operations

First, the circumstances that permission must be sought to operate a drone are: to operate a Category B drone; to operate a small drone without complying with any or all of the operational requirements applicable to small drones; to fly within a restricted flight area; and to carry any dangerous goods. The Commissioner of Police or the Secretary for Security will be consulted on the application for permission, and in certain circumstances where restricted flight areas are involved.

Second, Article 38 gives the Director the power to vary, suspend, or revoke a permit subject to specified conditions of reasonable suspicion.

3.1.3 Operation of Small Drones

Scope of Application

Under the territorial jurisdiction of the Small Unmanned Aircraft Order, the operation requirements for small drones apply only to the part of the Order that is flown in Hong Kong; they do not apply to flights within a residential enclosure; and the operation of flights for educational or research purposes under specified conditions applies only to offences relating to the operation of Division 3 of this Part.

A special relaxation for Category A-1 drones does not apply if the following three points are met: (1) at all times during the flight, the drone was operated in compliance with all operational requirements applicable to the drone (unless the flight was conducted entirely within the enclosure range); (2) at all times during the flight, the drone was not within a restricted flight area; or both were within a restricted flight area, but the flight was conducted entirely within the envelope; and (3) at any time during the flight, the drone was not carrying any dangerous goods.[35]

Provisions Relating to Operation

First, for the basic regulations, Article 11 regulates the airworthiness conditions for operating a small UAV, including that the UAV is registered, that the UAV is labelled and displayed clearly, that the insurance policy for the flight is valid, that it is capable of performing all safety system functions, that the pilot is a registered remotely piloted pilot and that the pilot's registration authorizes him to conduct the flight at the level of operation to which the flight belongs. The pilot must meet the conditions of permission granted in Article 37 for a Category B UAV. The second item of this article more specifically lists specific cases where the above airworthiness regulations do not apply. Particular attention is paid to the requirement to have a valid insurance policy for the flight of a small drone, which focuses on the

[35] The *Small Unmanned Aircraft Order*.

mandatory allocation of social risk and, although it increases the cost of a particular flight, it allows for a significant sharing of social risk and addresses the issue of drone infringement. In addition, Article 14 establishes the function of the safety system in accordance with the airworthiness requirements, reminds the obligation to store the information recorded by the safety system and stipulates the criminal liability. This security system is essentially a combination of the "electronic fence" function and the drone cloud system in the mainland of China.

Second, the Subdivision 2 stipulates: (1) Article 15 provides that this Division applies only to the extent that the flight is carried out entirely within the enclosure range; or is operated to carry out the flight in accordance with the permission granted under Article 37. It also stipulates criminal liability and penalties; and (2) the operation provides for 11 specific rules requiring respectively that the flight not be operated at the specified time on a particular day; that the drone be kept in view in the specified manner; that the altitude and speed of the flight not exceed the specified requirements; that the distance from third persons and other vehicles, vessels or buildings not exceed the specified distance; that the distance from the pilot not exceed the specified distance; that no person or thing be carried during the flight; and that no person or object be carried during the flight; that nothing is dropped from the drone during the flight (except for the specified purpose); that the pilot operates not more than the specified number of drones at any one time; and that the size of the drone does not exceed the specified size at all times during the flight. Particularly, Article 17 empowers the Director to specify any other requirements that the Director considers necessary in the interests of aviation safety or public safety, and the parameters specified in the above specific provisions are all determined by the Director.

Third, there are additional provisions for flights within the restricted area and for the carriage of dangerous goods. On the former, Article 18 requires that the flight may be operated only if permission is granted under Article 37 for the restricted area, except that permission is not required if the flight is carried out within the fully enclosed area. Criminal liability is also provided for in Article 18(3)(4). The restrictions on the power to designate specific restricted flight areas are regulated by Article 19. The carriage of dangerous goods by the latter is in principle prohibited, unless permission is sought from the Director.

3.1.4 Legal Provisions for Enforcement

This is the legal regulation of how law enforcement officers are to enforce the law under the Small Unmanned Aircraft Order, which is more detailed and specific and operational. First, it binds authorized persons by defining how to "appoint an authorised person" and how to produce identification documents to carry out their functions. "Secondly, Division 2 specifies the enforcement powers in relation to small drones and flight operations etc., enabling them to prevent or stop flight operations or interference with small drones etc., and to 'seize', detain, search and inspect small drones and their associated equipment and components. Again,

Division 3 empowers an authorized officer to require the production of certain documents and information. Division 4 states the offences of giving false or misleading information and obstructing or authorising an officer."[36] Table 1 is attached for reference.

3.2 Legal Provisions for Civil Drones in Macao

On 5 October 2021, the Civil Aviation Authority of the Macao Special Administrative Region (MSAR) completed the approval of the amendments to the Macao Air Navigation Regulations, which were published in the MSAR Official Gazette, Group I, Issue 40, Administrative Order No. 43/2021, in which Article 67 of Part IV and Part XVI "Requirements for the operation of drones" have made significant changes.

Article 67 of Part IV of the Macao Air Navigation Regulations provides the requirements for the restriction of drones in the Macao Air Traffic Control Zone: except indoors, no person shall be allowed to fly within the following guaranteed areas (a. the airspace within 1000 m of any airport or landing site; b. the flight path of an aircraft, i.e., the area defined as the perimeter of the polygon with the right-angled co-ordinate point as the apex in Table A below but excluding the area defined in Table B;[37] c. the airspace within 50 m of the headquarters of the MSAR Government, the Legislative Assembly Building, the Court of Final Appeal, the official residences of the Chief Executive and the principal officials of Macao, the institutions of the Central People's Government in Macao as defined in Administrative Regulation 22/2000, prisons and juvenile reformatories, the Kowloon Power Station and substations, the headquarters of the Macao Water Supply Company and the Kau Ou oil depot; d. the airspace of 22 historic buildings assessed as World Heritage Sites airspace; e. Prohibited or restricted flight areas established by the Civil Aviation Authority in accordance with paragraph 66 of these Regulations.), except with the written permission of the CAAC and subject to other conditions specified in that permission. In addition, Article 184 also imposes corresponding restrictions for the operation of drones with a total mass not exceeding 7 kg in the Macao Traffic and Navigation Control Area.[38]

Part XVI serves as a special chapter to stipulate " Requirements of UAV Operation." First, its regulations of applicability do not apply to drone operations indoors or underground with a total weight of no more than 25 kg. Meanwhile, its scope of application is very wide. If the application space is interpreted reversely by the legislative techniques of strictly negative inapplicable conditions, there is no suitable application space for medium, large and plant protection drones.

[36] Summary from the notes of the *Small Unmanned Aircraft Order*.

[37] For specific Tables A and B, please refer to the *Macao Air Navigation Regulations*.

[38] Civil Aviation Authority of the Macao Special Administrative Region (2021).

Table 1 Comparison of three categories' rules in Hong Kong[a]

Category	Category A1 aircraft	Category A2 aircraft	Category B aircraft
Basis of classification	A flight if its weight does not exceed 250 g at all times during the flight	A flight if its weight does not exceed 7 kg at all times during the flight but exceeds 250 g at any time during the flight	A flight if its weight exceeds 7 kg at any time during the flight, or other special situation
Registration and labelling regulations			
Small drones need to register and display labels	×	√	√
The minimum age of the person in charge of the small drone	×	18	18
Registration of remote-control driving			
Remote-control driver requires registration	×	√	√
Minimum age of remote-control driver	×	14	14
Training and assessment regulations			
Remote-control drivers need to receive training and assessment	×	×	√
Equipment regulations			
Basic regulations (flight record and suitable airspace identification function)	×	√	√
Operating regulations			
Operating regulations	√	√	√
Need to obtain a flight permit from the Civil Aviation Department in advance	Not applicable	Not applicable	√
Insurance regulations			
Purchasing third-party liability insurance for small drones	×	√	√
Minimum sum assured	×	HKD 5 million	HKD 10 million

[a]A Picture to Understand the New Rules of Drones in Hong Kong, https://view.inews.qq.com/a/20210817A006JO00?ivk_sa=1024320u. Accessed on 05 August 2022

Second, the hierarchical management of drones is divided into four levels according to weight: the first level of operation of drones with a total wight of more than 250 g must be affixed with the owner's name and phone label on the conspicuous position of the drone, and the relevant information must be easy to read. Except the above-mentioned regulations, the management also requires "notify the Civil Aviation Administration immediately before the operation starts." The notification must be made in written form through the Civil Aviation Administration's electronic platform or other means, with relevant materials attached. The above regulations apply to other grades at the same time, because the total weight exceeds 250 g. The second level operates drones are the drones with a total weight of no more than 7 kg. Article 184 summarizes ten conditions. Under general conditions, it is allowed as long as they meet: not flying within the boundary of any guarantee zone specified in Article 67; flying within the specified altitude limit; daytime operating; no carrying dangerous goods; no releasing any gas, liquid or solid; no dragging anything; no flying within 100 m of places where 100 people or more gather; drone operators should be on site and directly control the drones, the distance from the drone is no more than 100 m; the operation is within the line of sight; the flight can be carried out under safe conditions with reasonable confidence. Among the above regulations, there are the requirements of the distance between the operator and the UAV and the requirements for direct control. These regulations considerably limit the types of drones. The remotely controlled drones or drones with considerable autonomy cannot be defined and applied to these. Of course, the UAV which obtains authorization and operates in accordance with the drone activity permit does not need to be restricted by the above conditions; The third level is to operate a drone with a total weight of more than 7 kg but not more than 25 kg. Under the general conditions, it should not be operated unless the following conditions are met: (1) Operate in accordance with the permissions granted by the drone operator's license and the drone activity license. (2) The proposed drone activity has been underwritten by third-party insurance, and the minimum insured amount is 10 million patacas. This is an additional insurance clause for such drones, which is conducive to reasonable sharing of unpredictable risks; UAVs with a total operating mass of more than 25 kg at level IV are not allowed unless authorized in writing by the Civil Aviation Administration and any conditions established in accordance with the authorization.[39]

Furthermore, for the aforementioned required drone activity license, Article 187 specifies that the corresponding materials should be submitted and provided in the form and method required by the Civil Aviation Administration, and the Civil Aviation Administration must also make sure that the applicant has the ability to comply with the provisions of the permit or ensure the activities to be stipulated can be carried out safely.

Then, there are different age restrictions for drone operators to operate drones with a total weight of more than 250 g and more than 7 kg. Its pilots are prohibited

[39] Civil Aviation Authority (2021).

from using psychoactive substances to prevent the inability to operate the drone in a safe and appropriate manner. For its qualifications, there is no requirement to have a corresponding operating certificate, but in a general setting: know the performance specifications and operating restrictions of the drone; ensure that the drone is suitable for flight and its equipment meets the requirements of the Civil Aviation Administration; and it is reasonably certain that the flight can be carried out under safe conditions. For its responsibilities, they are required to abide by the performance specifications and operating restrictions of the UAV specified by the manufacturer; fly in the field of view unless authorized; and the obligation to notify the Civil Aviation Administration if there is a serious UAV accident.[40] Besides, Article 189 specifically defines the responsibilities of drone operator license holders.

Generally, the legal requirements for UAV in the newly revised *Macao Air Navigation Regulations* are sufficient to meet the status quo of drones in the Macao Special Administrative Region, as well as to adapt to the geographical and economic development of the Macao Special Administrative Region itself. It has the following features: First, hierarchical management, mainly for micro drones. As is known to all, Macao has a small land area and a large population. Intuitively speaking, the land area is only 32.9 km^2 while at the end of 2020, the total population is 683,200. The population density is extremely high. As a result, Macao is unable to broaden the application of large and medium-sized drones, and it is unnecessary to some extent. Therefore, Macao centralize to standardize micro and small drones, and the hierarchical management is consistent with the classification of drones in Hong Kong. Second, flight operations are strictly restricted. For specific flight operations such as flight altitude and flight, the specific coordinates of the chart are used to define. And it is necessary to implement direct control of drones which do not exceed 7 kg. For small drones, third-party liability insurance must be insured with a minimum insurance of 10 million patacas. Such requirements undoubtedly greatly restrict drone flight operations. Third, leniency combine with strictness. The regulation is generally loose for drones under 7 kg. The second level drones which exceed 250 g but not more than 7 kg, its flight activities that do not require special activity permits if they meet specific conditions. This is a post-review. Only for UAV activities over 25 kg, the prior permission mode is adopted, and high third-party liability insurance is required. Fourth, the unmanned aerial vehicle laws and regulations are unified but not perfect. The unified regulations provide a single chapter in the *Macao Air Navigation Regulations* to facilitate the reference and quotation of relevant regulations. However, the expression of driver qualifications is not clear enough, meaning, it does not require the driver to register, but uses an equivalent concept to show that he has the corresponding qualifications. The UAV also did not perform real-name registration. Instead, it made operation notifications and self-identified information before starting operation, which may be both inconvenient and effective.

[40] *Macao Air Navigation Regulations.*

4 The Similarities and Differences of Civil Drone Laws in the Three Regions

4.1 *Manage Objects*

4.1.1 Definition and Grading

In mainland China, civil AUV "refers to an aircraft that does not have an onboard pilot to control, has its own flight control system, and is engaged in non-military, police, and customs missions. It does not include aeronautical models, unmanned free balloons and tethered balloons."[41] Its definition emphasizes "unmanned" and "self-provided flight control system";[42] the Hong Kong Special Administrative Region defines it as "dynamically driven machines that are supported in the atmosphere by air reactions (except air reactions to the ground) that operate without a driver."[43] It is from the definition of the physical principle of the aircraft plus "the driver does not operate on the machine" to define; the definition of Macao Special Administrative Region ("refers to an aircraft that is not operated by a pilot") is defined from the perspective of "unmanned." We can see that in all cases, the characteristic of "unmanned" is the essence, and the definition of drones in the three regions is roughly the same. From the perspective of management objects, the differences are mainly focused on the hierarchical management of drones. The classification of civil drones currently in effect in China's mainland is divided into seven levels according to the type of weight. The I–IV level is doubled determined according to the weight of the empty aircraft and the total take-off weight. The plant protection drones and unmanned airships and type I and II UAVs that can operate beyond visual range which can reach 100 m can be distinguished according to the type. China's mainland places special emphasis on the relevant norms and operating rules of plant protection drones, and includes it as a key target. The classification management of drones in Macao is divided into four levels according to their weight. They are divided into 250 g, 7 kg, and 25 kg in weight. Hong Kong and Macao have the same classification, and they are specifically described as Class A plus, Class A, and Class B kind. Observing the development of our country's legislation, in the latest *Regulations on the Administration of Real-Name Registration of Civil Unmanned Aircraft*, it has been proposed to apply civil drones with a maximum take-off weight of 250 g or more (including 250 g). *The Interim Regulations on Flight Management (Draft for Solicitation of Comments)* clearly change my country's hierarchical management standards to follow the trend of internationally accepted standards.

---

[41] See The Ministry of Industry and Information Technology (2017).

[42] See The Ministry of Industry and Information Technology (2017).

[43] Transport and Housing Bureau, The Government of the Hong Kong Special Administrative Region of the People's Republic of China (2021).

It should be noted that different regions have different regulations that are not in the scope of application. Generally, they are not applicable to small civilian drones flying in "closed areas" because there is no need for supervision. The Hong Kong region places special emphasis on flying operations for educational or research purposes under certain conditions. In Macao, operations in closed areas and the weight exceeding 25 kg are managed. "This chapter applies to remote-controlled drone flight activities with open space outside the building" inappropriately exaggerated the scope of application.

4.1.2 Civil Drone Registration Management

The China's mainland implements a civil drone registration management system for the use of civilian drones exceeding 250 g, and conducts related operations online, which are reviewed by the Civil Aviation Administration; Hong Kong has also formulated special regulations to regulate drone registration and have a specific age limit. All natural persons are restricted to 18 years of age. The Director decides whether to register or not, and sets the validity period; there is no real-name registration system for drones in Macao, but there is a system of operation notice and self-identification of information before starting operation. Above all, the three regions generally require the weight of drone registration management to exceed 250 g. It is worth noting that Hong Kong has set age requirements, but its necessity remains to be investigated.

4.2 Manufacturing and Sales

Unfortunately, up to now, there are currently no effective legal regulations on the manufacture of drones in China's mainland. As for the sales field, there are only export controls for drones, and the types of drones listed are also civilian drones; for Hong Kong and Macao, due to the degree of development of related industries, there is no need to formulate airworthiness management for production and manufacturing of civil unmanned drones. But there is no special legislation on the sales of drones, and the *Small UAV Order* and the *Macao Air Navigation Regulations* have not stipulated the relevant regulations. Its application inspection system requires that the design, manufacture, and even modification of remotely controlled drones must apply for inspection. For imports, they may be approved by application for inspection. Only if the structure is simple and approved or announced by the Civil Aviation Administration, inspection or approval can be exempted. The *Management Rules for Remotely Controlled Unmanned Aerial Vehicles* further stipulates that the form inspection and entity inspection shall be classified according to different weights.

4.3 Personnel Qualification

The current law on effective driving qualifications in China's mainland, the *Regulations on the Administration of Civil Unmanned Aircraft Drivers*, is relatively complete and has recently been amended. Its hierarchical management is divided into three types of qualifications: (1) No license is required for micro-small UAVs, if necessary, the UAV cloud system can be archived; (2) Unmanned aerial vehicles operating in the separate airspace and type III-VII unmanned aerial vehicles operating in the fused airspace need to be implemented the dual mode of industry association management and bureau supervision; (3) The drivers of class XI and class XII UAVs operating in the fused airspace shall be managed by the local authority. Generally, the hierarchical sub-organizations conduct different supervision and management modes proportionally for their qualifications. Thus, supervision is loose. When the unmanned aerial vehicle cloud system is used independently, the management of trade associations fully respects the internal industry. Only in the highly dangerous situation, CAA supervises, and its supervisors vary according to necessity.

In Hong Kong, in principle, anyone who is over 14 years old can apply for registration with the Commissioner. Not only that, but it also distinguishes the registration level. The director must determine whether to grant the corresponding application level based on the relevant methods and the knowledge, experience, and ability of the person in combination with the results. The drone managers in Hong Kong are all "directors," and industry associations have no right to conduct qualification management. In addition, the permitted flight operations of a person at each level must be announced by the "Director" in an appropriate manner; Macao has different age restrictions for drone operators to operate unmanned aircraft with a total mass of more than 250 g and more than 7 kg. It does not require a corresponding operating license, but sets some requirements in general.

4.4 Flight Airspace

The flight airspace in China's mainland is distinguished into "converged airspace" and "isolated airspace," and civil unmanned aircraft are only allowed to fly in isolated airspace. In isolated airspace, the flight of civil drones within the airspace or in the above airspace of such as air routes, approaches (terminals) and airport control zones that have an impact on the operations are subject and constraints to the "Civil Unmanned Aircraft System Air Traffic Management Measures." Expect it meets special conditions, assessments must be made for the use of airspace. In principle, the designation of isolated airspace areas shall not be in densely flying areas, densely populated areas, key areas, and airspace around busy airports. Specifically, the horizontal distance between the boundary of the isolated airspace and the boundary of the airspace used by other aircraft should not be less than 10 km. The upper and lower limits of the isolated airspace shall not be less than 600 m below 8400 m (inclusive) in the airspace used by other aircraft, and shall not be less than 1200 m above 8400 m. In addition, the "Regulations on the Operation of Light and Small UAVs (Trial)" stipulates a "restricted area." Such regulations are not detailed. In the latest legislative development "Interim Regulations on Flight Management of Unmanned Aircraft (Draft for Review)," "By combining the horizontal protection range of important ground targets and limiting the height of the air, classify and manage the airspace for drone flight. Under the premise of clarifying the horizontal range, it is stipulated that the airspace suitable for micro-UAV flying below the true height of 50 m and the airspace suitable for light drones below the true height of 120 m" [*Interim Regulations on Flight Management of Unmanned Aircraft (Submission for Review Draft)*]—this is a more specific future legislative development that deserves further and more attention.

Hong Kong did not specifically delineate the flight airspace, but set up a restricted flight area model, and the "Small UAV Order" did not specify the scope. The "director," the "Secretary for Security," and the "director of police" were assigned the right to issue newspapers and publications to designate a certain area of Hong Kong as a restricted flight zone according to different conditions, and these three have the power to change, temporarily revoke and revoke the instructions made by the personnel. In Macao, the management of the flight airspace is relatively strict. It sets up a flight guarantee zone, and no one is allowed to fly in the flight guarantee zone without permission. The management of flight guarantee area is very specific. The first four categories specifically delineate the specific location of the guarantee area, and finally include the prohibited or restricted flight areas established by the Civil Aviation Administration in accordance with paragraph 66 of the regulations. Of course, the 66 authorized specifications do not specify a restricted flight area, which is an abstract baseline specification. It can be seen that there is no upper-level legal requirement for drones in China's mainland. The regulation of the *Civil Unmanned Aircraft System Air Traffic Management Measures* in which the flight airspace seems to be within a specific range has caused the author's confusion. In addition, what is the significance of the distinction between isolated airspace and converged airspace? If it only stipulates that it can fly in the isolated airspace, there is no further regulation or permission for converged airspace operations. In comparison, the regulations in the other two regions are all clear, or delineate general no-fly zones where other places can fly loosely. Or if the restricted flight zone is delineated by the relevant authority, it can be seen that the flight airspace is subject to specific airspace restrictions, and the restrictions can be blocked according to special applications. In fact, according to "*General Aviation Flight Control Regulations*" issued by the State Council and the Central Military Commission, the airspace allocation is in the charge of the Chinese Air Force. General aviation flights must be obtained through an application for "temporary airspace," otherwise they are not allowed to fly. Therefore, country's general aviation does not have immediately available flight space, and of course it cannot conduct autonomous flight at all. The complicated flight approval system cannot even meet the needs of our civil aviation flight, let alone include drones into the general aviation system for management.[44] In theory, recently it generally believed that UAV airspace access is a right rather than a privilege, and it should have general permission attributes.[45] Generally, the condition of administrative license is clearly stipulated by law, and the administrative agency shall approve the application of the counterparty within the statutory time. The administrative agency shall not refuse an application that meets the conditions. The same is true for drone airspace access permits. When drone users meet the statutory airspace access requirements, if the law does not impose special restrictions on flight activities, drones should not be prohibited from entering the airspace. In specific settings, when establishing the drone airspace access permit system, the principle of proportionality and risk balance should be followed to protect the freedom of citizens to the greatest extent.[46]

4.5 *Flight Operation*

For China's mainland, first, it is required to conduct a single aircraft inspection in the airworthiness review, not to conduct a single aircraft airworthiness review, and not to issue a standard airworthiness certificate but issue a Class I concessionary flight certificate. It is not necessary for Micro-UAVs in real operation. The second is that in the *Civil Unmanned Aircraft System Air Traffic Management Measures*, expect it meets specific conditions, the flight plan should be reviewed by the regional administration, and the review content includes the first point. For other applications outside the scope of its application, the *Regulations on the Operation of Small and Light UAVs (Trial)* adopts a hierarchical management method for the flight operations of micro and small civilian UAVs, applying electronic fences, UAV cloud systems and combining other methods for hierarchical management of different types of drones. But there are no very clear regulations on flight operations, and the regulations are very principle.

For Hong Kong, its regulations on the operation of unmanned aerial vehicles limit the application of civil unmanned aerial vehicles, that is, operation on unmanned aerial vehicles that do not need restrictions. In fact, the basic rules for operating a small unmanned aerial vehicle for flight operations are the simply listing conditions that need to be met and are easy to meet. It is worth noting that Class B UAVs require special permission and that all flights require specified insurance policy.

The fixation in Macao is clearer, requiring operation notifications and specific conditions for drones of different weight classes, which are detailed below. To operate a drone with a total mass of not more than 7 kg, Article 184 summarizes ten conditions. Finally, it is allowed to obtain authorization at discretion and operate

---

[44] Zhou and Pang (2019), pp. 85–94.

[45] Wang (2021a), pp. 111–123.

[46] Guo (2001), pp. 19–23.

according to the UAV activity license, which may not be limited by the above conditions—operate a UAV with a total mass of more than 7 kg but not more than 25 kg. Generally, it shall not be operated unless it is operated in accordance with the authority granted by the UAV operator's license and UAV activity license, and the proposed UAV activity has been insured by a third-party insurance—it is not allowed to operate UAVs with a total mass of more than 25 kg unless authorized in writing by the civil aviation authority and any conditions established in accordance with the authorization.

Generally, the three regions have set up corresponding flight operation regulations, and each has its own characteristics. They are a combination of management and release. The management of civil UAVs that meet specific conditions is relaxed, and the heavier UAVs are more strictly supervised. Specifically, (1) The requirements for the insurance of civil UAVs are listed in the three regions, but in China's mainland, according to the *Civil Aviation Law,* UAV operators should insure the ground third-party liability insurance for UAVs. Macao is the most clear. It requires compulsory insurance for specific levels of UAVs to determine the specific insurance amount, but it is suspected that whether the amount is too large to affect the development of civil UAVs. Hong Kong is the same. For class B UAVs, it is required to specify the policy without limiting the insured amount. (2) On special permission procedures for flights outside the restrictions, the mainland's provisions are quite limited and narrow. Its assessment and management is also aimed at the flight areas within its scope of application. In the provisions on the *Operation of Light and Small UAVs (Trial)*, it is clearly emphasized that it is not allowed to break through the airport obstacle control surface, flight restricted area, unapproved restricted area, and dangerous area. It seems that there is no special permission. (3) The special provisions on the carriage of dangerous goods are reflected in all regions except the mainland. This special provision should be paid attention to. For the potential risk of UAV, judging from the regulatory requirements of the mainland, it should be more appropriate to make a plan not to use transportation.

4.6 Force of Law

Judging from the analysis of relevant laws and regulations in each region mentioned above, there is no doubt that the civil UAV laws and regulations in the mainland area are messy and the level of effectiveness is too low. The Hong Kong officially implemented the "*Small UAV Order*" on June 1, 2022, which is subordinate to the *Civil Aviation Ordinance (Chapter 448)*. For the relevant areas of the common law system, the current laws are *the Basic Law* and its Annex III listed in the national laws implemented in the Hong Kong Special Administrative Region; the original laws before July 1, 1997 include common law and the law of equity, except those that violate the Basic Law or are amended by the legislature of the Hong Kong Special Administrative Region; and laws enacted by the legislature of the Hong Kong Special Administrative Region. *The Civil Aviation Ordinance* is a part of the

"Hong Kong Laws," and the "*Small Drone Order*" is a subsidiary legislation, which is enacted in accordance with delegated powers. This level of legal effect is quite high, and the relevant regulations are stipulated in a decree. The first group of the 40th issue of the "*Macao Special Administrative Region Bulletin*" published on October 5 by the Macao Special Administrative Region of China published Administrative Order No. 43/2021 to approve and amend the "*Macao Air Navigation Regulations,*" including Article 67 of Part IV and Part XVI "UAV Operation Requirements." It specifically standardizes the legal requirements for UAVs, which are also specialized legislation and are placed in aviation management legal regulations. It can be seen that the laws and regulations of the other two regions have been revised in recent years, and no separate laws have been set up, either as subsidiary legislation or directly revised as special chapters, but they are all systematic and have a considerable degree of legal effect. However, the laws and regulations related to drones in China's mainland are too scattered and are formulated by different agencies, and each department makes regulations for their specific jurisdictions; on the other hand, the legal effectiveness level is too low, and the main legal regulations currently in effect are basically "Departmental normative documents," a large number of drafts for solicitation of comments and trial regulations have not been officially Issued, even if they are issued, it is a bit more complicated than the current regulations.

5 Conclusion: The Future Direction of China's Civil Drone Legal System

5.1 Hierarchical Management

The current classification standards in our country are inconsistent with internationally accepted standards, and they distinguish the quality classification of take-off total weight and empty aircraft quality, while preferring higher standard. In fact, such a classification has no meaning in subsequent regulations. On the contrary, it causes inconvenience to the main body of production design, actual use, and specific supervision. According to the follow-up legislation and the "draft for comments," our country has a trend to change the current UAV grading standards and conform to the international mainstream. Although such regulations are considering the issue of the difference between the weight of civilian drones and the quality of unmanned aerial vehicles "for a specific flight." However, the registered air quality, the actual take-off weight, and the maximum total weight of a certain flight, the most important specific flight weight, are the sources of potential social dangers that lead to a certain drone flight. Even if we use strong supervision, we only need to regard the maximum weight as the quasi-grading to maximize the supervision risk. As for the so-called Chinese names in the draft, that is, the names of micro, small, medium, large, and Roman numerals constitute synonymous repetitions, it is unnecessary. And for different levels, generally, Category I drones that take-off with a total weight of less than 250 g do not need to be included in the scope of regulatory application. The key regulatory and legislative areas for civil drones should be Category II and III drones which are currently widely used in the civil market and has a large number.

5.2 Unified Legislation

At present, China's UAV regulation is basically composed of the policy opinions of State Council of the PRC, the regulatory documents of the Civil Aviation Administration of China, and the documents of the National Air Traffic Management Committee. As a result, the regulatory body is not clear. During the entire life cycle of drones, various departments have the willingness to participate in regulatory legislation. Further, the UAV legislation is messy. As it is known to all, the effect of the evaluation method lies not in the quantity of legislation but in the quality of legislation (Luan 2019). There is no need to formulate so many low-effect laws and regulations for civilian drones. However, these norms are mostly temporary and guiding, and relevant approval procedures, management regulations, and airworthiness standards are relatively lagging. Overall, the industry guidance and industry supervision are loose and messy, and a complete system of civil drone regulations and management systems have not yet been formed.[47] In view of the rapid development of the new generation of industrialization technology due to the epidemic and the 5G technology in recent years, the industry of civilian drones have also ushered in a blowout development. A large number of legislative drafts are delayed in the discussion, and may be enacted by another department before the official announcement. In the accumulation of considerable legislative experience and the legislative trend of various countries and regions around the world, China mainland region should speed up the formulation of a unified "civil drone" legal norms. On December 29, 2018, the seventh meeting of the Standing Committee of the 13th National People's Congress voted and passed the amendments to the *Civil Aviation Law of the People's Republic of China*, and President Xi Jinping signed Chairman Order No. 24 to promulgate it. The purpose of this amendment to *the Civil Aviation Law* is to implement the requirements of "delegation, control and service" reform and institutional reform, including legislative authorization for unmanned aircraft. "If the State Council and the Central Military Commission have other regulations on the management of unmanned aircraft, those regulations shall prevail." Obstacles have been cleared from the legislative level. The author does not agree with the provisional plan for placing civil and military drones together in the "*Interim Regulations on Flight Management of Unmanned Aircraft (Draft for Comment)*" of January 2018. Uniform legislation is not a hodgepodge, and there is a huge difference between civil and military use before the extraction of common factors.

[47] Luan (2017), pp. 32–37.

Under the current legal norms, it is entirely possible for the State Council to carry out administrative laws and regulations under the authorization of the *Civil Aviation Law*. Since the current legislation basically excludes the provision of single chapters in the law, the laws and regulations formulated by the State Council are also effective. Second only to the law. The same law can be applied, effectively avoiding the embarrassing situation that each department is in charge of separately formulating related regulations and the effectiveness of the regulations is too low.

5.3 Flight Operation

First, it should focus on clarifying the specific operating conditions and specifications for the hierarchical management of UAVs. The current operating specifications are not complete and specific enough, there are too few relevant specifications, and there are no regulations on "carrying dangerous goods." Second, the flight airspace needs to be further clarified. What is the direct relationship between restricted flight areas, converged airspace, isolated airspace, etc.? It also needs to be explored whether the special permission system for restricting flights within the flight zone should be adopted. Third, the flight system evaluation system is far less suitable for the application of civilian drones than the conditional permit system. my country's own administrative organs have a strong management mindset, and the concept that administrative organs are service-oriented institutions is not deep. Evaluation management gives the Civil Aviation Administration considerable discretion. Specific hierarchical management can adopt a more specific flight permit system for drones of different levels. Generally, if the formal requirements are met, they should be permitted to fly. The author believes that further specific regulations can be further specified to reach further refined management and relaxed regulations. There is no need to worry about the rules being specific and detailed, because the ones who read the rules are professional and qualified drivers, not ordinary citizens.

5.4 Risk Allocation

Civil unmanned aerial vehicle causing injury or damage to property can be interpreted through tort liability in essence (Shi et al. 2021). Civil drone laws can in fact clarify responsibilities among risks in the professional field, which may overlap with tort liability law, but such incidents are new and ambiguous, and the legal norms for civil drones should have the relevant content. On the one hand, it is to clarify the responsible person, change the problem of the lack of main responsibility at this stage now. The other is the insurance system for "civilian drones," that is, sharing the risk of the responsible person also solves the problem of "artificial intelligence" drone systems themselves. In other words, a more effective way is to delineate a liability insurance system to effectively hedge against the risk of potential infringement liability of civilian drones. At present, according to mainland, China's specific regulations, the compulsory insurance for which types of drones are regulated, and the compulsory insurance for flight regulations can all be discussed. Choosing the combination of release and management in the balance of interests also affects the distribution of risk burden. The rational use of insurance terms can maximize social interests and do not sacrifice the development of civil UAVs. Due to the big data risk assessment and premium calculation rules of the insurance system itself, it will also urge all parties to carefully produce and operate civil UAVs.

References

Journal Article

Deng X (2018) The legal obstacles and countermeasures for unmanned aircraft logistics development. J Beijing Jiaotong Univ (Soc Sci Ed) 17(01):136–142. https://doi.org/10.16797/j.cnki.11-5224/c.20180122.003

Gao G (2017) On China's regulatory regime for civil unmanned aerial vehicle. J Beijing Univ Aeronaut 30(50):28–36. https://doi.org/10.13766/j.bhsk.1008-2204.2017.0188

Guo D (2001) Principles and classification of administrative license. Polit Sci Law. https://doi.org/10.15984/j.cnki.1005-9512.2001.06.005

Huang X (2001) Study on the principle of proportionality of administration law. Sci Law (J Northw Univ Polit Sci Law). https://doi.org/10.16290/j.cnki.1674-5205.2001.01.009

Li Y (2019) Unmanned Aerial Vehicle Classification and Legal Regulations—Comment on the "Interim Regulations on the Flight Management of Unmanned Aircraft (Draft for Solicitation of Comments).". China J Appl Jurisprud 18(06):24–40

Luan S (2017) Thesis of legal regulation of unmanned aerial vehicle. J Nanjing Univ Aeronaut Astronaut (Soc Sci). https://doi.org/10.16297/j.nuaass.201701007

Wang X (2021a) Airspace access for UAV: legal analysis and system construction. Admin Law Rev 02:111–123

Wang X (2021b) Construction of a legal system for drone product management. J Beijing Univ Aeronaut 34(1):125–133. https://doi.org/10.13766/j.bhsk.1008-2204.2020.0369

Zhou C, Pang C (2019) A comparative analysis on statutory regulations on privacy violations by civilian drones. Leg Forum 06:85–94

Book

Luan S (2019) Research on legal regulation of civil unmanned aerial vehicle. Law Press China, Beijing

Shi Y, Zhu X, Guan X et al (2021) Basics of safe flight of civil UAVs. Tsinghua University Publishing House, Beijing

Song J (2019) Legal regulations and safe flight of unmanned aerial vehicle. China Machine Press, Beijing

Yang H, Hao X (2017) Principles and examples of aviation law. Law Press China, Beijing

Online Document (No DOI Available)

Airworthiness and Safety WeChat Team of Airworthiness Technology Research and Management Center of Civil Aviation Administration and Ministry of Industry and Information Technology (2014) An Analysis of the Current Status of Civil Drones Airworthiness Management in China. http://news.carnoc.com/list/293/293102.html. Accessed on 01 August 2022

Askci Consulting Co., Ltd (2019) 2019 China industrial drone industry market outlook research report. https://wk.askci.com/details/d147c5447bc1481284bf44584a1eecc9/. Accessed on 01 August 2022

Benowitz D (2021) How has 2021 changed the drone industry? https://droneanalyst.com/2021/09/14/how-has-2021-changed-the-drone-industry. Accessed on 01 August 2022

Civil Aviation Administration of China (2018) Interpretation | Measures for the Administration of Civil Unmanned Aerial Vehicle Operational Flight Activities (Provisional). http://www.caac.gov.cn/XXGK/XXGK/ZCJD/201804/t20180409_56264.html. Accessed on 04 August 2022

Civil Aviation Authority (2021) Civil Aviation Authority of Macao SAR. https://www.aacm.gov.mo/index.php?r=law/fly-activity&id=103&pageid=198. Accessed on 05 August 2022

Civil Aviation Authority of the Macao Special Administrative Region (2021) Unmanned Aircraft Flight Chart. https://www.aacm.gov.mo/uploads/202110/pdf/16334099365534.pdf. Accessed on 05 August 2022

Lin X (2021) Nanhu Public Security Bureau Introduces Technological Police Equipment Drone Air Patrol Guarding Safe Nanning. http://www.szuavia.org/news_cen.php?cid=26&id=5163. Accessed on 01 August 2022

Lushan (2021) Stable recovery in industrial production and continuous improvement in enterprise efficiency. http://www.stats.gov.cn/ztjc/zthd/lhfw/2021/lh_sjjd/202102/t20210219_1813636.html. Accessed on 10 October 2021

The Ministry of Industry and Information Technology (2017) Guidance on promoting and regulating the development of civil drone manufacturing. https://www.miit.gov.cn/jgsj/zbys/gzdt/art/2020/art_73aee6b161b54841a7480b3da01f6234.html. Accessed on 10 October 2021

Transport and Housing Bureau, The Government of the Hong Kong Special Administrative Region of the People's Republic of China (2021) Small unmanned aircraft order to be Gazetted on Friday. https://www.thb.gov.hk/sc/psp/pressreleases/transport/air/2021/20210714a.htm. Accessed on 17 October 2021

Wang T (2021) "Wild Birds Trapped by Sticky Nets Regain Freedom" as 'Operation on Clearing the Nets' is Patrolled by Powerful Drones. http://www.szuavia.org/news_cen.php?cid=26&id=5176. Accessed on 01 August 2022

Xu S (2021) Digital wisdom empowerment: Jiangsu electricity inspection opens autonomous drone mode. https://www.chinanews.com.cn/ny/2021/10-06/9580741.shtml. Accessed on 01 August 2022

Other

Schroth L, Bödecker H, Radovic M (2020) Drone market report 2020–2025. Drone Industry Insights, Hamburg

Part II
Ships: Maritime Automated/Autonomous Transportation Systems

The Intersection Between Law and Technology in Maritime Law

Aybüke Naz Durmuş

Abstract Maritime stakeholders have been benefiting from recent technological developments particularly in autonomous maritime vehicles (AMVs) which are used in numerous activities, not only by private maritime sector but also by States and criminals. The concept of maritime safety and security has started to evolve due to such utilisation. Although AMVs may be vulnerable in certain aspects and used by perpetrators, they present several opportunities to enhance maritime security since they considerably reduce human error and risks to human life, can perform missions more precisely and efficiently, and can carry out tasks beyond human capability. Furthermore, these vehicles can monitor and detect illegal activities occurring at sea and allow States to respond swiftly. Thus, States use AMVs for intelligence and surveillance, mine countermeasures, identification, inspections and protection of ports and critical facilities. This shift brings several challenges and opportunities. This chapter attempts to examine the intersection between law and technology in maritime law by addressing the evolution of maritime security, the terminology and the legal status of AMVs, the international legal framework, AMVs in enhancing and threatening maritime security and law enforcement by tackling how AMVs correspond to the current legal framework in the light of these recent developments.

Keywords Autonomous maritime vehicles · Autonomous ships · Maritime law · Maritime security · Cyber security

---

I would like to thank the Republic of Türkiye Ministry of National Education for providing me a fund for my studies. The opinions and views expressed in this chapter are those of the author.

A. N. Durmuş (✉)
University of Reading, Reading, UK
e-mail: a.n.durmus@pgr.reading.ac.uk

K. Noussia, M. Channon (eds.), *The Regulation of Automated and Autonomous Transport*, https://doi.org/10.1007/978-3-031-32356-0_5

1 Introduction

The recent years have seen remarkable developments in technology, and maritime sector is no exception. To achieve safer, more efficient, and environmentally friendly shipping, maritime sector keeps making progress in the integration of technology to maritime industry by way of autonomous systems and vehicles. Maritime stakeholders have been benefiting from various recent technological advancements especially in autonomous maritime vehicles (AMVs) which are already being used in various maritime activities, not only by the shipping sector but also by States and the criminals. The concept of maritime safety and security has started to evolve because of these new technological advancements which provide new potentials and challenges.

AMVs have the possibility to threaten maritime security as well as the potential to transform the maritime world to a safer and more secure version. States gradually utilise AMVs which can provide many opportunities to enhance maritime security with their various sizes, shapes, capabilities, and roles since this type of technology reduces human error and risks to human life especially in conducting dangerous missions, allows to carry out missions in a more precise way, improves the performance and outcomes of the missions, and enables to perform tasks beyond human capability by covering larger areas at high speed. Furthermore, equipped with numerous sensors, radars, cameras, and identification systems, they can monitor and detect harmful substances and illegal activities occurring at sea, record video images, collect and send data to the relevant authorities, and therefore, provide a swift and efficient response. In this regard, AMVs are used for intelligence and surveillance, coastal patrols, searching for illegal activities, fishing boats, underwater weapons, smugglers, and many other threats to security. Given that States keep trying to find new ways to enhance their maritime security, AMVs can be the new instrument to provide that.

On the other hand, maritime security may be threatened and disrupted by the utilisation of new technology given that criminals use autonomous vehicles to commit offences since such vehicles enable them to carry out activities with high capacity without the risk to human life along with the risk of getting caught and prosecuted. Moreover, autonomous vehicles may be more vulnerable to cyber incidents such as cyberattacks, malware, ransomware, data breaches, and manipulation of data all of which have the possibility to give rise to the loss of operational control, service disruptions, system failures and downtimes, financial loss, and a variety of criminal activities. These threats and incidents could affect international shipping companies, vessels, ports, oil and gas facilities, and even States.

It is apparent from the approaches of States along with the budgets they provide for these vehicles that autonomous maritime vehicles will become increasingly prominent for various practices in the future which will require a new perspective for maritime security and international cooperation. This new shift towards a more technologically developed phase brings several challenges and opportunities to both commercial shipping and maritime security. This chapter examines the intersection between law and technology in maritime law by addressing the role that technology may play in enhancing and threatening maritime security and law enforcement and attempts to explain how AMVs correspond to the current maritime security legal framework along with the approaches considered by the international maritime community in the light of these recent developments. In this framework, the first section of this chapter provides an introduction to maritime security, the distinction between maritime security and maritime safety, and the evolution of maritime security. Section 2 addresses autonomous maritime vehicles by examining terminology, legal status of an AMV as a ship, legal status of an AMV in naval warfare, non-ship autonomous maritime vehicles, and the degrees of autonomy. Section 3 tackles the legal framework in relation to AMVs, addresses the legal framework under UNCLOS,[1] the approach of IMO[2] on MASS, and the proposal of the United Kingdom for legislative change. Section 4 evaluates the use of AMVs as instruments for enhancing maritime security by analysing intelligence and surveillance, right of visit, illicit traffic in narcotic drugs and psychotropic substances, migrant smuggling, and damage to marine environment. Section 5 considers the use of AMVs as instruments to threaten maritime security by focusing on cyber security issues and crimes and terrorism committed with the use of AMVs. Finally, the chapter concludes that AMVs can be accommodated within the existing international legal framework depending on the characteristics of the particular situation, whereas there are some gaps which may require further interpretations and amendments.

1.1 Introduction to Maritime Security

Maritime security does not have a uniform and universally accepted definition,[3] and there is not an international consensus over the meaning of it.[4] Much of the disagreement over the definition derives from the various interpretations of 'security' which may become a group of activities including legislative, judicial, executive, military and police actions all of which can be constructed to react to the need for order and protection from both internal and external threats.[5] The term maritime security has been attempted to be defined numerous times. Maritime security may vary in meaning based on the context and the users, and at its narrowest conception,

---

[1]The United Nations Convention on the Law of the Sea (UNCLOS).

[2]International Maritime Organization (IMO).

[3]Kraska and Pedrozo (2013), p. 1; United Nations General Assembly, Report of the Secretary-General, Oceans and the Law of the Sea (10 March 2008), UN Doc A/63/63 (UN A/63/63), para. 39, http://daccess-ods.un.org/access.nsf/Get?Open&DS=A/63/63&Lang=E. Accessed on 27 July 2022.

[4]Bueger (2015), p. 160.

[5]Klein (2011), p. 2.

it involves the protection from threats to a State's territorial integrity.[6] By some authors, maritime security is regarded '*as a stable order of the oceans subject to the rule of law at sea*'.[7] According to Hawkes, maritime security is '*those measures employed by owners, operators and administrators of vessels, port facilities, off-shore installations, and other marine organizations or establishments to protect against seizure, sabotage, piracy, pilferage, annoyance, or surprise. It can also be considered as embracing all measures taken to prevent hostile interference with lawful operations*'.[8] Another definition refers to maritime security as '*the protection of a state's land and maritime territory, infrastructure, economy, environment and society from certain harmful acts occurring at, or from the sea*'.[9] Jones makes the definition of maritime security as '*the state of a shipping company/vessel/crew/port, being or feeling secure or the safety of a shipping company/vessel/crew/port against such threats as terrorism, piracy, and other criminal activities*'.[10] Klein identifies maritime security as an inclusive interest considering the common interest in battling against a group of maritime security threats.[11]

Although maritime security is not precisely defined and may change depending on the context and numerous interests of States, it is evident that the term incorporates being secure from maritime threats. These include piracy, armed robbery, terrorism, hijacking vessels, illicit trafficking of narcotic drugs and psychotropic substances, illicit trafficking of arms and weapons of mass destruction, smuggling and trafficking of people by sea, as well as intentional and unlawful damage to the maritime environment such as illegal dumping, the discharge of pollutants from vessels, illegal, unreported and unregulated (IUU) fishing, and depletion of natural resources.[12]

The considerable aspect of maritime security is that each of the interests of States differs contingent upon the context and the use of maritime environment. Consequently, a coastal State's and a landlocked State's interests in the use of maritime space and interpretations in respect of maritime security are inevitably different. In accordance with this, maritime security has been tackled with numerous approaches subject to various perceptions of States regarding their interests. Furthermore, in the wake of remarkable technological developments, maritime security threats have been changing and evolving especially with the introduction of several new autonomous maritime vehicles and cyber-security threats.

A disruption in maritime security does not merely have an effect on the relevant parties but also has the potential to affect the world. Maritime threats have the possibility to impact the entire global economy as a disruption in the supply chain

[6] UN A/63/63, para. 39.
[7] Kraska and Pedrozo (2013), p. 1.
[8] Hawkes (1989), p. 9.
[9] Klein et al. (2009), p. 8.
[10] Jones (2012), p. 1.
[11] Klein (2011), p. 3.
[12] UN A/63/63, paras. 63, 72, 82, 89, 98, 107.

can have severe consequences. Even when a maritime crime is committed against a single ship, the impact of such crime may be well beyond the direct victims and may have an influence over the entire maritime security of people, marine environment, and maritime industry. Given that approximately 80% of global trade by volume and over 70% of global trade by value are carried by sea, and such trade is managed by ports worldwide, it goes without saying that maritime transport is the backbone of the international trade along with the global economy.[13] Maritime transport and its associated activities possess an immense overall effect on the global economy, influencing, either directly or indirectly, a vast majority of industries.[14] An interruption to the operations of a single shipping company may result in a halt in its entire business which may result in a global supply chain disruption. For example, suicide bombing of the M/V Limburg in 2002, one small incident, had the potential to have a long-term widespread effect. The attack led to a temporary halt of international shipping business in the Gulf, a 48 cent per barrel hike in the Brent crude oil price, a 93% drop in container terminal as a consequence of the tripling of war risks premiums for ships calling at Aden, giving rise to an estimated loss of $3.8 million a month in port revenues to the Yemeni economy.[15] As one of the recent disruptions in maritime transport, although not resulting from a maritime security threat, the grounding of the container ship Ever Given led to the blockage of Suez Canal and caused the interruption of traffic, delays, rerouting of vessels, and increase of freight rates.

Maritime security threats have the potential to endanger human life, global economy, and maritime environment. It is a realistic assumption that the maritime environment will remain to be vulnerable to maritime threats due to its unique character, expanse, lack of regulation, and widespread significance to the global trade.[16] Considering that the sea is the least regulated domain, it is of crucial importance for the international community to develop a range of mechanisms to prevent and respond to the threats towards maritime security and safety which require multi-jurisdictional, multi-national, multi-pronged, and multi-agency reaction.[17] The developments in technology allow the threats to be more connected and not confined to national boundaries. The interconnected, advanced, and evolving nature of these threats necessitates a new vision of collective security, and they must be handled at all levels.[18] It must be emphasised that autonomous maritime vehicles, when used by States for maritime security and law enforcement purposes, provide an exceptional opportunity to deal with the threats to maritime security.

[13] United Nations Conference on Trade and Development (UNCTAD) (2018).

[14] Fratila et al. (2021), p. 3.

[15] Chalk (2008b), pp. 23–24.

[16] Chalk (2008a), p. 127.

[17] Richardson (2004), p. ix.

[18] UN A/63/63, para. 40.

1.2 The Distinction Between Maritime Security and Maritime Safety

Although the words safety and security are synonymous in the vast majority of situations, a distinction has been made between maritime safety and maritime security in the shipping sector.[19] Such distinction can be fundamental for the reason that each indicates protection against different kinds of threats towards life and property at sea.[20] Although safety and security are distinguished from each other in the English language, the other languages usually have the same word for both safety and security which makes it hard to separate these two concepts.[21] This situation has caused some complications in the formation of international regulations. One particularly salient example is the revision of Chapter XI of the SOLAS.[22] The revised chapter is comprised of two parts. Part I is titled as 'special measures to enhance maritime safety' which explains the requirements on authorisation of recognised organisations, enhanced surveys, ship identification number scheme, and port state control on operational obligations. Part II is titled as 'special measures to enhance maritime security' which is comprised of security measures. It lays down the ISPS Code,[23] ship security alert system, requirements for port facilities, port facility security obligations for contracting governments, and the control of ships in port. The English language makes a distinction between safety and security semantically; however, many other languages do not possess the same separation including Spanish and French. '*Seguridad*' word in Spanish and '*securité*' word in French refer to both safety and security.[24] In the revision process, this problem resolved in the Spanish language by ascertaining *seguridad marítima* for maritime safety and *protección marítima* for maritime security whereas in the French language by establishing *securité maritime* for maritime safety and *sûreté maritime* for maritime security.[25]

In this regard, maritime safety can be defined as '*those measures employed by owners, operators, and administrators of vessels, port facilities, offshore installations, and other marine organizations or establishments to prevent or minimize the occurrence of mishaps or incidents at sea that may be caused by substandard ships, unqualified crew, or operator error*'.[26] The distinction between maritime security and maritime safety may become essential since each of them indicates protection against different matters. While maritime safety-related threats consist of accidents

[19] Mejia (2003), p. 154.

[20] Mejia (2004), p. 317.

[21] Mejia (2003), p. 154.

[22] The International Convention for the Safety of Life at Sea (SOLAS).

[23] The International Ship and Port Facility Security Code (ISPS Code).

[24] Mejia (2003), p. 154.

[25] Ibid, p. 154.

[26] Mejia (2002), p. 28.

caused by unsafe ships and ship operations, maritime security-related threats involve crimes committed by people against the vessel, crew, passengers, or cargo.[27] The fundamental difference between these two concepts is caused by the motivation behind the incidents and threats. Maritime security incident and maritime safety accident can be distinguished as while the former is intentional and focuses on injury to individuals and/or damage to property for political motivations, or on theft in the case of piracy, the latter is unintentional and does not emanate from a political reason.[28] Although the motivations behind them are differed from each other, maritime safety accidents and maritime security incidents may have the same outcome, namely damage to property and harm to human life.[29] For example, the difference between maritime security and maritime safety can be observed in two of the conventions of IMO, namely SOLAS and SUA Convention.[30] While the former pertains to safety at sea and is designed to protect people from accidents caused by unsafe ships, the latter deals with security at sea and is designed to protect people against the attacks of other people with criminal purpose.[31]

Although maritime safety and maritime security are two distinct concepts, they are not mutually exclusive and inharmonious with each other, rather they complement each other.[32] The legal regimes for maritime safety and maritime security have common and mutually reinforcing objectives.[33] A secure maritime space is undoubtedly a safer one; and a maritime regime which prioritises safety is less vulnerable to criminal activities and further threats to maritime security.[34] Consequently, attempts to improve either maritime safety of maritime security have cascading effects upon the conduct and regulation of other activities in the oceans, and both of them require collaborative efforts at all levels to enhance their success and deal with new challenges.[35]

International Maritime Organization, who is responsible for the safety and security of shipping as the specialised agency of the United Nations, has brought several regulations and conventions to enhance maritime security and safety. Under IMO, the Maritime Safety Committee (MSC) handles all matters relevant to maritime safety and maritime security which fall within the responsibility of IMO. This duty contains making updates to the SOLAS Convention and relevant codes,

[27] Mejia (2004), p. 317.

[28] Talley (2008), p. 1.

[29] Talley (2008), p. 1.

[30] The Convention for the Suppression of Unlawful Acts against the Safety of Maritime Navigation (SUA Convention).

[31] Mejia (2003), p. 154.

[32] Ibid, p. 156.

[33] UN A/63/63, para. 36.

[34] Ibid, para. 36.

[35] Ibid, para 36.

and managing human element matters such as amending the STCW Convention[36] in respect of training and certification of seafarers.[37] The current agenda of the MSC consists of comprehensive matters including but not limited to goal-based standards, piracy and armed robbery against ships, autonomous vessels, e-navigation, cyber security, and the modernisation of the Global Maritime Distress and Safety System.[38] The mission of IMO is stated as '*to promote safe, secure, environmentally sound, efficient and sustainable shipping through cooperation. This will be accomplished by adopting the highest practicable standards of maritime safety and security, efficiency of navigation and prevention and control of pollution from ships, as well as through consideration of the related legal matters and effective implementation of IMO's instruments with a view to their universal and uniform application*'.[39] IMO's approach towards maritime security and safety and the wide range of issues tackled by the Maritime Safety Committee reveal the strong connection between maritime security and maritime safety.

1.3 The Evolution of Maritime Security

Twenty-first century has seen remarkable developments in technology, and maritime sector is no exception. The new technological innovations involve enhanced analytics, communications technology, port-call optimisation, on-board sensors, big data, blockchains, autonomous ships and vehicles.[40] Maritime stakeholders have been benefiting from various recent technological advancements especially in automated and autonomous systems and vehicles. COVID-19 pandemic has accelerated the pre-existing digitalisation along with the environmental sustainability trends. During the pandemic, technological developments have facilitated reducing physical contact and interaction, kept ships moving, ports open and cross-border trade flowing, and allowed shipping sector to continue its operations while also stimulating online consumer spending and an expansion in e-commerce.[41] Industry guidelines keep getting updated to provide useful guidance for shipowners and operators on procedures to maintain the security of IT systems. These guidelines embrace a cyber-risk management perspective and consider IMO requirements along with other

[36]The International Convention on Standards of Training, Certification and Watchkeeping for Seafarers (STCW).

[37]IMO, Maritime Safety Committee, https://www.imo.org/en/MediaCentre/MeetingSummaries/Pages/MSC-Default.aspx. Accessed on 27 July 2022.

[38]Ibid.

[39]Ibid.

[40]United Nations Conference on Trade and Development (UNCTAD), Review of Maritime Transport 2021, p. xxiii, available at: https://unctad.org/system/files/official-document/rmt2021_en_0.pdf. Accessed on 27 July 2022.

[41]Ibid, pp. xvi, xxiii.

pertinent guidelines.[42] The significance of new technologies is increasingly acknowledged by the transport operators, customs officials, and port workers, not only for enhancing efficiency but also for preserving business stability in the event of disruption.[43]

One particularly critical development in technology is autonomous systems and vehicles from which the maritime stakeholders have been progressively benefiting. States have been increasingly utilising autonomous aerial, surface, and underwater vehicles for a variety of activities such as intelligence, surveillance, and detecting illegal activities occurring at sea such as smuggling, illegal fishing, and marine pollution.

The vital obstacle of ensuring and enhancing maritime security and safety has been the difficulties on surveillance, prevention and prosecution of crimes, and enforcement problems given the detection and practicability issues along with the costs. The advancements in respect of autonomous technology introduce several opportunities to tackle the existing problems in maritime security and enforcement. The utilisation of autonomous vehicles allows States to have better surveillance of maritime activities, provides opportunities to interfere and prevent crimes, and offers faster response to maritime security threats.

While these innovations possess numerous advantages, they also bring various new threats and vulnerabilities to the table. This technology does not only attract innocent maritime stakeholders but also appeals to the criminals, and consequently, crimes committed at sea have evolved. According to an average person, the word piracy visualises the images of Blackbeard, Jack Sparrow, eye patches, peg legs, cutlasses, and parrots.[44] In the twenty-first century, pirates do not dress as characters from movies, and rocket-propelled grenades and automatic rifles have taken the place of cannons and cutlasses.[45] With the recent developments in technology, grenades and rifles are being replaced by hackers and autonomous systems and vehicles. Now, shipping companies face ransomware attacks which have the capacity to cripple the entire operation of them. Autonomous vehicles can be used in various criminal activities such as carrying out attacks to people, ships, ports, and offshore installations. Additionally, since autonomous technology may be more vulnerable to cyber incidents, perpetrators can hijack other vehicles and systems which belong to numerous shipping companies and even States to manipulate them to achieve their goals.

'Safe, secure and efficient shipping on clean oceans' has been the slogan of IMO, and the recent developments can be employed as a very effective instrument to be one step closer to this aim. The evolving advancements are tackled by IMO which pursues to integrate new technologies to its regulatory framework. Additionally, IMO aims to balance the benefits obtained from new technologies against safety and

---

[42] Ibid, p. xxii.

[43] Ibid, p. xxiii.

[44] Mejia (2003), p. 157.

[45] Chamberlain (2008), p. 37.

security concerns, the effect on the environment, on the facilitation of international trade and on personnel, and the possible costs to the industry.[46]

The concept of maritime safety and security has started to evolve as a consequence of these new technological innovations which provide new potentials and challenges. While autonomous maritime vehicles are increasingly developed and deployed, there is a lack of clarity as to the correct application of existing IMO instruments (such as conventions, codes, guidelines, and recommendations) to these systems and vehicles.[47] This evolving nature of autonomous vehicles and its impact on maritime safety and security necessitate an analysis of their position in the current international legal framework.

This chapter attempts to look at the significant intersection between law and technology in the light of maritime security, while also bearing in mind that although technology is evolving, autonomous vehicles are still required to operate within the current international legal framework which can accommodate them to a certain extent, and to examine the terminology and legal framework pertinent to autonomous maritime vehicles and the use of them to enhance maritime security and law enforcement. The chapter then tackles the other direction, namely how the new advancements may be utilised to threaten maritime security.

Given that autonomous maritime vehicles are utilised not only by the stakeholders of the private maritime sector but also by States and criminals, such utilisation necessitates a discussion in respect of autonomous maritime vehicles and the international legal framework in relation to maritime security. Such aspect and how autonomous maritime vehicles may be used to improve or threaten maritime security have not been considerably explored in the literature. Moreover, as far as the author is aware, the most recent developments in the area such as International Maritime Organization's Regulatory Scoping Exercise on Maritime Autonomous Surface Ships (MASS) and the United Kingdom's proposal for legislative change have not been examined in relation to this subject. Consequently, this chapter attempts to shed some light on these issues by addressing the most recent developments.

[46] IMO, Autonomous Shipping, https://www.imo.org/en/MediaCentre/HotTopics/Pages/Autonomous-shipping.aspx. Accessed on 27 July 2022.

[47] IMO, Outcome of the Regulatory Scoping Exercise for the use of Maritime Autonomous Surface Ships (MASS), MSC.1/Circ.1638 (3 June 2021), para. 2.2, available via https://wwwcdn.imo.org/localresources/en/MediaCentre/HotTopics/Documents/MSC.1-Circ.1638%20-%20Outcome%20Of%20The%20Regulatory%20Scoping%20ExerciseFor%20The%20Use%20Of%20Maritime%20Autonomous%20Surface%20Ships...%20(Secretariat).pdf. Accessed on 27 July 2022.

2 Autonomous Maritime Vehicles: Terminology and Legal Status

Defining and characterising AMVs do not simply constitute a practical exercise and do not merely have technical consequences, considering that various legal rights and obligations attach to certain classification of vehicles. Such practical consequences on the status of AMVs are of significant concern to maritime stakeholders, considering that AMVs can be utilised not only as an asset which can be employed to enhance maritime security but also as a threat instrument which is more and more exploited for criminal activities.[48] Given the rapid developments in these technologies, and the lack of such advancements when the majority of the international legal instruments were established, the legal status of AMVs will ultimately depend on the interpretation of the international instruments and customary international law which may lead to new understandings in respect of AMVs.

2.1 *Terminology*

In examining unmanned systems and vehicles, several terms have been expressed based on the autonomy degrees, depending on the position of the vehicle (below, on, or above the water surface) and whether it is utilised in combats.[49] These include Unmanned Vehicle (UV), Marine Unmanned Vehicle (MUV), Unmanned Surface Vehicle (USV), Unmanned Aerial Vehicle (UAV), Unmanned Underwater Vehicle (UUV), Remotely Operated Vehicle (ROV), Maritime Autonomous Vehicle (MAV), Maritime Autonomous Surface Ship (MASS), Unmanned Maritime (or Marine) System (UMS) and Unmanned Maritime Vehicle (UMV).[50]

A distinction between Autonomous Maritime Systems (AMS) and Autonomous Maritime Vehicles (AMV) may be of necessity. AMS is the broader term which incorporates not only autonomous maritime vehicles but also the additional components such as the programmes, infrastructure, sensors, and payloads that are required for AMVs to undertake their duties.[51]

In military, the term 'unmanned' is used pervasively. For example, in the United States, the focal point is the capability of the vehicle rather than the mode of present operation, i.e. although it is manned during a period of time, a surface vehicle is considered as an unmanned surface vehicle if it has the capability of unmanned operation.[52] A distinction can also be made between the terms 'unmanned' and

[48] Allen (2018a), p. 491.

[49] Klein (2019), p. 248.

[50] Ibid, p. 248.

[51] Allen (2018a), p. 484.

[52] Ibid, p. 485. U.S. Navy also distinguishes the unmanned vehicles based on their sizes such as Large Unmanned Surface Vehicles (LUSVs), Medium Unmanned Surface Vehicles (MUSVs), and

'autonomous'. While unmanned refers to the lack of human presence on board, the term autonomous denotes the decision-making process of the system. In this framework, the term 'unmanned' can incorporate all degrees of autonomy given that in remote-controlled vehicles, the decision-making and operation are conducted by humans rather than the autonomous decision-making process of the vehicle itself.

IMO uses the terms autonomous shipping and Maritime Autonomous Surface Ships (MASS) to cover all degrees of autonomy including ships with automated processes and decision support where there are seafarers are on board to operate and control the systems, remotely controlled ships with or without seafarers on board, and fully autonomous ships.[53] This chapter also adopts IMO's glossary and uses autonomous term to cover all degrees of autonomy. For this chapter, the term Autonomous Maritime Vehicle (AMV) is used to address vehicles in every degree of autonomy. As it will be examined in the following sections, it must be emphasised that autonomous maritime vehicles incorporate not only ships but also non-ship vehicles. IMO handles and regulates autonomous shipping by using the term Maritime Autonomous Surface Ships (MASS), and various governments have already adopted the term MASS to proceed with their regulations. Within this context, in this chapter, although the term autonomous maritime vehicle (AMV) is used in general, the term MASS is used in particular parts to refer to autonomous ships, the instruments of IMO, and the regulations adopted by some States.

AMVs can be defined as vehicles with technological advancements which allow them to operate independently from human interaction to a varying degree. AMVs present themselves in various sizes (from small to extra large), in numerous capabilities (such as surveillance, data collection, and scientific research), and may be utilised for a wide range of purposes. A distinction can also be made whether they are underwater, surface, or aerial vehicles.[54] Surface vehicle can be defined as one that displaces water in an inoperative state and operates by close continuous contact with the water surface.[55] AMVs include not only surface vehicles but also underwater vehicles which can operate below the water surface along with aerial vehicles that can operate above water surface. In practice, military operations consist of the deployment of AMVs from a surface vessel, aircraft, submarine, or shore facility, and when AMVs are deployed, they may remain under the control of the deploying craft or shore facility, or they may operate independently.[56]

Extra-large Unmanned Undersea Vehicles (XLUUVs): O'Rourke R (Updated February 17, 2022), Navy large unmanned surface and undersea vehicles: background and issues for congress, congressional research service report, available via: https://crsreports.congress.gov/product/pdf/R/R45757/47. Accessed on 27 July 2022.

[53] MSC.1/Circ.1638, para. 3.4.

[54] Klein et al. (2020), p. 720; Dinstein and Dahl (2020), p. 43; Klein (2019), p. 248.

[55] U.S. Navy, The Navy Unmanned Surface Vehicle (USV) Master Plan (23 July 2007), available via: https://www.hsdl.org/?abstract&did=479083. Accessed on 27 July 2022.

[56] Allen (2018a), p. 484.

Evaluating AMVs in the context of international legal framework necessitates an explanation and a distinction of terminology since international instruments distinguish ship/vessel, boat, equipment, installation, and aircraft at sea.[57]

2.2 *Legal Status of an AMV: Ship or Vehicle*

Defining a surface AMV as a ship or a vessel under UNCLOS and relevant customary international law is a threshold issue which concerns the application of several international instruments[58] considering that identifying the autonomous maritime vehicles' legal status is a prerequisite for not only their regulation but also their protection.[59] One of the most essential questions on AMVs is that whether ships without seafarers on board are considered as ships or vessels in the context of the legal framework. Identifying them as ships has vital consequences. If AMVs are regarded as ships or vessels in the context of UNCLOS, it indicates that they are subject to the exact rules of the law of the sea as ordinary manned ships, meaning that AMVs and their flag States are subject to the same obligations to comply with international rules and regulations; additionally, they have the same navigational rights, such as innocent and transit passage, as ordinary ships.[60]

Although 'ship' and 'vessel' terms are expressed in an interchangeable way in UNCLOS, none of the two is defined in it.[61] Article 91 asserts that the conditions for the grant the nationality to ships, for the registration of ships in its territory, and for the right to fly its flag shall be fixed by every State, and there must be a genuine link between the ship and the State. This article indicates the criticality of the flag State's national law for the relevant definitions.[62]

A number of international instruments define ship or vessel. For example, Rule 3 (a) of COLREGs[63] states that '*the word "vessel" includes every description of water craft, including non-displacement craft and seaplanes, used or capable of being used as a means of transportation on water*'; MARPOL[64] Article 2(4) declares that '*"ship" means a vessel of any type whatsoever operating in the marine environment and includes hydrofoil boats, air-cushion vehicles, submersibles, floating craft and*

[57] Klein et al. (2020), p. 721.

[58] Allen (2018a), p. 480.

[59] Chang et al. (2020), p. 1.

[60] CMI International Working Group Position Paper on Unmanned Ships and the International Regulatory Framework, p. 3, available via: https://comitemaritime.org/wp-content/uploads/2018/05/CMI-Position-Paper-on-Unmanned-Ships.pdf. Accessed on 27 July 2022.

[61] Zhu and Xing (2021), p. 129; CMI Paper, p. 1.

[62] CMI Paper, p. 3.

[63] The Convention on the International Regulations for Preventing Collisions at Sea (COLREGs).

[64] The International Convention for the Prevention of Pollution from Ships (MARPOL).

fixed or floating platforms'; ship is defined in OILPOL[65] as '*any sea-going vessel of any type whatsoever, including floating craft, whether self-propelled or towed by another vessel, making a sea voyage*'; under Article 1 of the SUA Convention, '"*ship*" *means a vessel of any type whatsoever not permanently attached to the sea-bed, including dynamically supported craft, submersibles, or any other floating craft*'; and SALCON[66] defines 'vessel' in Article 1(b) as '*any ship or craft, or any structure capable of navigation*'.

The foregoing provisions reveal that there is not a single definition of 'ship' or 'vessel', and this itself creates an ambiguity on the legal status of AMVs as ships. It must be noted that these treaties are designed for specific purposes, and therefore, their approaches in tackling the definition of ship inevitably vary to satisfy such purposes. The most appropriate approach to address this problem perhaps would depend on two inquiries. The first inquiry involves the context and application in question, focusing on the purpose of classification and the consequences of inclusion or exclusion of an AMV thereof.[67] The second inquiry requires the application of the purpose-based definition to the design and individual AMV's operational qualities.[68] Considering the variety of AMV designs and modes of operation, one specific characterisation which would be applicable to the entire class of AMVs is not the most appropriate solution.[69]

It has been submitted that although there is not an agreed definition of 'ship' or 'vessel', a ship would normally be required to have navigation capabilities,[70] meaning being able to have controlled movement on the surface of water,[71] along with transportation capabilities[72] meaning that a ship should be able to carry passengers, cargoes, weapons, systems, or other particular objects to maintain specific purposes such as commercial, military, public service, and scientific objectives.[73] Although there are certain AMVs that fulfil such conditions, particular aspects of AMVs may make them fall outside the scope of ships given that the existing AMVs are mainly in small size and international regulations are only applicable to the ships which are exceeding certain minimum tonnages,[74] and the lack of transportation capabilities also has the ability to exclude such AMVs from the scope of ships.[75] In this regard, the legal status of autonomous maritime vehicles also depends on the users considering that the users of AMVs incorporate commercial

[65] The International Convention for the Prevention of Pollution of the Sea by Oil (OILPOL).

[66] The International Convention on Salvage (SALCON).

[67] Allen (2018a), p. 493.

[68] Ibid, p. 493.

[69] Ibid, p. 493.

[70] Veal et al. (2019), p. 25; Xing and Zhu (2021), p. 127.

[71] Veal et al. (2019), p. 25.

[72] Xing and Zhu (2021), p. 127; ibid., p. 26.

[73] Xing and Zhu (2021), p. 127.

[74] Veal et al. (2019), p. 25.

[75] Ibid, p. 25; Xing and Zhu (2021), p. 127.

and scientific communities along with militaries.[76] Consequently, the size, capabilities, characteristics, and users have direct impacts on evaluating a particular AMV as a ship or not.[77] Nevertheless, due to the pace of new technology, more and more complex and large systems which allow AMVs to have the capabilities of navigation and transportation are developed, and this brings the question whether such AMVs without crew on board may possess the legal status of a ship/vessel.[78]

It must be emphasised that the existing international instruments which incorporate definitions of the term ship do not refer to crewing, and similarly in national regulations, the ship definition is typically independent of whether the ship is manned or unmanned.[79] According to CMI, given the nature of the activities performed by large, cargo-carrying, self-propelled, and commercially operated unmanned ships, they are most likely to be considered as ships/vessels because of their size, functions, and features, and it would not be justified that two ships, manned and unmanned, carrying out analogous tasks under similar dangers being not subject to the exact rules and regulations which have been intended to deal with those dangers.[80]

CMI defines unmanned ships as '*those which are capable of controlled movement on the water in the absence of any onboard crew*'[81] and states that although there are more methods of modes of operation, a binary distinction between two modes is important for regulatory compliance. In this framework, control is carried out primarily in two ways. First, control can be executed by way of remote-control where there is a remote controller on shore who controls the operation of the ship.[82] Secondly, the ship may be autonomously controlled where the ship is pre-programmed prior to its operation, and subsequently, carries out a predetermined nautical course with no human interaction of any kind.[83]

IMO defines Maritime Autonomous Surface Ships (MASS) '*as a ship which, to a varying degree, can operate independent of human interaction*'.[84] MASS includes ships with different automation levels, from partially automated systems which assist the crew to fully autonomous systems that can undertake all features of the operation of a ship without the necessity for human intervention.[85] Thus, given that IMO regulates MASS, it can be deduced that there is an inclination towards categorising at least some types of AMVs as ships.

[76] Showalter and Manley (2009), p. 1.

[77] For non-ship AMVs, see Sect. 2.4.

[78] Xing and Zhu (2021), p. 125.

[79] CMI Paper, p. 3.

[80] Ibid, p. 3.

[81] Ibid, p. 1.

[82] Ibid, p. 1.

[83] Ibid, p. 1.

[84] MSC.1/Circ.1638, para. 3.3.

[85] Ibid, para. 2.1.

In this framework, when evaluating whether an AMV is a ship or a vessel, it should be analysed within the context of the related legal instrument, and then it shall be decided whether the rules of a particular instrument will be applicable. It is also of crucial significance to bear in mind that considering the lack of consensus in the international instruments and the impact of the flag State's national law established in Article 91 of UNCLOS, the legal status of a specific AMV and whether such vehicle constitutes a ship/vessel shall be determined in accordance with the national laws of States.[86] Moreover, identifying AMVs as ships may become customary international law in the future since the general practice of States and *opinion juris* may develop and evolve in the light of the rapid technological advancements.

2.3 *Legal Status of an AMV in Naval Warfare*

2.3.1 AMV As a Warship

Although this chapter mainly focuses on the operations of AMVs during peacetime, the legal status of AMVs as warships and auxiliary vessels are briefly mentioned under this section to provide a broader understanding in respect of the legal status of autonomous maritime vehicles.

If an AMV is identified as a ship, then another question arises in the international legal framework, namely whether an autonomous ship can be considered as a warship. Identification of a vessel as a warship has numerous consequences in the law of the sea. UNCLOS establishes that warships and other government ships operated for non-commercial purposes possess immunity.[87] Identifying AMVs as warships gain special significance in respect of navigational rights along with belligerent rights in an armed conflict.

Article 29 of UNCLOS defines warship as '…*a ship belonging to the armed forces of a State bearing the external marks distinguishing such ships of its nationality, under the command of an officer duly commissioned by the government of the State and whose name appears in the appropriate service list or its equivalent, and manned by a crew which is under regular armed forces discipline*'.[88] As to belonging to the armed forces of a State and bearing the distinguishing external marks, AMVs can easily facilitate these requirements. However, two particular expressions stand out regarding autonomous ships in this description, namely 'under the

---

[86] Chang et al. (2020), pp. 2–3; Veal et al. (2019), p. 29.

[87] UNCLOS Articles 32, 95, 96.

[88] This definition emanates from Article 8 of 1958 Geneva Convention on the High Seas in which warship is defined as '*a ship belonging to the naval forces of a State and bearing the external marks distinguishing warships of its nationality, under the command of an officer duly commissioned by the government and whose name appears in the Navy List, and manned by a crew who are under regular naval discipline*' which is derived from the 1907 Hague Convention VII relating to the Conversion of Merchant Ships into War-Ships: Nasu and Letts (2020), p. 85.

command of an officer' and 'manned by a crew which is under regular armed forces discipline'.

Some authors have submitted that AMVs could not be qualified as warships since such identification would necessitate extending the concept of 'under the command of an officer' to incorporate remote control, and furthermore, the plain text on 'manned by a crew' would make it highly unlikely to consider unmanned AMVs as warships.[89] On the contrary, it has been expressed that 'under the command' focuses on navigational safety and interaction with other vessels, the authority, responsibility, control, direction, and coordination all of which can be accomplished remotely; and what concerns 'command' the most is the capacity to take action, rather than a physical presence.[90] According to some authors, there is no evidence which reflects that the parties intended to preclude the categorisation of AMVs as warships when warship was defined in conventions, instead, the matter was not even discussed since technology then did not incorporate AMVs.[91] The author of this chapter submits that in accordance with the degrees of autonomy established by IMO,[92] certain autonomous ships may be qualified as warships especially the ones with the autonomy level one, namely *'ship with automated processes and decision support'* and level two, namely *'remotely controlled ship with seafarers on board'*. These degrees of autonomy can provide 'manned by a crew' requirement, and human control is key in their operations. However, if the vessel is fully autonomous and there is no human involvement, it gives rise to the question of whether an autonomous system or AI can qualify 'under the command of an officer' requirement. In the current legal frame, it is highly unlikely to evaluate fully autonomous vessels as warships since there is no officer who commands the ship, and when there are no seafarers on board, manning by a crew obligation likewise fails.

It is noteworthy that even if an AMV is not considered in the category of warships in the context of international law, it may still possess immunity as 'other government ships operated for non-commercial purposes' under Article 32 of UNCLOS.

The legal status of AMVs will inevitably depend on the approaches of States in the qualification of such vehicles. Although the legal status of AMVs as warships is not established under the current legal framework, either a treaty or customary international law which reflects the general practice of States may resolve the ambiguity in the future.

2.3.2 AMV As an Auxiliary

If AMVs are not categorised as warships, an alternative has been proposed on their classification under the law of naval warfare, namely auxiliary vessels. According to

[89] Schmitt and Goddard (2016), p. 579; McLaughlin (2014), p. 238.

[90] Klein et al. (2020), p. 724.

[91] Nasu and Letts (2020), p. 85.

[92] See Sect. 2.5 of this chapter.

this view, given that an AMV can fall under the category of a ship, it can be an auxiliary vessel[93] which is defined as '*a vessel, other than a warship, that is owned by or under the exclusive control of the armed forces of a State and used for the time being on government non-commercial service*' in Section 13 (h) of the San Remo Manual.[94] Considering that this definition does not involve command or crew requirements, but rather being owned by or under the control of the armed forces and used for non-commercial service, it has been stated that it is less challenging for AMVs to meet these requirements of the auxiliary vessel status.[95] However, auxiliary vessels do not possess the equal belligerent rights vested in warships. A number of international instruments provide a few belligerent rights auxiliary vessels may have such as carrying troops and replenishing warships,[96] and there are various tasks which auxiliaries cannot exercise such as attacking submarines and engaging in attack–hunting,[97] however, there remains substantial uncertainty on the scope of the rights and obligations of auxiliaries which ultimately depends on State practice.[98]

2.4 AMV As a Non-Ship Vehicle

In the light of the variety in AMV designs and operational modes, it is highly impossible and imprudent to make a single classification which applies to the entire category of autonomous vehicles.[99] AMVs differentiate in size, capabilities, operational limits, and purposes. As a result, naturally, they fall within the scope of several categories under numerous international instruments. Consequently, the problem of categorising an AMV which is not considered as a ship/vessel arises. Various terms are used to address AMVs such as objects,[100] systems, auxiliaries,[101] military devices,[102] autonomous and unmanned submersible apparatus,[103] and marine equipment.[104] There are particular types of maritime vehicles and items identified and

[93] Klein et al. (2020), p. 722.

[94] San Remo Manual on International Law Applicable to Armed Conflicts at Sea, 12 June 1994.

[95] Klein et al. (2020), pp. 724–725.

[96] San Remo Manual, Section IV Rule 60 (b).

[97] Klein et al. (2020), p. 725.

[98] Ibid, p. 725.

[99] Allen (2018a), p. 493.

[100] Ibid.

[101] Klein et al. (2020), p. 722.

[102] UNCLOS Article 19 (2) (f); Vallejo (2015), p. 414.

[103] Department for Transport, Future of transport regulatory review: maritime autonomy and remote operations (28 September 2021), available via: https://www.gov.uk/government/consultations/future-of-transport-regulatory-review-maritime-autonomy-and-remote-operations/future-of-transport-regulatory-review-maritime-autonomy-and-remote-operations. Accessed on 27 July 2022.

[104] Ibid.

regulated under UNCLOS such as ships/vessels,[105] boats,[106] installations and structures,[107] artificial islands,[108] devices,[109] submarine cables and pipelines,[110] scientific research installations, or equipment.[111] The identification of AMVs gains significance especially with regard to the provisions in the international instruments which provide certain rights and obligations based on the qualification of the vehicle. If a particular AMV is considered within the scope of such provisions, then it will be subject to certain rights and obligations attached to its characterisation.

It is also proposed that AMVs should be categorised as 'military devices'.[112] The author of this view claims that the ambiguity of the legal status of AMVs becomes particularly concerning in the event of an accident involving such vehicles, and it takes years to make amendments to the international instruments.[113] Therefore, Vallejo enunciates that Section 845 of Title 18 of the United States Code[114] in which it is stated that '*military device includes, but is not restricted to, shells, bombs, projectiles, mines, missiles, rockets, shaped charges, grenades, perforators, and similar devices lawfully manufactured exclusively for military or police purposes*' can be used as a starting point given the characteristics and purposes of the vehicles and the fact that these are owned by militaries.[115] The author of this chapter respectfully differs from this view on the grounds that the term 'military device' is not widely used in the international law and such use may complicate the legal status rather than clarify the terminology. Similarly, to identify an AMV which falls outside the scope of a vessel, the term 'object' is used by some authors.[116] This definition then again does not correspond to the classifications identified by the international law and may be perceived as a very broad term which overlooks the specific qualities of AMVs.

With regard to aerial autonomous vehicles, more commonly known as drones, they fall under the category of aircraft. Such identification necessitates aerial AMVs to be subject to international aviation law and overflight rules.[117] The overflight of aerial vehicles is of importance in the assessment of the legality of its passage over

[105] See Sect. 2.2.

[106] UNCLOS Articles 110–111.

[107] UNCLOS Articles 56, 60, 79, 80, 208, 214, 246.

[108] UNCLOS Articles 11, 56, 60, 79, 80, 208, 214, 246.

[109] UNCLOS Articles 19, 145,194, 209.

[110] UNCLOS Articles 21, 58, 79, 87, 112–115, 145, 297.

[111] UNCLOS Articles 258–263.

[112] Vallejo (2015), p. 414.

[113] Ibid, pp. 415–416.

[114] 18 U.S.C. § 845 (c) (3) (2006).

[115] Vallejo (2015), pp. 414–416.

[116] Allen (2018a), p. 513.

[117] Klein (2019), p. 251.

the territorial sea of a coastal State or over an international straight in respect of the transit passage regime.[118]

Therefore, the legal status of an AMV depends on its features and how it can be accommodated within the international legal instrument it has connections to. For example, float indicates '*an autonomous vehicle used for collection of ... data ... and floating passively at a preprogrammed pressure level until at predetermined time intervals rising to the ocean surface to broadcast its position and, as the case may be, collected data to a satellite*'.[119] Glider is an autonomous underwater vehicle which uses wave and solar energy to move and may be equipped with sensors to monitor currents, temperature, salinity and other ocean conditions, and also used for data collection purposes.[120] Considering that gliders do not require human assistance, or need very little assistance, during the course of travelling, they can be quite advantageous in the utilisation of collecting data in further locations, in a safe manner, and at a lower cost.[121] They can also be utilised for underwater battlespace in high-threat areas to detect underwater and surface threats, and they may be capable of sending crucial information assisting submarine hunting operations.[122] There are particular conditions which make categorising these AMVs as ships highly questionable. First issue is navigation. Ships are either used in navigation or suitable for such use. On the other hand, although floats have movement in water, they follow currents with no opportunity to change their own course; consequently, since they do not navigate, floats cannot be considered as ships.[123] In terms of gliders, they may possess tools of navigation on board which can allow them to reach destinations, however, gliders lose this ability in the event of strong currents.[124] The second point is the capability of transportation which is frequently intermixed with navigation or size of a ship.[125] It may be argued that a ship must be in a certain size to be capable of the transportation of humans or goods, and given the small size of gliders, they do not have that capability.[126] The final point is the propulsion issue of which gliders

[118] Ibid, p. 251.

[119] Intergovernmental Oceanographic Commission of UNESCO, Draft [Practical] Guidelines of IOC, Within the Context of UNCLOS, for the Collection of Oceanographic Data By Specific Means, Seventh Meeting of the Advisory Body of Experts on the Law of the Sea (IOC/ABE-LOS VII) (Libreville, Gabon 19–23 March 2007), available via: https://unesdoc.unesco.org/ark:/48223/pf0000218771. Accessed on 27 July 2022.

[120] National Ocean Service National Oceanic and Atmospheric Administration, U.S. Department of Commerce, What is an ocean glider?, https://oceanservice.noaa.gov/facts/ocean-gliders.html. Accessed on 27 July 2022.

[121] Ibid.

[122] Naval Technology (2020), Royal Navy to test underwater gliders in North Atlantic, https://www.naval-technology.com/news/royal-navy-to-test-underwater-gliders-in-north-atlantic/. Accessed on 27 July 2022.

[123] Bork et al. (2008), pp. 308–309.

[124] Ibid.

[125] Ibid.

[126] Ibid.

merely possess passive version and cannot move freely in any direction. Unlike a ship which has the capability of following a course not depending on currents, gliders heavily rely on the environment for movement, and in the event of an encounter with strong currents, they cannot stick to the programmed course.[127]

To conclude, the foregoing points make it difficult to consider some of the AMVs such as floats and gliders as ships in the context of international law, however, they may be regarded as installations, structures, devices, or equipment in the scope of UNCLOS.[128] The main challenge is that the terms 'ship', 'vessel', 'device', 'equipment' and 'installations' do not have precise and internationally accepted definitions. It can be assumed that these AMVs do not correspond to the definitions of ship or vessel and can be regarded as equipment instead of installations under the marine scientific research provisions of UNCLOS.[129]

2.5 *The Degrees of Autonomy*

AMVs vary in degrees of autonomy starting from the ones with the most human involvement to the fully autonomous ones. The differentiation of the levels of autonomy and human involvement has crucial consequences particularly in respect of the characterisation of the vehicle and determining liability.[130]

Lloyd's Register (LR) has established Unmanned Marine Systems Code[131] in which six levels of autonomy are identified ranging from manual to fully autonomous. Autonomy Level (AL) 0 refers to manual where there is no autonomous function. All action and decision-making are carried out manually (n.b. systems may possess a level of autonomy with human in/or loop). In other words, humans control the actions. AL1 level represents on-board decision support where every action is taken by a human operator, however, there is a decision support tool which can produce options or influence the actions chosen. The systems on board provide data. AL 2 means on and off-board decision support where a human operator takes every action, however, the decision support tool can give options or affect the actions chosen. The systems on or off-board may provide data. AL3 refers to active human in the loop where the decisions and actions are carried out with human supervision. The systems on or off-board may provide data. AL 4 denotes human on the loop, operator/supervisory where the decisions and actions are carried out autonomously together with human supervision. High impact decisions are executed in a way providing the opportunity to human operators to intervene and over-ride. AL5 refers

[127] Ibid.

[128] Ibid.

[129] Ibid.

[130] Klein et al. (2020), p. 721.

[131] The Lloyd's Register Unmanned Marine Systems Code, available via: https://www.lr.org/en/unmanned-code/. Accessed on 27 July 2022.

to fully autonomous in which supervised operation rarely occurs and the system fully makes and actions the decisions. AL 6 denotes fully autonomous where there is an unsupervised operation in which the decisions are completely made and actioned by the system in the course of the operation.[132] According to the relevant code, various levels of autonomy can be used at the same time. Several systems at different levels may be combined to create a complex system. Similarly, a higher level of autonomous system may make use of a lower level of autonomous system as a part of its reversionary control.[133]

The Maritime Safety Committee, under IMO, has conducted the regulatory scoping exercise (RSE) for the use of Maritime Autonomous Surface Ships (MASS) which delivers the examination of the existing regulatory framework in connection with MASS operations, provides advice to the relevant parties in respect of identifying, selecting and deciding on the future work on MASS.[134] RSE is a significant first step which paves the way to dedicated discussions to ensure that regulations will keep up with the technological advancements.[135]

IMO has developed four degrees of autonomy in respect of MASS operations. These categories of the degrees of autonomy may be analogously applicable to the other autonomous maritime vehicles to the extent that they correspond to. It is noteworthy that the list which reflects the levels of autonomy does not provide a hierarchical order, and MASS can operate not only at one but also more than one degree of autonomy for the duration of one single voyage.[136] IMO has also acknowledged that references to the degrees of autonomy in the RSE document refer merely to the definitions examined within RSE and do not preclude prospective future definitions which shall be considered at a further stage.[137] This approach indicates that IMO accepts the evolving nature of these technological developments and there may be additional categories of autonomy in the future.

Degree One: 'Ship with Automated Processes and Decision Support' This level of autonomy is the one with the most human intervention and where there are seafarers on board operating and controlling the systems and functions of the ship. Although various operations may be automated and every so often be unsupervised, it comes along with seafarers on board who are ready to take control.[138]

Degree Two: 'Remotely Controlled Ship with Seafarers on Board' In this degree of autonomy, while the ship is controlled and operated from a different location,

[132] Ibid, Section 4.

[133] Ibid, Section 4.

[134] MSC.1/Circ.1638, para 1.2.

[135] IMO, Autonomous ships: regulatory scoping exercise completed, https://www.imo.org/en/MediaCentre/PressBriefings/pages/MASSRSE2021.aspx. Accessed on 27 July 2022.

[136] MSC.1/Circ.1638 para. 3.5.

[137] Ibid, para. 3.13.

[138] Ibid, para. 3.4.

there are seafarers on board to take control and operate the systems and functions of the ship.[139]

Degree Three: 'Remotely Controlled Ship Without Seafarers on Board' This includes the ships which do not have seafarers on board and are controlled and operated from a different location.[140]

Degree Four: 'Fully Autonomous Ship' This level of autonomy is the highest degree. The ship's operating system is capable of making decisions and determining actions by itself.[141]

AMVs can operate at various degrees of autonomy. Based on the development of technology, more categories may be developed in the future. At the moment, IMO tackles the levels of autonomy with these four categories, and the majority of the international community and States adopt this categorisation. The levels of autonomy play a significant role in the regulation of AMVs, and therefore, the most prudent approach perhaps would be internationally adopting these levels of autonomy established by IMO for a unified method.

3 Autonomous Maritime Vehicles: Legal Framework

The increasing utilisation of AMVs requires an analysis in respect of their position in the existing legal framework. One should always keep in mind that international treaties should be interpreted '*in good faith in accordance with the ordinary meaning to be given to the terms of the treaty in their context and in the light of its object and purpose*'.[142]

In this context, UNCLOS, IMO's approach on MASS, and the United Kingdom's proposal for legislative change shall be addressed in the following sections.

3.1 Legal Framework Under UNCLOS

As one might expect, given that a vast majority of the conventions and instruments relevant to the law of the sea have been established many years ago, they do not possess the same pace as the technological developments, and it would be nonsensical to assume that the existing rules perfectly align with the novelties presented by such developments. There are various gaps on AMVs in the context of the

[139] Ibid, para. 3.4.

[140] Ibid, p. 4.

[141] Ibid, p. 4.

[142] Article 31 (1) of the Vienna Convention on the Law of Treaties.

international legal framework and UNCLOS. Addressing all of them is beyond the scope of this chapter, however, it may be beneficial to point out some of the challenging issues.

3.1.1 Gaps in UNCLOS

Various possible challenges present themselves within the context of the legal framework of UNCLOS. Addressing all these challenges is beyond the scope of this chapter, however, some articles of UNCLOS are referred under this section. There are several compatibility matters with UNCLOS and the operations of autonomous ships.[143] Although UNCLOS does not contain provisions which prohibit the operation and classification of autonomous vehicles,[144] a number of articles presents various challenges for autonomous vehicles especially concerning autonomous ships.[145] One particularly salient example of a challenging provision is Article 94 of UNCLOS. Article 94 establishes the duties of the flag state which includes taking measures for ships which are necessary to ensure safety at sea including the manning of ships. This may constitute problems especially for fully autonomous ships. Article 94 (4) also states that such measures shall include that '*each ship is in the charge of a master and officers who possess appropriate qualifications, in particular in seamanship, navigation, communications and marine engineering, and that the crew is appropriate in qualification and numbers for the type, size, machinery and equipment of the ship*'[146] and '*the master, officers and, to the extent appropriate, the crew are fully conversant with and required to observe the applicable international regulations concerning the safety of life at sea, the prevention of collisions, the prevention, reduction and control of marine pollution, and the maintenance of communications by radio*'.[147] Article 94 of UNCLOS presumes that the vessel shall be crewed, and such presumption creates problems on the regulation of autonomous ships by flag States.[148] The obligations of the flag States as established by Article 94 present several challenges concerning autonomous ships such as the definition of the terms master and crew, the qualifications of the master and crew, and the regulation of autonomous ships where the master and crew are not on board.[149] Considering that, in practice, the ship's Master has the authority and the responsibility to implement the flag State's regulations meaning that the flag State controls the ship via the Master of the vessel, the exercisability of such control in the

[143] Ringbom (2019), p. 161.
[144] Parker (2021), p. 34.
[145] Ibid, p. 34.
[146] UNCLOS Article 94 (4) (b).
[147] UNCLOS Article 94 (4) (c).
[148] Parker (2021), p. 34.
[149] Ibid, p. 35.

event of autonomous ships may become problematic.[150] Traditionally, the master and the crew of a vessel have been appropriately qualified mariners who carry out their duties on board.[151] The main problem with Article 94 of UNCLOS is that whether such role is restricted to onboard personnel or whether programmers and remote controllers may be perceived as the master and the crew.[152] It is noteworthy that the master and the crew are not defined in UNCLOS, and customary international law does not provide direction in the definitions all of which make it difficult to reach clarity.[153] However, such provisions of Article 94 should not be interpreted as excluding autonomous ships. Article 94(5) regulates that in taking such measures, '*each State is required to conform to generally accepted international regulations, procedures and practices and to take any steps which may be necessary to secure their observance*'. In this context, Article 94 should be interpreted by bearing in mind the other relevant regulations.[154] This wording suggests that UNCLOS has referred to generally accepted international rules that are not only abstract but also keep changing and being established somewhere else, such as at IMO, rather than precisely forming the obligations of the flag States.[155] By doing so, UNCLOS has prevented the requirements to be frozen at a particular technical degree or at a certain time.[156] The identification and examination of the other related international regulations is at the heart of determining whether the operation of an autonomous ship that has no crew on board would violate manning obligations under UNCLOS.[157] Consequently, when the international community wishes for autonomous ships to be part of such international legal framework, the wording of Article 94 should not be seen as precluding technological advancements in the shipping sector,[158] and IMO can regulate autonomous ships in this regard[159] given that the competent international organisation in UNCLOS denotes IMO and the regulations of IMO can be used in the interpretation of the provisions of UNCLOS.[160] In this context, specific safety standards which include the manning requirements shall be established by IMO instruments.[161] Subsequently, IMO's proactive role in the clarification of the Article 94 may alleviate legal concerns on autonomous ships.[162]

150 Zhu and Xing (2021), p. 137.
151 Veal et al. (2019), p. 36.
152 Ibid, p. 36.
153 Ibid, p. 36.
154 Zhu and Xing (2021), p. 133.
155 Ringbom (2019), p. 161.
156 Ibid, p. 161.
157 Prasetya (2020), p. 108.
158 Ringbom (2019), pp. 161–162.
159 Ibid, p. 161.
160 Zhu and Xing (2021), p. 134. For a detailed analysis of the relationship between UNCLOS and IMO, see Sect. 3.1.2.
161 Nawrot and Peplowska-Dabrowska (2019), p. 247.
162 Zhu and Xing (2021), p. 134.

There are various other gaps in UNCLOS as well. Under Article 211, in which pollution from vessels is regulated, States shall require the vessel's master to provide information. Article 217 (3) demands States to ensure that vessels carry the relevant on-board certificates which are necessary in accordance with the international instruments, and vessels are required to be inspected by States for verification of such certificates. This again creates problems on how these certificates will be carried and verified in a ship that do not have seafarers on board.

Another important challenge is seaworthiness. States are required to take the necessary measures to ensure safety which involves the construction, equipment, and seaworthiness of ships[163] and they have to take measures concerning seaworthiness of vessels to avoid pollution.[164] This reveals further problems on seaworthiness of autonomous ships which may require further regulations since the rules created by existing instruments and technological advancements do not have the same pace.

These rules bring several challenges particularly in respect of fully autonomous ships which do not have seafarers on board and when the decision-making process is conducted by the autonomous systems, all of which makes it questionable how AMVs fit in the existing international legal framework and how well they can be accommodated within it.

3.1.2 UNCLOS and IMO

UNCLOS is considered as 'the constitution of the oceans'[165] which is commonly acknowledged as the general legal framework within which every activity in the oceans and seas must be performed.[166] Due to its constitutional nature, framework characteristic which involves fundamental norms that set out guiding principles and a general legal frame instead of particularised rules similar to statutes, and finally living nature which grows and adapts to changing conditions similar to a living organism, it has been submitted that UNCLOS may be interpreted in an evolutionary manner depending on the circumstance.[167] The law of the sea is dynamic rather than static, and technology has significantly progressed since 1982. Accordingly, adhering and giving positive effect to such dynamism can be accomplished by way of interpretation and construction of the articles.[168]

[163] UNCLOS Article 94 (3) (a).

[164] UNCLOS Article 219.

[165] Koh TTB, 'A Constitution for the Oceans' Remarks by Tommy T.B. Koh of Singapore, President of the Third United Nations Conference on the Law of the Sea, available via: http://www.un.org/Depts/los/convention_agreements/texts/koh_english.pdf. Accessed on 27 July 2022.

[166] Sekimizu (2014).

[167] Petrig (2020), pp. 114–117.

[168] Lucky (2015).

Notwithstanding that IMO is clearly mentioned in merely one of the articles of UNCLOS, namely Article 2 of Annex VIII, the Convention's various provisions address to a competent international organisation with respect to the adoption of rules and standards on maritime safety, navigation, and preventing and controlling marine pollution.[169] In this scope, the competent international organisation, when it is used in a singular form in the convention, denotes IMO.[170]

Given the 'framework convention' nature of UNCLOS, a vast majority of its provisions can only be implemented by way of special operative regulations within additional international agreements.[171] This situation reveals itself in various UNCLOS provisions which bring requirements to States to 'give effect to', 'implement', 'take account of' or 'conform to' the related international rules and standards established by the competent international organisation, namely IMO.[172] It can be observed that the provisions containing such requirements establish an obligation on Parties to UNCLOS to apply the rules and standards of IMO. The particular type of such application is based on the interpretations of Parties to the expressions of UNCLOS articles which provide requirements relevant to IMO provisions.[173] Such application also depends on the types of IMO instruments, whether they are the recommendations adopted by the IMO Assembly, the IMO Maritime Safety Committee (MSC), and the IMO Marine Environment Protection Committee (MEPC), or whether the rules and standards exist in IMO treaties.[174] The former involves resolutions adopted by Member States that incorporate recommendations concerning the implementation of technical rules and standards which IMO treaties do not incorporate. Such resolutions are typically adopted through consensus and therefore indicate IMO Member States' global agreement which necessitates an expectation from Parties to UNCLOS as to conform to such rules and standards.[175] On IMO treaty instruments, the obligations determined by UNCLOS on compliance with IMO rules and regulations should be evaluated in relation to each treaty's particular operative features which relate to how the rules and standards regulate substantial matters along with how procedural rules govern the interrelations between flag, coastal, and port State.[176]

In this framework, the overlapping of possible conflict between IMO and UNCLOS has been prevented by way of including various provisions to IMO

[169] IMO, Implications of the United Nations Convention on the Law of The Sea for the International Maritime Organization, Study by the Secretariat of the International Maritime Organization (30 January 2014), LEG/MISC.8, p. 7, available via: https://www.kuestenpatent-kroatien.at/LEG%20MISC%208-1.pdf. Accessed on 27 July 2022.

[170] Ibid, p. 7.

[171] Ibid, p. 8.

[172] Ibid, p. 8.

[173] Ibid, p. 10.

[174] Ibid, p. 10.

[175] Ibid, p. 10.

[176] Ibid. p. 10.

instruments enunciating that they should not be interpreted as prejudicing the development and codification of the law of the sea in UNCLOS, or incorporating clauses stating that these treaties should not prejudice States' claims and legal views regarding the law of the sea along with the nature and scope of jurisdiction of the coastal and flag State.[177] Consequently, it has been ensured that the regulatory activities of IMO do not overlap with the developments in the law of the sea, and the legal certainty has been improved.[178]

To conclude, the legal framework of autonomous maritime vehicles and systems will primarily be contingent on IMO's approach along with the flag and coastal States' practices.[179] Every instrument of IMO operates within the legal framework of UNCLOS.[180] What allows the legal framework's evolution, effective response, and flexibility to the new developments and problems is the dynamic relationship between UNCLOS and IMO.[181] Subsequently, it can be assumed that IMO will continue to elaborate the ambiguities on AMVs, and therefore, attempt to facilitate the integration of autonomous maritime vehicles and systems into the international legal framework in its further efforts.

3.2 IMO's Approach Regarding MASS

3.2.1 In General

As anticipated, IMO has started to tackle the regulation of autonomous ships. IMO's Strategic Plan (2018–2023) involved a vital Strategic Direction to 'integrate new and advancing technologies in the regulatory framework' which comprises of balancing the benefits resulting from new technologies against concerns for safety and security, the possible costs to the industry, the impact on the environment, on international trade and on personnel.[182] Maritime Safety Committee (MSC), in the 98th session which took place in 2017, put scoping exercise on autonomous vessels on its agenda. This was in the structure of a scoping exercise to ascertain how safe, secure, and environmentally sound operation of Maritime Autonomous Surface Ships (MASS) might be established in the instruments of IMO.[183] It was acknowledged by MSC that IMO's role should be proactive and leading considering the new technological developments in respect of the introduction of the commercially operated

[177] Ibid, p. 11.

[178] Ibid, p. 11.

[179] Zhu and Xing (2021), p. 152.

[180] Beckman and Sun (2017), p. 235.

[181] Ibid, p. 236.

[182] IMO, Autonomous Shipping, https://www.imo.org/en/MediaCentre/HotTopics/Pages/Autonomous-shipping.aspx. Accessed on 27 July 2022.

[183] IMO, Maritime Safety Committee (MSC) 98th session 7–16 June 2017, https://www.imo.org/en/MediaCentre/MeetingSummaries/Pages/MSC-98th-session.aspx. Accessed on 27 July 2022.

autonomous/unmanned ships. In this framework, scoping exercise was determined as a starting point to tackle the issue.[184]

In June 2019, MSC approved 'interim guidelines for Maritime Autonomous Surface Ships (MASS) trials'. It has been established that these guidelines should be applied when performing trials in respect of MASS-related systems and infrastructure with the aim of ensuring that the manner in which trials are conducted supports at least the same level of security, safety, and environment protection as provided by the related instruments.[185]

Maritime Safety Committee conducted the regulatory scoping exercise (RSE)[186] for the use of MASS. The aim of RSE was ascertaining '*how safe, secure and environmentally sound MASS operations might be addressed in IMO instruments*'[187] and the objective was assessing the extent to which the existing regulatory framework might be affected to deal with the operation of MASS.[188] Scoping exercise addressed the instruments[189] such as safety and maritime security (SOLAS), training of seafarers and fishers (STCW, STCW-F), collision regulations (COLREG), search and rescue (SAR),[190] loading and stability (Load Lines),[191] tonnage measurement (Tonnage Convention),[192] special trade passenger ship instruments (SPACE STP, STP),[193] and Safe Containers (CSC[194]).[195]

IMO has revealed the instruments under the responsibility of the Maritime Safety Committee where the potential gaps and themes have been detected.[196] Various provisions have been determined for each instrument and degree of autonomy. The potential gaps and themes have been established in accordance with appendix 2 of the IMO document[197] in which the results of the Regulatory Scoping Exercise at instrument level are displayed. According to appendix 2, the application of currently drafted IMO instruments is divided into four categories. Category A is 'applied to

[184] Ibid.

[185] IMO, Interim Guidelines for MASS Trials, MSC.1/Circ.1604 (14 June 2019), p. 1, available via: https://wwwcdn.imo.org/localresources/en/MediaCentre/HotTopics/Documents/MSC.1-Circ.1604%20-%20Interim%20Guidelines%20For%20Mass%20Trials%20(Secretariat).pdf. Accessed on 27 July 2022.

[186] MSC.1/Circ.1638.

[187] Ibid, para. 3.1.

[188] Ibid, para. 3.2.

[189] IMO, Autonomous Shipping, https://www.imo.org/en/MediaCentre/HotTopics/Pages/Autonomous-shipping.aspx. Accessed on 27 July 2022.

[190] The International Convention on Maritime Search and Rescue (SAR).

[191] The International Convention on Load Lines (Load Lines).

[192] The International Convention on Tonnage Measurement of Ships (Tonnage Convention).

[193] The Special Trade Passenger Ships Agreement (STP).

[194] The International Convention for Safe Containers (CSC)

[195] IMO, Autonomous Shipping, https://www.imo.org/en/MediaCentre/HotTopics/Pages/Autonomous-shipping.aspx. Accessed on 27 July 2022.

[196] MSC.1/Circ.1638, para. 5.1.

[197] MSC.1/Circ.1638.

MASS and prevented MASS operations'; category B is 'applied to MASS and did not prevent MASS operations and required no actions'; category C is 'applied to MASS and did not prevent MASS operations but might need to be amended or clarified, and/or might contain gaps'; and category D 'had no application to MASS operations'.[198] Furthermore, there is also a categorisation with regard to the most appropriate ways of addressing MASS operations. Category I is 'equivalences as provided for by the instruments or developing interpretations', and/or category II is 'amending existing instruments', and/or category III is 'developing a new instrument' or category IV is none of these.[199] A number of conventions can accommodate MASS as drafted, while some of them require further interpretations and amendments to tackle the potential gaps and themes created by MASS operations.[200]

It should be emphasised that although UNCLOS was not considered as part of the RSE given that it is not a convention of IMO, MASS would be required to operate in the legal framework established by UNCLOS, and accordingly, IMO will need to consider UNCLOS in its future work on MASS, especially in the event of developing a separate instrument which regulates the operations of MASS.[201]

It is noteworthy that each of the potential gaps is of different nature while several of them are critical and fundamental problems that have the potential to influence on tackling MASS operations, some of them relate to technical qualities.[202] These gaps include but are not limited to terminology, meaning of the terms master, crew or responsible person, remote operator as a seafarer, remote control station/centre, search and rescue, provisions requiring actions by personnel (fire, spillage cargo management, onboard maintenance), connectivity, and cyber security with relevance to instruments such as SOLAS Convention and mandatory codes under SOLAS (for example, International Safety Management (ISM), Ship and Port Facility Security (ISPS), Fire Safety Systems (FSS), Bulk Chemical (IBC), Grain, and Polar), COLREG, the Convention on Safe Containers (CSC), the International Convention on Maritime Search and Rescue (SAR), Load Lines Convention and 1988 Protocol, STCW Convention and Code, the Code of Safe Practice for Cargo Stowage and Securing (CSS Code), and IMO Instruments Implementation Code.[203]

3.2.2 High-Priority Issues

IMO has differentiated high-priority issues given that certain potential gaps are at the heart of introducing MASS operations safely and effectively into the existing

[198] MSC.1/Circ.1638, appendix 2.

[199] MSC.1/Circ.1638, appendix 2, p. 18.

[200] IMO, Legal Committee, 108th session (LEG 108), 26–30 July 2021, https://www.imo.org/en/MediaCentre/MeetingSummaries/Pages/LEG-108th-.aspx. Accessed on 27 July 2022.

[201] Ibid.

[202] MSC.1/Circ.1638, paras. 5.2–5.3.

[203] Ibid, para. 5.2, appendix 2.

regulatory framework and some of the issues may necessitate further policy decisions.[204]

Meaning of the Terms Master, Crew or Responsible Person It has been established that several instruments are comprised of a potential gap in respect of the meaning of the terms master, crew or responsible person which need further clarification. One particularly salient gap is concerning the master's definition, role, and responsibility in autonomy degrees three and four where the ship is controlled on shore.[205]

Remote-Control Station/Centre It is also possible that MASS is operated through a remote-control station or centre which is not only a new concept to IMO instruments but also a potential gap in various instruments. IMO has observed that the remote-control station's or centre's functional and operational requirements are needed to be referred.[206]

Remote Operator as Seafarer Another potential gap in numerous instruments is the potential designation of a remote operator as seafarer which creates complicated problems such as responsibility, qualifications, and the role of remote operator as seafarer.[207]

Terminology A further high priority issue is terminology. The introduction of various new technologies requires the identification of new definitions along with updates to some of the existing terminology.[208]

3.2.3 Priorities for Further Work

Considering the complicated and extensive output of the regulatory scoping exercise, IMO has established numerous priorities for further work noting that determined priorities are non-exhaustive.[209] A variety of priorities for further work have been recognised. In this framework, it has been observed that the principal high-priority matters require review of the terminology and definition, consideration of the development of a new instrument, and examination of high-priority common gaps and themes, however, it should be emphasised that the specified priorities are non-exhaustive.[210]

[204] Ibid, para. 5.4.
[205] Ibid, para. 5.5.
[206] Ibid, para. 5.6.
[207] Ibid, para. 5.7.
[208] Ibid, para. 5.8.
[209] IMSC.1/Circ.1638, para. 6.1.
[210] Ibid, para. 6.1.

Development of a New Instrument IMO wishes to establish the most appropriate ways to tackle MASS operations. Having acknowledged that the vast majority of the common potential gaps and themes expand to numerous instruments, IMO believes that a holistic approach by way of a new instrument such as MASS Code may be more preferable given that separately addressing every single instrument or chapters of SOLAS can cause inconsistencies and confusion along with developing possible barriers in respect of applying the existing regulations to the conventional ships.[211] Consequently, rather than making amendments to the individual instruments, a MASS instrument can be made mandatory through amending a current IMO convention, for instance SOLAS.[212]

IMO also indicates that for the purpose of facilitating MASS operation at an early stage, the establishment of interim guidelines designed for MASS may be of value to ensure safe, secure, and environmentally friendly operations of MASS.[213]

It has been revealed that amending instruments or creating a new instrument necessitates an agreement on terminology. One of the questions that requires focus is the re-evaluation of the autonomy degrees. The further work may also include developing a glossary.[214]

Emphasis has been made on the issue that a number of the potential gaps and themes are considered as high-priority issues which influence various IMO instruments and may require a policy decision prior to tackling individual instruments.[215]

Amending the Existing Instruments IMO further tackled the position in the event of the decision to make amendments to the existing instruments instead of developing a new one and proposed an order of priorities to address the instruments.[216] High-priority instruments are required to be addressed before the other issues and involve the instruments that include the common potential gaps and themes such as terminology, meaning of the terms master, crew or responsible person, remote operator as a seafarer, remote control station/centre, provisions containing manual operations, alarms to the bridge, certificates and manuals on board, provisions requiring actions by personnel (fire, spillage cargo management, onboard maintenance, etc.), connectivity, cyber security, watchkeeping, implication of MASS in SAR, information to be available on board and required for the safe operation.[217] Medium-priority involves the instruments, which are outside of the scope of

[211] Ibid, para. 6.2.

[212] Ibid, para. 6.2.

[213] Ibid, para 6.3.

[214] Ibid, para. 6.4.

[215] Ibid, para. 6.5.

[216] Ibid, para. 6.6.

[217] Ibid, Section 5. According to this, the high-priority instruments are as follows: COLREG, SOLAS chapters II-1, II-2, III, IV, V, VI, VII, IX, XI-1 and XI-2, 1966 LL Convention and 1988 Protocol, 1979 SAR Convention, STCW Convention and Code; STCW-F Convention, FSS Code, IMSBC Code, IMDG Code, TONNAGE 1969, IBC Code and IGC Code: MSC.1/Circ.1638 para. 6.7.1.

high-priority, that require considering the effect of the use of MASS.[218] It is noteworthy that nearly every medium-priority instrument has been determined to be addressed through making amendments to the individual instruments.[219] Low-priority contains the instruments which do not require a substantial action for the use of MASS.[220] However, this situation can change and some of these low-priority instruments may require further considerations in the future with regard to the developments in technologies.[221]

The approach of IMO and States will depend on the nature of each international instrument given that while a number of them can accommodate AMVs within their existing legal scope, several of them will inevitably necessitate new interpretations, amendments, or instruments. For example, especially on the provisions for systems and appliances that require manual operations, particularly fire-fighting, it has been expressed that it would be better to develop new instruments such as a new code for or a new chapter in SOLAS instead of amending them individually since the identical potential gaps and themes can present themselves in multiple instruments.[222]

3.2.4 Legal Committee of IMO

Following the initiation of the regulatory scoping exercise by MSC, Legal Committee of IMO (LEG) likewise commenced a parallel exercise for conventions emanating from the Legal Committee with the aim of assessing the extent to which the existing regulatory framework may be impacted to address operations of MASS.[223] Volunteer Member States, interested Non-governmental Organizations and Intergovernmental Organizations took part in the completion of the work.[224] Legal Committee has approved the RSE's outcome, and MASS has been deemed to be accommodated within the existing regulatory framework of the conventions of the Legal Committee without the necessity for major adjustments or requirement for a new instrument.[225] It has been enunciated that a number of conventions might accommodate MASS as the way they were drafted, whereas some of them might

[218] Ibid, para. 6.6. Accordingly, medium-priority instruments are as follows: SOLAS chapter XII, CSS Code, Grain Code, INF Code, Casualty Investigation Code, III Code, 2008 Intact Stability Code and Standards for owners' inspection and maintenance of bulk carrier hatch covers.

[219] Ibid, para. 6.8.3.

[220] Ibid, para. 6.6. These include SOLAS chapters XIII and XIV, CSC Code, Polar Code, ISM Code, LSA Code, ISPS Code, FTP Code, RO Code, and ESP Code.

[221] Ibid, para. 6.9.3.

[222] Ibid.

[223] IMO, Legal Committee, 108th session (LEG 108), 26–30 July 2021, https://www.imo.org/en/MediaCentre/MeetingSummaries/Pages/LEG-108th-.aspx. Accessed on 27 July 2022.

[224] Ibid.

[225] Ibid.

necessitate supplementary interpretations or amendments to tackle the potential gaps and themes established by the RSE of which the most significant ones are coherent definitions of MASS, the role and responsibility of the master and operator, issues on liability and carriage of certificates.[226]

It has been concluded that the role and responsibilities of the master and the remote operator are high-priority issues which need be tackled as a foundation for any additional work, a number of particular legal terms such as fault, negligence, and intention require deliberation of the damage caused by autonomous technology, and these issues can be addressed jointly between the committees to facilitate both technical and legal aspects, while also considering the different purposes and functions of the conventions.[227] The Committee has also underlined that although UNCLOS was not addressed in the RSE since it is not a convention of IMO, MASS would be required to operate within the legal framework of UNCLOS, and consequently, UNCLOS will need to be examined in the future work of IMO, especially in the event of the development of a new instrument regulating the operations of MASS.[228]

3.3 *The United Kingdom's Proposal for Legislative Change*

The United Kingdom has conducted 'the future of transport regulatory review' which aims to address the transport regulation areas which are deemed to be outdated, barrier to innovation, or designed without considering the recent technologies and business models.[229] The review comprises of maritime autonomy and remote operations. This review has been conducted due to the '*need to ensure that UK law keeps pace to ensure the safe, secure, and environmentally sound operation of remotely operated and autonomous vessels*' since autonomous and remote operations technology keeps getting introduced to the marine environment.[230]

The proposal aims at preparing the domestic legal framework for upcoming changes in international law[231] and expresses that there is a requirement of government intervention which is designed as a comprehensive regulatory framework that will be able to assist the manufacturers and operators in the light of the rapid

---

[226] Ibid.

[227] Ibid.

[228] Ibid.

[229] Department for Transport, Future of transport regulatory review: maritime autonomy and remote operations (28 September 2021), available via: https://www.gov.uk/government/consultations/future-of-transport-regulatory-review-maritime-autonomy-and-remote-operations/future-of-transport-regulatory-review-maritime-autonomy-and-remote-operations. Accessed on 27 July 2022.

[230] Ibid.

[231] Ibid.

advancement and operation of MASS and to make certain that such evolving nature is sufficiently accommodated in the UK legislation.[232]

The proposal is comprised of four primary elements. The first one is identifying key definitions and roles in the operation of remotely operated and autonomous vessels while focusing on enabling flexibility to develop suitable definitions or allowing the amendments of the existing definitions as international law and the MASS industry progress.[233] It has been proposed that there should be either an entity or a person who could be held responsible and accountable for a remotely operated or an autonomous vessel all the time including in case of an emergency or accident, by highlighting that, having discussed with the industry and operators, further details of the relevant requirements would be established in the secondary legislation.[234] The second element is to ensure that the Maritime Coastguard Agency (MCA) can make regulations concerning MASS of any size, including crafts which are not traditionally deemed as ships. The third one is granting the MCA additional powers to make regulations for Remote Operation Centres (ROCs) to ensure remotely operated or autonomous vessels' safe operation and management.[235] Finally, the fourth element is to ensure that the MCA, Department for Transport, ports and harbours possess appropriate powers to regulate various attributes of MASS and ROCs such as health, safety, security and the environmental matters.[236]

The report proposes several definitions and clarifications of terms and roles in the operation of remotely operated and autonomous vessels in primary and secondary legislation.[237] In the consultation, remotely operated vessels are defined as '*vessels where there is a human element involved in the control or operation of the vessel, but that human element is not located onboard the vessel; or to a vessel that carries crew but some functions of the vessel are controlled from a location remote from the vessel*' and autonomous vessels are defined as '*vessels that are capable of decision-making and operating without human input*'.[238] Having made this differentiation, MASS is established more comprehensively. MASS includes '*every description of vessel or craft used in navigation that can for any part of its voyage, fully or in part navigate or operate autonomously or through remote operations*' and remote operations denotes '*controlling the functioning of an operation on a MASS from a different place or location from that MASS*'.[239] The definition of MASS has been proposed to be applicable to every vessel and craft notwithstanding the size including small crafts which are not regarded as ships because they are outside the scope of

[232] Ibid.
[233] Ibid.
[234] Ibid.
[235] Ibid.
[236] Ibid.
[237] Ibid.
[238] Ibid.
[239] Ibid.

the Merchant Shipping Act 1995 (MSA 1995).[240] This approach clearly differs from the above-mentioned distinctions, given that this proposal's perspective is to bring every AMV together under the same roof. The proposal indicates that the UK's forthcoming regulation on MASS is expected to be more comprehensive and inclusive towards the terminology and the legal framework concerning autonomous maritime vehicles.

In respect of the degrees of autonomy, three levels are stated, namely 'remotely operated vessels or craft that have no persons on board', 'remotely operated vessels or craft that may have persons on board (for example, crew, personnel and/or passengers)', and finally, 'vessels or craft operating fully autonomously (currently no distinction as to whether persons are on board or not)'.[241] As it can be observed, the degrees of autonomy established in the United Kingdom's proposal are slightly different than the degrees established by IMO. The proposal refers to remotely operated ships with and without people on board and fully autonomous ships. These levels are consisted with the degrees two, three, and four determined by IMO.[242] However, the proposal does not refer to IMO's degree one, namely '*ship with automated processes and decision support*'. Thus, it may be deduced that the United Kingdom's proposal tackles the levels of the autonomy in a more confined manner than IMO. The master of MASS is defined as '*a person (except a pilot) having command or charge of a MASS*' who is not required to be onboard focusing on the roles and responsibilities rather than physical location in relation to a vessel or vessel's type such as manned or unmanned by anticipating the legislation to make sure that every responsibility is enforceable against a master who is not on board while not changing the requirements of the need for the master to be onboard on a non-MASS vessel.[243] Remote Operator indicates '*every person, including a MASS master, who is employed or engaged in any capacity to undertake remote operations of a MASS*' and Remote Operation Centre (ROC) '*is a place or location from where at least one Remote Operator is operating a MASS*'.[244]

In this context, amending the existing legal framework to take powers in primary legislation to regulate every MASS irrespective of size has been proposed by supporting this view with the reasons that such approach would ensure that the innovative and developing MASS sector is properly regulated and supported, would provide a unified approach to tackle maritime operations and regulatory oversight between MASS and non-MASS, would make sure that every vessel is built, surveyed, operated, and inspected to ensure that they do not give rise to damage to maritime users, human health, environment, property, and resources, would let the UK have an active and learned position in the forthcoming international discussions in respect of the MASS regulations and IMO instruments, would allow the UK legal

[240] Ibid.
[241] Ibid.
[242] See Sect. 2.5.
[243] Ibid.
[244] Ibid.

framework to be more prepared for upcoming changes in international law.[245] However, it has been noted that developing such legal framework at the moment possesses the risk of diverging from the international standards when they will be established in the future, for instance, in determining the definitions of MASS; and it has been indicated that MCA is needed to be given the authority to make definitions and clarifications of the terms and roles for the MASS operation, along with those defined in the primary legislation, so that sufficient flexibility is provided for the changes in secondary legislation while the MASS industry progresses and the international legal framework develops.[246] It has been noted that to make certain there is uniformity in the industry and to establish safety, security, and the protection of marine environment, it is pursued that MCA has the authority to apply its existing statutory responsibilities such as survey, inspection, certification, and enforcement on either MASS with UK flag or operations of MASS in the UK waters.[247]

In respect of 'autonomous and unmanned submersible apparatus' which are widely used for defence and marine research, it has been acknowledged that these will potentially be utilised in a wider and more varied scope, and therefore, there is a need to establish powers to regulate them in a way coherent with their manned counterparts through secondary legislations as these concepts develop.[248] As to the marine equipment attached to a vessel, the general system is found to be applicable for MASS and ROCs while also acknowledging that the relevant regulations and standards may be required to be amended to incorporate new kinds of equipment, particularly, there may be a necessity to tackle software systems and algorithms independently from the hardware they may be used with.[249]

3.4 Concluding Remarks

Sections 2 and 3 of this chapter attempted to shed some light on the terminology and the legal status of autonomous maritime vehicles. In this regard, these sections addressed the different terminology and definitions used for autonomous maritime vehicles and evaluated their legal status as ships/vessels, warships, auxiliaries, along with the non-ship vehicles which are comprised of various other vehicles. AMVs were also examined based on their degrees of autonomy. Additionally, AMVs were addressed under a number of legal frameworks, namely UNCLOS, IMO, and the United Kingdom.

A vast majority of the international conventions and instruments were established many years ago, and the drafters did not have autonomous vehicles in their minds.

[245] Ibid.

[246] Ibid.

[247] Ibid.

[248] Ibid.

[249] Ibid.

Consequently, international instruments do not perfectly align with the novelties presented by such vehicles. These rules bring several challenges particularly in respect of fully autonomous vehicles which do not have seafarers on board and when the decision-making process is conducted by the autonomous systems. Therefore, it is difficult to establish how AMVs fit in the existing international legal framework, however, they still operate within it. While particular legal instruments can accommodate these vehicles, some of them require further interpretations and amendments to tackle the potential gaps as identified by IMO in its regulatory scoping exercise.

It is also of crucial importance to categorise AMVs in their appropriate status given that numerous rights and obligations attach to each of these vehicles. IMO and several States have already initiated the process to regulate autonomous maritime vehicles, and it is safe to assume that these issues will be addressed more thoroughly in the future. However, in the meantime, one must always bear in mind that they still operate in the existing legal framework.

4 AMVs As Instruments to Enhance Maritime Security

As technology has progressed, several new AI technologies and autonomous vehicles have been developed in various sizes, shapes, capabilities, and roles. The wide range utilisation of AMVs necessitates evaluating their position in the framework of maritime security. With their diverse capabilities, AMVs present many opportunities for States to enhance their maritime security and law enforcement. Several countries are increasingly utilising and investing in AMVs, and this situation reveals itself in their budget plannings. For instance, the US navy requested a budget for $434.1 million for research and development for three types of large unmanned surface and undersea vehicles along with their enabling technologies,[250] and Australia has invested up to $1.3 billion in a new unmanned aircraft systems development programme to improve situational awareness across maritime domain.[251]

Autonomous vehicles have a wide variety of advantages and present many opportunities. First of all, they substantially reduce risks to human life especially in conducting dangerous missions. Moreover, they can carry out their missions in a more precise way, therefore, improve the performance and outcomes of the missions along with reducing the adverse effects of human error. AMVs can perform tasks beyond human capabilities by covering larger areas at a high speed in rough weather conditions. Furthermore, supported by numerous sensors, radars, cameras, and identification systems, they can monitor and detect harmful substances and illegal activities occurring at sea, record video images, collect and send data to the relevant authorities, and therefore, provide a swift and efficient response. By virtue of their

[250] O'Rourke (2022), p. 1.

[251] Australian Government Department of Defence (2020).

sophisticated components, they can be utilised in detecting activities, other vehicles, human movements, weapons, and changes in marine environment which would otherwise be highly difficult for other manned vehicles. AMVs not only provide adaptability, survivability, persistence, and lowered risk to human life, but also are ideal for dull, dirty, or dangerous missions.[252] In this framework, dull missions include long-duration tasks which may be inappropriate for humans to conduct such as surveillance missions that require lengthy observations.[253] Dirty missions are the ones that may cause exposure to hazardous conditions such as biological, chemical, and nuclear detection operations.[254] Dangerous missions are with high risk and inherently dangerous,[255] such as disabling explosive devices and mine sweeping operations.[256]

In the light of the benefits provided by them, utilisation of autonomous maritime vehicles, whether surface, underwater, or aerial, presents many realistic opportunities for States to enhance their maritime security and law enforcement. Countries progressively use AMVs to improve their maritime security. US Navy aims to utilise autonomous underwater vehicles in intelligence, surveillance, and reconnaissance, anti-submarine warfare, mine countermeasures, inspection and identification, information operations oceanography, communication and navigation network node, payload delivery, and time critical strike.[257] The United States also employs aerial autonomous vehicles to monitor piracy threats occurring on high seas[258] and suspicious activities taking place at sea such as trafficking of illicit drugs.[259] Singapore has new AMVs equipped with collision detection and avoidance systems to complement manned ships in support of maritime security.[260] In Türkiye, AMV ULAQ is made of advanced composite material and possesses cruising range of over 400 kilometres, over 70 kilometres speed per hour, advanced day and night vision, autonomy, crypto and electronic warfare protected communication infrastructure, and will be used in intelligence, surveillance and reconnaissance, armed escort and

[252] US Department of Defense, Unmanned Systems Integrated Roadmap FY2013–2038 (2014), p. 20, available via: https://www.hsdl.org/?abstract&did=747559. Accessed on 27 July 2022.

[253] FY2013–2038, p. 20.

[254] Ibid, p. 20.

[255] Ibid, p. 20.

[256] Pedrozo (2020), p. 218.

[257] US Department of the Navy, Navy Unmanned Undersea Vehicle (UUV) Master Plan (9 November 2004) accessible via: https://www.hsdl.org/?abstract&did=708654. Accessed on 27 July 2022.

[258] Burgers and Romaniuk (2017).

[259] For instance, an aerial AMV detected a suspicious go-fast boat which led to the confiscation of illicit drugs: 'Fire Scout Scores First-Ever Drug Bust' (2010), https://www.defencetalk.com/fire-scout-scores-first-ever-drug-bust-25563/. Accessed on 27 July 2022.

[260] Vavasseur (2021).

force protection, surface warfare and strategic facility security, guarding sea borders, ports, naval bases, and critical facilities.[261]

Underwater, surface, and aerial AMVs produce a comprehensive interactive system, generally with the assistance of humans rather than full autonomy, that can allow carrying out challenging tasks successfully. States may benefit from AMVs in enhancing their maritime security and law enforcement abilities particularly in detecting crimes occurring at sea.[262] The rapid development of technology and various applications of AMVs require a further analysis in respect of how they may enhance maritime security and law enforcement along with how they may be accommodated in the existing international legal framework. In this context, this section of the chapter attempts to tackle the intersection between the use of AMVs and a number of maritime security-related issues. It must be emphasised that although AMVs keep improving, changing, and becoming subject to a variety of use, they still operate within the international legal framework. One particularly challenging problem in this regard is that the existing inadequacies in the legal framework of maritime security becomes more prominent by the operation of AMVs as the functionality of new technology may be hindered by the operation of law.[263] States always pursue finding new ways to enhance their maritime security and law enforcement, and the latest advancements in technology and various autonomous vehicles are the new instruments to provide that purpose. It must, however, always be kept in mind that AMVs are also subject to the international law of the sea.

It has been explained in the first section of this chapter[264] that there is not a single precise definition of maritime security, and the term encompasses being secure from maritime threats and incidents. Accordingly, the following sections attempt to tackle the use of AMVs in connection with enhancing maritime security in respect of intelligence and surveillance, right of visit, illicit traffic in narcotic drugs and psychotropic substances, migrant smuggling, and damage to marine environment.

4.1 Intelligence and Surveillance

AMVs present great opportunities for surveillance and intelligence activities particularly when interacting with other systems. The Command, Control, Communications, Computers, Intelligence, Surveillance and Reconnaissance oriented systems can be utilised to connect to autonomous maritime vehicles to enhance their

[261] 'The First Indigenous and State of the Art Armed Unmanned Surface Vehicle-AUSV', https://www.ulaq.global/. Accessed on 27 July 2022.

[262] Klein (2019), p. 247.

[263] Ibid, pp. 247–248.

[264] See Sect. 1.1.

functions by providing over-the-horizon mapping/targeting and expanding the line-of-sight in every direction.[265]

The United Kingdom's Royal Navy has declared that ships will be equipped with artificial intelligence systems with the aim of better detecting the incoming threats and assessing combat scenarios.[266] The idea is that AI can potentially process data more swiftly than humans, can assess and confirm potential threats, and make informed decisions; but instead of taking the place of humans, they are designed to enhance their situational awareness.[267]

AMVs, especially in relation to different aerial and land-based systems, are of crucial significance and potential to conduct surveillance activities, intelligence and data gathering, interpreting data, acting based on the interpretation or sending their findings to the relevant authorities to take the necessary actions. Such intelligence and surveillance activities can enable detecting criminal activities without delay, and by doing so, facilitate immediate response.

There are different navigational regimes under UNCLOS applicable to ships and aircrafts, namely innocent passage, transit passage, and archipelagic sea lanes passage all of which give rise to different rights and obligations, for example, exercising or practicing with weapons of any kind[268] and launching, landing or taking on board of any military device[269] may be considered against the scope of innocent passage. Although there are certainties to some extent that unmanned aerial vehicles are considered as aircraft and possess same navigational rights as their manned counterparts,[270] and when an AMV is part of this category, the rights and obligations applicable to vessels attach to such AMVs as well. As to AMVs which cannot be identified as vessels, UNCLOS and customary international law do not provide a passage regime for non-ships, and it is yet to be seen whether non-ship AMVs will be able to benefit from such rights. As explained in the previous sections of this chapter,[271] AMVs outside the categorisation of ships or vessels are subject to different rules such as floats and gliders being subject to marine scientific research rules when they are used for such purpose. AMVs such as floats and gliders can also be utilised for surveillance and collecting data for defence or operational purposes.[272] Ships may possess freedom of navigation rights and various immunities; however, the full extent of the rights and obligations attach to AMVs that are not

---

[265] Mukherjee (2018), Securing the Maritime Commons: The Role of Artificial Intelligence in Naval Operations 7 [Observer Research Foundation (ORF)], ORF Occasional Paper No. 159, available via: https://www.orfonline.org/research/42497-a-i-in-naval-operations-exploring-possibilities-debating-ethics/. Accessed on 27 July 2022.

[266] Moore-Colyer (2016).

[267] Ibid.

[268] UNCLOS Article 19 (b).

[269] UNCLOS Article 19 (2) (f).

[270] At least in the US: see Norris (2013), p. 40.

[271] See Sect. 2.4.

[272] Klein (2019), p. 267.

identified as ships is of ambiguity. There is a view that non-ship objects and devices do not possess the rights of innocent passage, transit passage, and archipelagic sea lanes passage.[273] On the contrary, it has been submitted that AMVs may similarly enjoy the freedom of navigation rights on the high seas and in exclusive economic zones, since such freedom belongs to States instead of attaching to ships or vessels.[274]

One must always keep in mind that such activities of AMVs may be permitted in different levels depending on where the surveillance activity takes place. There is no doubt that coastal States can exercise surveillance activities within their own territorial seas.[275] However, other States cannot carry out surveillance activities in the territorial sea of a coastal State since it is contradictory to the right of innocent passage regulated under Article 19 of UNCLOS as acts aimed at gathering information to the prejudice of the coastal State's defence or security,[276] launching, landing or taking on board of any aircraft[277] or any military device,[278] and carrying out research and survey activities[279] are deemed prejudicial to the peace, good order or security of the coastal State. It is also noteworthy that since the coastal State has sovereignty in its territorial sea and the right of innocent passage can be enjoyed by ships and not aircrafts, the consent of the coastal State is necessary for any overflight.[280] Accordingly, surveillance carried out by an AMV, whether it is underwater, surface, or aerial, can give rise to the violation of the right of innocent passage given that the lack of a master, commander, and crew on board does not make a difference in evaluating whether the vessel's activities constitute the breach of innocent passage.[281]

Under Article 33 of UNCLOS, in the contiguous zone, the coastal State may exercise the control required to '*prevent violation of its customs, fiscal, immigration or sanitary laws and regulations within its territory or territorial sea*'. Accordingly, AMVs can be of significant use in respect of the costal State's law enforcement to ensure the protection of its maritime security and compliance with domestic laws and regulations.

On the Exclusive Economic Zone (EEZ), there remains significant controversy on the military activities conducted by another state in the coastal State's EEZ,[282] and accordingly, on the legality of surveillance activities by AMVs.[283] This debate is

[273] Norris (2013), p. 40.

[274] Klein (2019), p. 268.

[275] McLaughlin and Klein (2021), p. 400; UNCLOS Article 2.

[276] UNCLOS Article 19 (2) (c).

[277] UNCLOS Article 19 (e).

[278] UNCLOS Article 19 (f).

[279] UNCLOS Article 19 (j).

[280] McLaughlin and Klein (2021), p. 400.

[281] Klein (2019), p. 269.

[282] McLaughlin and Klein (2021), p. 401.

[283] Klein (2019), p. 270.

beyond the scope of this work, however, it is noteworthy that under Article 58 of UNCLOS, all States enjoy the freedom of navigation, overflight, the laying of submarine cables and pipelines, and other internationally lawful uses of the sea relevant to these freedoms, like those in relation to the operation of ships, aircraft, submarine cables and pipelines, and compatible with the other provisions of the Convention, however, when States exercise their rights and duties in the EEZ, they must have regard to the coastal State's rights and duties, and comply with the coastal State's laws and regulations. Therefore, these activities require consideration on the type of research, whether such research is part of a military activity, or whether the consent of the coastal State is required if research is performed in its EEZ.[284]

As to the high seas, Article 87 of UNCLOS provides that all States have freedom of navigation, overflight, laying submarine cables and pipelines, constructing artificial islands and other installations, fishing, and scientific research, however, these freedoms shall be carried out with due regard for other States' interests in their exercise of the freedom of the high seas.[285] In this regard, AMVs may operate surveillance activities on the high seas.

4.2 *Right of Visit*

Depending on their autonomy levels and capabilities, AMVs can have the right of visit. In this regard, an AMV may have the right of visit under Article 110 of UNCLOS which provides that a warship may board to a foreign ship on the high seas, unless the ship possesses full immunity under Articles 95 and 96, if there is a reasonable ground for suspecting that the ship is engaged in piracy, slave trade, or unauthorised broadcasting; or the ship does not have nationality, or although flying a foreign flag or refusing to show its flag, the ship is of the same nationality as the warship.[286] In these circumstances, the warship can proceed for the verification of the ship's right to fly its flag, and for that, warship may send a boat to the suspected ship and check for documents, and if there remains suspicion after checking the documents, it may proceed to an additional examination on board.[287] These provisions are also applicable to '*any other duly authorized ships or aircraft clearly marked and identifiable as being on government service*'.[288] Accordingly, the right of visit may be exercised by an AMV which is not a warship.[289] It must be noted, however, that since AMVs may lack some of the capabilities to conduct physical inspection, their utilisation may be restricted to visual assessment.

[284] Ibid, p. 270.
[285] UNCLOS Article 87.
[286] UNCLOS Article 110 (1).
[287] UNCLOS Article 110 (2).
[288] UNCLOS Article 110 (5).
[289] Klein (2019), p. 256.

4.3 *Illicit Traffic in Narcotic Drugs and Psychotropic Substances*

Article 108 of UNCLOS regulates illicit traffic in narcotic drugs and psychotropic substances and asserts that every State shall cooperate in the suppression of such illicit traffic committed by ships on the high seas in contradiction of international conventions.[290] This requirement for cooperation is regulated in more detail in the UN Drug Convention.[291] In this framework, UN Drug Convention, whose purpose is promoting cooperation between the Parties so that the different aspects of illicit traffic with an international element may be addressed more efficiently, brings the Parties an obligation to take necessary measures involving legislative and administrative ones.[292] The cooperation obligation of the Parties is emphasised in Article 17 by expressing that the Parties shall fully cooperate to suppress illicit traffic by sea in accordance with the international law of the sea.[293] Article 17 also brings an obligation to the Parties to provide assistance '*within the means available to them*'.[294] If there is a suspicion that a vessel exercising freedom of navigation is involved in illicit traffic of drugs, a Party may request authorisation from the flag State to take appropriate measures against the vessel. Such authorisation may include boarding and searching the vessel and taking the appropriate action in respect of the vessel, people and cargo on board if evidence concerning illicit traffic is detected.[295] Any type of such action shall be performed '*only by warships or military aircraft, or other ships or aircraft clearly marked and identifiable as being on government service and authorized to that effect*'.[296]

AMVs have a great potential to be widely utilised to assist in the prevention and detection of drug trafficking.[297] In this context, an AMV may have the right of visit in accordance with UN Drug Convention.[298] AMVs are also subject to the same obligations such as requesting authorisation from the flag State to take appropriate measures,[299] '*not to endanger the safety of life at sea, the security of the vessel and the cargo or to prejudice the commercial and legal interests of the flag State or any other interested State*',[300] swiftly providing information to the flag State regarding

[290] UNCLOS Article 108 (1).

[291] The United Nations Convention against Illicit Traffic in Narcotic Drugs and Psychotropic Substances, 1988. (UN Drug Convention).

[292] UN Drug Convention Article 1.

[293] UN Drug Convention Article 17 (1).

[294] UN Drug Convention Article 17 (2).

[295] UN Drug Convention Article 17 (3) (4).

[296] UN Drug Convention Article 17 (10).

[297] See also McLaughlin and Klein (2021).

[298] Klein (2019), p. 259.

[299] UN Drug Convention Article 17 (3).

[300] UN Drug Convention Article 17 (5).

the results of that action,[301] not interfering with or affecting the rights and obligations and the exercise of jurisdiction of the coastal States pursuant to the international law of the sea.[302]

The US Coast Guard uses drones, which provide real-time imagery, data, target illumination, and communications relay to the vessels of the Coast Guard, in surveillance, detection, and classification operations.[303] These unmanned ariel vehicles facilitated the confiscation of nearly 11 tons of illicit drugs.[304] Mekong River, placed between Laos, Myanmar, and Thailand, have tremendously attracted drug traffickers during the COVID-19 pandemic which resulted in maritime law enforcement and security agencies' request on advice and assistance from the United Nations Office on Drugs and Crime to set up advanced technology and training solutions such as providing aerial surveillance equipment to advance reconnaissance operations with the assistance of drone technology equipped with high-resolution and infrared cameras which will enhance both day and night patrols.[305] In addition to aerial vehicles that support maritime security, other AMVs such as floats and gliders may also be used to detect activities and collect data. In this context, AMVs may be able to detect drug traffickers, migrant smugglers, or illegal fishing activities, classify them, and report their locations, routes, speeds to Coast Guard command centres[306] from where maritime security and law enforcement activities can be further progressed. AMVs can be utilised to detect suspicious activities at sea, track the subject of such activity, collect data which may be used in the prosecution process, and finally alert the law enforcement authorities to take the necessary actions to prevent or halt such activities. It is noteworthy that although AMVs may be used in surveillance purposes in a very effective way, they possess inadequacies in some of the law enforcement issues considering their inability to board and search the suspect vessel, carry out arrests, and seizure of evidence all of which may be required for prosecution.[307]

---

[301] UN Drug Convention Article 17 (8).

[302] UN Drug Convention Article 17 (11).

[303] Jarocki (2018).

[304] Ibid.

[305] United Nations Office on Drugs and Crime (UNODC), UNODC provides a technological edge to address drug trafficking on the Mekong River, https://www.unodc.org/roseap/2021/09/drug-trafficking-mekong-river/story.html. Accessed on 27 July 2022.

[306] Keller (2014).

[307] Klein (2019), p. 259.

4.4 Migrant Smuggling

The pervasive use of unseaworthy vessels in migrant smuggling repeatedly give rise to situations of distress requiring search and rescue operations,[308] therefore, migrant smuggling and law enforcement process against these operations often interact with search and rescue obligations.[309]

In preventing migrant smuggling, from the perspective of law enforcement, States exercise authorities as the coastal States over the territorial seas and contiguous zones of them, and the Migrant Smuggling Protocol (the Protocol Against the Smuggling of Migrants by Land, Sea and Air) provides the rights and obligations for beyond the contiguous zone.[310] Article 8 of the relevant protocol regulates the right of visit in the event of the suspicion of smuggling of migrants. The use of AMVs in this context give rise to similar challenges mentioned in drug smuggling[311] in respect of boarding, searching, checking documents, and gathering evidence along with arresting smugglers, and assisting the needs of migrants on board.[312]

At the time, in respect of migrant smuggling, the primary utilisation of AMVs is in the form of autonomous aerial vehicles which are used for detecting migrants. In this context, AMVs may detect migrant smuggling activities and provide such information to close by warships or government authorities to take the required measures such as boarding.[313] AMVs possess the potential to provide a swifter detection of such illegal activities.

There is a duty to rescue those in peril at sea under maritime law and custom, and this duty should not be neglected by law enforcement officers to suppress migrant smuggling.[314] Smugglers often use ill-conditioned unseaworthy vessels to transport migrants that cause them to sink, and therefore, possess danger to the lives of migrants. There are various international instruments which bring obligations to States with regard to search and rescue operations. Article 98 of UNCLOS regulates duty to render assistance, by providing that States shall require the master of a ship flying its flag to render assistance to people found at sea in risk of being lost, and to proceed with all possible speed to the rescue of people in distress.[315] This gives rise

[308] McLaughlin and Klein (2021), p. 413.

[309] Klein (2019), p. 263.

[310] Klein (2019), p. 263.

[311] See Sect. 4.3.

[312] Klein (2019), p. 263.

[313] McLaughlin and Klein (2021), p. 413.

[314] United Nations Office on Drugs and Crime (UNODC), Legislative Guide for the Implementation of the Protocol Against the Smuggling of Migrants by Land, Sea and Air, Supplementing the United Nations Convention Against Transnational Organized Crime, para. 100, available via: https://www.unodc.org/pdf/crime/legislative_guides/04%20Legislative%20guide_Smuggling%20of%20Migrants%20Protocol.pdf. Accessed on 27 July 2022.

[315] UNCLOS Article 98 (1) (a), (b). SOLAS Chapter V also provides a similar obligation for the master of a vessel by stating that '*The master of a ship at sea which is in a position to be able to provide assistance on receiving a signal from any source that persons are in distress at sea, is*

to the question of whether an AMV can render such assistance especially if it is fully autonomous and do not have a master on board. As also acknowledged as a high priority issue by IMO in its regulatory scoping exercise,[316] the master of an AMV is still of ambiguity. Duty to rescue obligations also prompts challenging problems such as who the master of an AMV is, and if there is no master, whether there remains a duty to rescue[317] which, at the moment, do not have clear answers. If master of a vessel is perceived as someone who is in charge of the vessel, remote operator of an AMV may be identified as the master, and therefore would have the duty of rescue.[318] However, when it comes to fully autonomous vessels, since the term master becomes vague, it would be highly unlikely to define the certain scope of such duty. SAR Convention defines search as '*an operation, normally co-ordinated by a rescue coordination centre or rescue sub-centre, using available personnel and facilities to locate persons in distress*',[319] and describes rescue as '*an operation to retrieve persons in distress, provide for their initial medical or other needs, and deliver them to a place of safety*'.[320] In accordance with these provisions, it can be deducted that there must be a human element for such operations to succeed. There are also various practical challenges and limitations on search and rescue operations which AMVs may exercise, given that victims of such distress almost always need medical supplies, water, and food.[321]

In this context, it is clear that AMVs possess limitations as to the search and rescue operations, however, they may be of help in respect of detecting the people in distress and they can act in accordance 'with all possible speed' given that they may cover larger areas and proceed faster than manned vehicles. UNCLOS also declares that States '*shall promote the establishment, operation and maintenance of an adequate and effective search and rescue service regarding safety on and over the sea*'.[322] Whether AMVs can provide adequate and effective search and rescue service will indeed depend on their structure and capabilities. While they may assist in the detection of people in distress, it seems unlikely that they can solitarily provide search and rescue operations, and accordingly, the best possible solution to reach the aim of adequate and effective search and rescue would be the coordination between AMVs and the traditional manned vehicles and vessels with human element. AMVs can be of assistance to States to enhance their maritime security by detecting illegal activities, alarming them to take action, and fulfil their duties to render assistance in the event of a distress situation.

---

bound to proceed with all speed to their assistance, if possible informing them or the search and rescue service that the ship is doing so': SOLAS Chapter V Regulation 33 (1).

[316] MSC.1/Circ.1638.

[317] Pritchett (2015), p. 208.

[318] Ibid, p. 209.

[319] SAR Annex, Chapter 1, 1.3.1.

[320] Ibid, 1.3.2.

[321] Pritchett (2015), p. 209.

[322] UNCLOS Article 98 (2).

4.5 Damage to Marine Environment

The illegal disposal, dumping, and discharge of hazardous materials and wastes constitute an increasing threat to maritime security.[323] The adverse effects of this problem intensify when those illegal activities occur in or off the coastlines of financially struggling States which are not adequately and appropriately equipped to deal with such damage.[324] Damage to marine environment does not merely affect the coastal States, but rather has a much wider long-term influence on the overall environment, and the illicit movement and dumping of toxic and dangerous products and wastes have numerous detrimental effects on the enjoyment of several human rights.[325]

Under Article 210 of UNCLOS, States are obligated to adopt laws and regulations along with other necessary measures to prevent, reduce, and control pollution of the marine environment as a result of dumping.[326] Dumping in the coastal State's territorial sea, EEZ, and continental shelf shall not be conducted without the express prior approval of the coastal State which has the right to permit, control, and regulate such dumping.[327]

Immediate and swift measures are required to be taken to diminish the adverse effects of such activities which are threatening to marine environment. AMVs can be utilised to adopt fast, effective, safe, and environmentally friendly measures to reduce the impact of marine pollution.[328] AMVs, such as floats and gliders, equipped with various sensors to monitor ocean conditions may identify changes in the ocean environment, analyse the data, and send notifications to the relevant authorities to take action; moreover, they can even assist in the cleaning process. The main struggle in responding to marine pollution is identifying and cleaning the affected areas which can be overcome with AMVs that can monitor and detect the zones impacted by the pollution and clean them by implementing a cooperative and coordinated response.[329] Such utilisation can allow States to identify the problems, take measures immediately, and conduct cleaning activities efficiently.

[323] Calley et al. (2016), p. 515.

[324] Ibid, p. 515.

[325] UN Human Rights Council Resolution, Mandate of the Special Rapporteur on the implications for human rights of the environmentally sound management and disposal of hazardous substances and waste (Adopted 29 September 2011), 18/11, A/HRC/18/L.6, available via: https://documents-dds-ny.un.org/doc/RESOLUTION/LTD/G11/164/65/PDF/G1116465.pdf?OpenElement. Accessed on 27 July 2022.

[326] UNCLOS Article 210 (1)(2).

[327] UNCLOS Article 210 (5).

[328] Bella et al. (2021).

[329] Ibid, p. 1.

5 AMVs As Instruments to Threaten Maritime Security

Although AMVs have a variety of advantages, they also possess some disadvantages. Initial investment and research and development costs of these new technologies and vehicles are extremely high, not only for the vehicles but also for setting up onshore operations for them to perform properly,[330] which makes it very difficult to use them pervasively and in large quantities. The lack of crew also presents disadvantages especially when there is a need for immediate human response and interaction, along with the repair issues since the moving parts' maintenance could be quite difficult during long voyages and breakdowns may give rise to considerable delays.[331]

Concerning maritime security, the most challenging disadvantages of AMVs are cyber security issues along with their utilisation in criminal activities and terrorism. States and innocent maritime stakeholders are not the only ones benefiting from the utilisation of autonomous maritime vehicles, criminals also use these vehicles to accomplish their objectives. The lack of human element can make AMVs more vulnerable to cyber incidents. Furthermore, these vehicles are already being utilised by perpetrators in several criminal activities.

5.1 Cyber Security Issues

The integration of systems and vehicles with autonomous technology may make them more vulnerable to cyber risks and threats than their manned counterparts given the lack of human on board to take control and immediately react to such incidents. Cyber risks constitute a substantial risk to maritime security and require States to take the appropriate steps for cyber risk management.

IMO defines maritime cyber risk as '*a measure of the extent to which a technology asset could be threatened by a potential circumstance or event, which may result in shipping-related operational, safety or security failures as a consequence of information or systems being corrupted, lost or compromised*'.[332] In the recent years, there have been cyber incidents involving several international shipping companies, and these incidents reveal that the victims of a potential cyber incident may be the vessels, ports, oil and gas companies, and even States.

Although the term cyber incident incorporates cyber threats, it also involves various other events which may be on account of both an intentional and an unintentional act. In this regard, the sources of cyber incidents are cyber security incident which has an effect on the availability and integrity of systems, an

[330] O Brian (2018).

[331] Ibid.

[332] IMO, Maritime Cyber Risk, https://www.imo.org/en/OurWork/Security/Pages/Cyber-security.aspx. Accessed on 27 July 2022.

unintentional system failure during software maintenance, loss of or manipulation of data which may be of critical significance for the operation of AMVs, failure of a system as a result of software crashes, phishing which constitutes one of the most common attacks that may lead to the loss of sensitive data, the introduction of malware to the systems, and even the loss of operational control[333] all of which may give rise to service disruptions, system failures, loss of confidential data, financial loss, and criminal activities.

Given the susceptibility of autonomous systems and vehicles to cyber incidents, perpetrators may access their systems, confidential data, and even take control of the vehicles to carry out offences. Threat actors include accidental actors who do not have malicious motive, however, cause unintended harm, activists who act for revenge, media attention, disruption of operations and reputational damage, criminals who may act for financial gain or disrupting maritime security, and States, State sponsored organisations and terrorists acting for political or ideological aims.[334]

The perpetrator's motivation and objectives may be accessing sensitive or confidential data, manipulating cargo manifests, crew or passenger/visitor lists, stow plans or loading lists which can be utilised for transporting illegal cargo or enabling thefts, closing operational systems, facilitating a variety of crimes such as piracy, theft and fraud, disrupting the operation of systems, and demanding a ransom for the data or release of the vehicle.[335]

Maritime security may as well be under the threat of cyber incidents which also brings various responsibilities to States to manage cyber risks and to take the appropriate measures to ensure maritime security. In this regard, cyber risk management means '*the process of identifying, analysing, assessing and communicating a cyber-related risk and accepting, avoiding, transferring or mitigating it to an acceptable level, considering costs and benefits of actions taken to stakeholders*'[336] and maritime stakeholders should ensure that they can manage cyber risks.

IMO has provided guidelines in respect of recommendations on maritime cyber risk management to protect shipping from existing and emerging cyber threats and vulnerabilities.[337] In this framework, effective cyber risk management should consider not only the malicious actions but also benign actions both of which expose or exploit vulnerabilities in information or operational technology.[338] IMO has expressed numerous functional elements which need to be properly incorporated

---

[333] The Guidelines on Cyber Security on board Ships (version 4) issued by ICS, IUMI, BIMCO, OCIMF, INTERTANKO, INTERCARGO, InterManager, WSC and SYBAss, p. 3, available via: https://wwwcdn.imo.org/localresources/en/OurWork/Security/Documents/ANNEX%20Guidelines%20on%20Cyber%20Security%20Onboard%20Ships%20v.4.pdf. Accessed on 27 July 2022.

[334] Ibid, p. 12.

[335] Ibid, p. 15.

[336] IMO, Maritime Cyber Risk, https://www.imo.org/en/OurWork/Security/Pages/Cyber-security.aspx. Accessed on 27 July 2022.

[337] Ibid.

[338] IMO, Guidelines on Maritime Cyber Risk Management, MSC-FAL.1/Circ.3 (5 July 2017), para. 2.1.4, available via: https://wwwcdn.imo.org/localresources/en/OurWork/Security/Documents/

in a risk management framework, namely identifying the systems, assets, data, and capabilities that may pose risks when disrupted, protecting against a cyber incident and ensuring the continuity of shipping operations, detecting cyber incidents in a timely manner, responding to cyber incidents in a resilient way, and finally, recovery measures to restore systems after a cyber incident.[339]

5.2 *Crimes and Terrorism Committed with AMVs*

Like every other item in the world, AMVs are not merely used by innocent maritime stakeholders and States, but also utilised by criminals who threaten maritime security. Innocent maritime stakeholders no longer deal with eye patched pirates, they deal with criminals who use the state-of-the-art technology. Autonomous vehicles and systems can be used in a variety of criminal activities such as carrying out attacks to people, ships, ports, offshore installations, and pipelines.[340] Perpetrators may use their own AMVs to commit crimes or may take over the control of other AMVs which belong to States or other innocent maritime stakeholders to carry out their illegal activities.

AMVs are already being used by criminals to conduct their illegal activities one of which is drug smuggling. Autonomous vehicles offer smuggling transportation to be more efficient in terms of space and speed, as they are harder to detect as opposed to manned vehicles, and the lack of people on board makes it more difficult to prosecute the crime since there is nobody who may give information to the authorities.[341] Perpetrators do not risk their own lives by committing crimes with AMVs, and it is considerably trickier to detect who committed the crime and from where it was perpetrated, furthermore, given that the vast majority of the defensive measures are equipped to prevent surface threats rather than subsurface ones, detecting especially the underwater AMVs is considerably more difficult for States.[342] This again brings the question of how to deal with AMVs in the context of international law. For instance, if the legal status of the AMV used in drug smuggling is a ship, there is a necessity to determine whether it is registered to a State which holds exclusive jurisdiction over the AMV in such case the boarding could only be carried out under Article 17 of the UN Drug Convention,[343] or whether it is a vessel without nationality in which case, although the convention does not provide the right of visit,

MSC-FAL.1-Circ.3%20-%20Guidelines%20On%20Maritime%20Cyber%20Risk%20Management%20(Secretariat).pdf. Accessed on 27 July 2022.

[339] Ibid, para. 3.5.

[340] Petrig (2020), p. 108.

[341] Allen (2018b).

[342] Petrig (2020), pp. 109–110.

[343] Klein (2019), p. 260.

Parties shall give assistance to suppress the vessel's use for drug smuggling.[344] If the AMV used in drug smuggling is a non-ship vehicle, States can seize the drug loaded AMV on the high seas arguing that since it is not a vessel, it does not possess the possible protections or assumptions which might otherwise appear in the stateless vessels' treatment; and if the AMV's ownership is not known and could not be determined via any markings, there would not be a violation of the freedoms of the high seas.[345]

AMVs are capable of carrying out notorious attacks, especially when they are equipped with explosive payloads, to damage and destroy other vessels, ports, and offshore installations.[346] For example, Houthi rebels attacked a Saudi warship by employing a remote-controlled boat in January 2017 and made an attempt to explode a Saudi Aramco fuel terminal and distribution station by utilising a high-speed boat equipped with explosives.[347] Oil disruptions have the potential to affect the world economy and supply chains as it can be seen from the Abqaiq–Khurais attack where Houthi rebels attacked the world's biggest oil processing facility in Saudi Arabia that is crucial to global supply of energy[348] by using several unmanned aerial vehicles that hit two key oil installations, damaging facilities where most of the country's crude output is processed and giving rise to the risk of a world oil supply disruption,[349] nearly 15% increase in oil prices,[350] and 2.3% plummet in Saudi Arabia's stock market.[351]

These incidents clearly reveal that AMVs are and will be used to commit crimes and terrorism which may have grave adverse effects on vessels, structures, human life, States which are the victims of these offences, and the wider global community. One of the most significant international instruments in this regard is 1988 Convention for the Suppression of Unlawful Acts against the Safety of Maritime Navigation and its 2005 Protocol (SUA Convention). First of all, ship is defined as '*a vessel of any type whatsoever not permanently attached to the sea-bed, including dynamically supported craft, submersibles, or any other floating craft*'.[352] In this context, the use of AMVs in committing crimes along with the acts committed against them would be covered in the SUA Convention which makes the convention well positioned to deal with threats to or posed by AMVs.[353] Article 3, which provides a list of offences, indicates that the use of AMVs in this regard can be accommodated by

[344] Ibid, p. 260.

[345] Ibid, p. 260.

[346] Petrig (2020), p. 108.

[347] Perper (2018).

[348] Associated Press (2019).

[349] Hubbard et al. (2019).

[350] Kearney (2019).

[351] Turak (2019).

[352] SUA Article 1 (1) (a).

[353] Kraska (2017), pp. 13–14.

SUA Convention.[354] Furthermore, it has been argued that cybercrimes could be accommodated within the convention.[355] Article 3 declares that there is an offence if a person unlawfully and intentionally seizes or controls a ship by force, threat or intimidation; if that act is likely to endanger the safe navigation of that ship, performs an act of violence against a person on board a ship, destroys or damages a ship or her cargo, places a device or substance that is likely to harm the ship, seriously damages or destroys maritime navigational facilities or seriously interferes with the operation of them; or harms any person in relation to these offences. AMVs have the potential to fit into these descriptions and to be accommodated in the legal framework of SUA Convention.

When we analyse the international legal framework, we must always bear in mind that they have been introduced with the underlying assumption that there is a human physically present on board, and the scenario where an autonomous vehicle is utilised to commit crimes at sea was not on the radar of the drafters.[356] It must, however, be noted that using AMVs to commit crimes does not indicate that there is no involvement of humans by any means, but the crime's nature shifts where the involvement of the offender is remote instead of proximate in respect of both time and geography.[357] If the criminals use AMVs, although they act in real time, they act physically at a distance from where the actual crime is committed whether it is from the dry land, the territorial sea, the exclusive economic zone or the high seas.[358] If the offenders perpetrate the crime with an autonomous system, their involvement is remote not only geographically but also as to time since launching the system and the actual time of the crime may be significantly different.[359] This evolving nature of crimes conflicts with the proximity assumption of UNCLOS provisions dealing with maritime crimes.[360] For example, the definition of piracy in Article 100 of UNCLOS consists of acts committed by the crew or the passengers which necessitates the determination of whether a remote operator can be the perpetrator of this crime.[361] Some authors are of the view that maritime crime provisions of UNCLOS cannot be interpreted in an evolutionary way considering that these provisions present specificity, precision, and completeness; and they are similar to criminal codes rather than constitutions.[362] It is therefore stated that the attributes presented to justify the evolutionary interpretation of UNCLOS do not exist in the context of maritime

[354] Klein (2019), p. 262.

[355] Kraska (2017), p. 16, Klein (2019), p. 262.

[356] Petrig (2020), p. 112.

[357] Ibid, p. 112.

[358] Ibid, p. 113.

[359] Ibid, p. 113.

[360] Ibid, p. 113.

[361] It has been submitted that since the piracy definition of UNCLOS involves acts committed by the crew or the passengers of a private ship or a private aircraft, a cyberattack from a land-based location on an AMV would not be considered piracy: Pritchett (2015), p. 215.

[362] Petrig (2020), p. 119.

crime provisions;[363] and the hybrid and interdisciplinary nature of UNCLOS, namely governing a plurality of relationships and having different provisions in respect of international institutions, marine scientific research, and maritime security, do not allow for an evolutionary interpretation of maritime crime provisions[364] given that the definition of criminal offences necessitates the principle of certainty.[365] On the contrary, it has been argued that regard must be given to the ordinary meaning of the treaty provision in its context and taking into consideration its object and purpose, consequently, by referring to the International Court of Justice's position that every treaty is '*interpreted and applied within the framework of the entire legal system prevailing at the time of the interpretation*', it has been stated that although we may be encountering disruptive technology, 'we are not facing disruptive international law'.[366] The main issue here derives from the terminology and how AMVs should be treated in this scope. The international legal framework on maritime crimes involving AMVs may only provide a solution to the extent of the inclusivity of their ship/vessel definitions and depending on the wide range of offences embracing the variety of the means which AMVs could be utilised in maritime crimes.[367]

It is noteworthy that international law does not operate as a national law. There is not a single government working towards the benefit of a single State and it does not have the ability to conduct swift examinations and evaluations and rapidly implement the regulations. International law requires an adoption at the international level along with its implementation.[368] It is a necessity for States to integrate their commitments to IMO into effective national action which involves domestic rules and regulations, interagency resources and authorities, and structures for collaboration.[369] When the international instruments were drafted, the status of technology was quite different than that of today, therefore, it would be unwise to expect every provision of these instruments can accommodate AMVs. However, AMVs still operate within this framework, so long as they can fit into the definitions and scope. The extent of the international legal framework and the evaluation of AMVs in the context of crimes will ultimately depend on the approach of States to these new threats and the definitions of new crimes along with the procedural and logistical mechanisms such as designating a capable authority to enable international collaboration for maritime security.[370]

363 Ibid, p. 120.
364 Ibid, pp. 121–123.
365 Ibid, p. 130.
366 Klein (2019), p. 247.
367 Klein (2019), pp. 262–263.
368 Kraska (2017), p. 20.
369 Ibid, p. 20.
370 Ibid, p. 20.

6 Conclusion

The wide range utilisation of AMVs necessitates evaluating their position in the legal framework and their connection to maritime security. This chapter attempted to show the intersection between law and technology in the context of maritime law and how AMVs can enhance maritime security also bearing in mind that they have the potential to hinder it.

The recent years have seen astonishing developments in technology and autonomous maritime vehicles which are already being used in a variety of activities by States. With their diverse sizes, shapes, capabilities, and roles, AMVs present many opportunities for States to enhance their maritime security and law enforcement given that they considerably reduce human error and risks to human life, they can perform their missions more precisely, and carry out tasks beyond human capabilities. Moreover, they can monitor and detect illegal activities occurring at sea and allow a swift and efficient response. Thus, States can utilise them in intelligence and surveillance, mine countermeasures, identification, inspections and in the protection of vessels, ports, installations, and facilities. On the other hand, maritime security may be threatened and disrupted by the utilisation of autonomous maritime vehicles considering the cyber security risks and the use of AMVs by criminals.

The increasing utilisation of AMVs requires a discussion on where they currently stand in the international legal framework. Yes, it is a new technology, the drafters of the international instruments did not have in their minds that such technology would be available at some day, and the vast majority of the current international law instruments do not perfectly align with AMVs. However, it must always be remembered that AMVs still operate within the existing international legal framework. Klein enunciates that the existing principles in international law can be applicable to AMVs in various situations and interpreting the treaties in accordance with their ordinary meanings in context may offer some opportunity to address circumstances on a case-by-case basis contingent upon the operations of AMVs.[371] Kraska declares that although the utilisation of AMVs is in its early stages, longstanding international law regime already offers a basic legal framework, and the existing international law can apply either directly or by analogy to autonomous vehicles with the required customisation to make them suitable for the individual circumstances.[372]

It is clear that there are gaps in the existing legal framework which is why IMO and multiple States such as the United Kingdom are already tackling the issue and trying to make the necessary amendments and changes. One particularly challenging aspect of the matter has been the legal status of AMVs. Considering the numerous types of AMVs, one simple categorisation does not provide an appropriate and sufficient answer, and the legal status of them along with the rights and obligations attach to them are required to be evaluated individually in every one of the circumstances considering the purpose of the relevant legal instrument. IMO will keep

[371] Klein (2019), p. 271.

[372] Kraska (2010), p. 64.

working on the matter with its different committees to tackle the gaps and provide solutions in relation to the use of AMVs which may be equivalences, developing interpretations, amending existing instruments, or developing a new instrument. The practice of States will also be able to shed some light on the issue while AMVs are increasingly utilised by them.

References

Allen CH (2018a) Determining the legal status of unmanned maritime vehicles: formalism vs functionalism. J Marit Law Commer 49:477–514

Allen C (2018b) 20,000 drones under the sea. https://www.usni.org/magazines/proceedings/2018/july/20000-drones-under-sea. Accessed on 27 July 2022

Associated Press (2019) Major Saudi Arabia oil facilities hit by Houthi drone strikes. https://www.theguardian.com/world/2019/sep/14/major-saudi-arabia-oil-facilities-hit-by-drone-strikes. Accessed on 27 July 2022

Australian Government Department of Defence (2020) Morrison Government invests in new Unmanned Aircraft Systems development program. https://www.minister.defence.gov.au/minister/lreynolds/media-releases/morrison-government-invests-new-unmanned-aircraft-systems. Accessed on 27 July 2022

Beckman R, Sun Z (2017) The relationship between UNCLOS and IMO instruments. Asia Pac J Ocean Law Policy 2:201–246

Bella S, Belalem G, Belbachir A et al (2021) HMDCS-UV: a concept study of hybrid monitoring, detection and cleaning system for unmanned vehicles. J Intell Robot Syst 102:44

Bork K, Karstensen J, Visbeck M, Zimmermann A (2008) The legal regulation of floats and gliders – in quest of a new regime? Ocean Dev Int Law 39:298–328

Bueger C (2015) What is maritime security? Mar Policy 53:159–164. https://doi.org/10.1016/j.marpol.2014.12.005

Burgers T, Romaniuk SN (2017) Countering the 'pirates' paradise': unmanned systems and marine security in Southeast Asia. https://thediplomat.com/2017/10/countering-the-pirates-paradise-unmanned-systems-and-marine-security-in-southeast-asia/. Accessed on 27 July 2022

Calley D, Hulme K, Ong DM (2016) New marine security threats. In: Attard DJ, Fitzmaurice M, Martinez N, Hamza R (eds) The IMLI manual on international maritime law: volume III: marine environmental law and international maritime security law. OUP, Oxford, pp 511–540

Chalk P (2008a) Maritime terrorism: threat to container ships, cruise liners, and passenger ferries. In: Herbert-Burns R, Bateman S, Lehr P (eds) Lloyd's MIU handbook of maritime security, 1st edn. Auerbach Publications, Boca Raton, pp 117–132

Chalk P (2008b) The maritime dimension of international security: terrorism, piracy, and challenges for the United States. Available via: https://www.rand.org/pubs/monographs/MG697.html Accessed on 27 July 2022

Chamberlain G (2008) Bloodthirsty and beyond the law: rise of the gun-toting pirates terrorising the seas. The Sunday Telegraph (25 May 2008), p. 37

Chang Y, Zhang C, Wang N (2020) The international legal status of the unmanned maritime vehicles. Mar Policy 113:1–7

CMI International Working Group Position Paper on Unmanned Ships and the International Regulatory Framework, available via: https://comitemaritime.org/wp-content/uploads/2018/05/CMI-Position-Paper-on-Unmanned-Ships.pdf. Accessed on 27 July 2022

Department for Transport, Future of transport regulatory review: maritime autonomy and remote operations (28 September 2021) Available via: https://www.gov.uk/government/consultations/future-of-transport-regulatory-review-maritime-autonomy-and-remote-operations/future-of-

transport-regulatory-review-maritime-autonomy-and-remote-operations. Accessed on 27 July 2022
Dinstein Y, Dahl AW (2020) Section IV: unmanned maritime systems. In: Dinstein Y, Dahl AW (eds) Oslo manual on select topics of the law of armed conflict. Springer, Cham, pp 43–53
Fratila A, Gavril IA, Nita SC, Hrebenciuc A (2021) The importance of maritime transport for economic growth in the European Union: a panel data analysis. Sustainability 13:7961. https://doi.org/10.3390/su13147961
Hawkes KG (1989) Maritime security. Cornell Maritime Press, Centreville
Hubbard B, Karasz P, Reed S (2019) Two major Saudi oil installations hit by drone strike, and U.S. blames Iran. https://www.nytimes.com/2019/09/14/world/middleeast/saudi-arabia-refineries-drone-attack.html. Accessed on 27 July 2022
IMO, Implications of the United Nations Convention on the Law of The Sea for the International Maritime Organization, Study by the Secretariat of the International Maritime Organization (30 January 2014), LEG/MISC.8, available via: https://www.kuestenpatent-kroatien.at/LEG%20MISC%208-1.pdf. Accessed on 27 July 2022
IMO, Guidelines on Maritime Cyber Risk Management, MSC-FAL.1/Circ.3 (5 July 2017) Available via: https://wwwcdn.imo.org/localresources/en/OurWork/Security/Documents/MSC-FAL.1-Circ.3%20-%20Guidelines%20On%20Maritime%20Cyber%20Risk%20Management%20(Secretariat).pdf. Accessed on 27 July 2022
IMO, Legal Committee 108th session (LEG 108), 26–30 July 2021., https://www.imo.org/en/MediaCentre/MeetingSummaries/Pages/LEG-108th-.aspx. Accessed on 27 July 2022
IMO, Autonomous Shipping., https://www.imo.org/en/MediaCentre/HotTopics/Pages/Autonomous-shipping.aspx. Accessed on 27 July 2022
IMO, Autonomous ships: regulatory scoping exercise completed, https://www.imo.org/en/MediaCentre/PressBriefings/pages/MASSRSE2021.aspx. Accessed 27 on July 2022
IMO, Maritime Cyber Risk., https://www.imo.org/en/OurWork/Security/Pages/Cyber-security.aspx. Accessed on 27 July 2022
IMO, Maritime Safety Committee., https://www.imo.org/en/MediaCentre/MeetingSummaries/Pages/MSC-Default.aspx. Accessed on 27 July 2022
IMO, Outcome of the Regulatory Scoping Exercise for the use of Maritime Autonomous Surface Ships (MASS), MSC.1/Circ.1638 (3 June 2021) Available via https://wwwcdn.imo.org/localresources/en/MediaCentre/HotTopics/Documents/MSC.1-Circ.1638%20-%20Outcome%20Of%20The%20Regulatory%20Scoping%20ExerciseFor%20The%20Use%20Of%20Maritime%20Autonomous%20Surface%20Ships...%20(Secretariat).pdf. Accessed on 27 July 2022
Intergovernmental Oceanographic Commission (of UNESCO), Draft [Practical] Guidelines of IOC, Within the Context of UNCLOS, for the Collection of Oceanographic Data By Specific Means, Seventh Meeting of the Advisory Body of Experts on the Law of the Sea (IOC/ABE-LOS VII), (Libreville, Gabon 19–23 March 2007) Available via: https://unesdoc.unesco.org/ark:/48223/pf0000218771. Accessed on 27 July 2022
Jarocki AC (2018) With 11 tons of seized drugs, US Coast Guard asks for more ScanEagle drones. https://www.defensenews.com/unmanned/2018/06/08/with-11-tons-of-seized-drugs-us-coast-guard-asks-for-more-scaneagle-drones/. Accessed on 27 July 2022
Jones S (2012) Maritime security: a practical guide, 2nd edn. The Nautical Institute
Kearney L (2019) Oil jumps nearly 15% in record trading after attack on Saudi facilities. https://www.reuters.com/article/us-global-oil/oil-prices-surge-15-after-attack-on-saudi-facilities-hits-global-supply-idUSKBN1W00UG. Accessed on 27 July 2022
Keller J (2014) Coast Guard looks for unmanned ocean vehicles to hunt smugglers, poachers, and icebergs. https://www.militaryaerospace.com/unmanned/article/16719196/coast-guard-looks-for-unmanned-ocean-vehicles-to-hunt-smugglers-poachers-and-icebergs. Accessed on 27 July 2022
Klein N (2011) Maritime security and the law of the sea. Oxford University Press, Oxford
Klein N (2019) Maritime autonomous vehicles within the international law framework to enhance maritime security. Int Law Stud 95:244–271

Klein N, Mossop J, Rothwell DR (2009) Australia, New Zealand and maritime security. In: Klein N, Mossop J, Rothwell DR (eds) Maritime security: international law and policy perspectives from Australia and New Zealand, 1st edn. Routledge, London, pp 1–21
Klein N, Guilfoyle D, Karim MS, Mclaughlin R (2020) Maritime autonomous vehicles: new frontiers in the law of the sea. Int Comp Law Q 69:719–734
Koh TTB, "A Constitution for the Oceans" Remarks by Tommy T.B. Koh of Singapore, President of the Third United Nations Conference on the Law of the Sea. Available via: http://www.un.org/Depts/los/convention_agreements/texts/koh_english.pdf. Accessed on 27 July 2022
Kraska J (2010) The law of unmanned naval systems in war and peace. J Ocean Technol 5:44–68
Kraska J (2017) Effective implementation of the 2005 convention for the suppression of unlawful acts against the safety of maritime navigation. Nav War Coll Rev 70:Article 3
Kraska J, Pedrozo R (2013) International maritime security law. Martinus Nijhoff Publishers, Leiden
Lucky A (2015) International Tribunal for the Law of the Sea, Request for an Advisory Opinion submitted by the Sub-Regional Fisheries Commission (SRFC) (Request for Advisory Opinion submitted to the Tribunal). Available via: https://www.itlos.org/fileadmin/itlos/documents/cases/case_no.21/advisory_opinion_published/2015_21_SO_Luck-E.pdf. Accessed on 27 July 2022
McLaughlin R (2014) Unmanned naval vehicles and the law of naval warfare. In: Nasu H, McLaughlin R (eds) New technologies and the law of armed conflict. Hague, T.M.C. Asser Press, pp 229–246
McLaughlin R, Klein N (2021) Maritime autonomous vehicles and drug trafficking by sea: some legal issues. Int J Mar Coast Law 36:389–418. https://doi.org/10.1163/15718085-bja10061
Mejia MQ Jr (2002) Defining maritime violence and maritime security. In: Mukherjee PK, Mejia MQ Jr, Gauci GM (eds) Maritime violence and other security issues at sea: proceedings of the symposium on maritime violence and other security issues at sea, 26–30 August 2002. WMU Publications, Malmö
Mejia MQ Jr (2003) Maritime gerrymandering: dilemmas in defining piracy, terrorism and other acts of maritime violence. J Int Commer Law 2:153–175
Mejia MQ Jr (2004) Selected issues of law and ergonomics in maritime security. J Int Marit Law 10: 316–326
Moore-Colyer R (2016) Royal Navy will use AI to detect threats in maritime combat. https://www.silicon.co.uk/e-innovation/ai-to-aid-navy-in-detecting-combat-threats-199266?inf_by=5a1ea147681db800768b46d8. Accessed on 27 July 2022
Mukherjee T (2018) Securing the maritime commons: the role of artificial intelligence in naval operations. Available via: https://www.orfonline.org/research/42497-a-i-in-naval-operations-exploring-possibilities-debating-ethics/. Accessed on 27 July 2022
Nasu H, Letts D (2020) The legal characterization of lethal autonomous maritime systems: warship, torpedo, or naval mine? Int Law Stud 96:79–97
National Ocean Service National Oceanic and Atmospheric Administration U.S. Department of Commerce, What is an ocean glider?, https://oceanservice.noaa.gov/facts/ocean-gliders.html. Accessed on 27 July 2022
Naval Technology (2020) Royal Navy to test underwater gliders in North Atlantic. https://www.naval-technology.com/news/royal-navy-to-test-underwater-gliders-in-north-atlantic/. Accessed on 27 July 2022
Nawrot J, Peplowska-Dabrowska Z (2019) Revolution or evolution? challenges posed by autonomous vessels for national and international legal framework. Comp Law Rev 25:239–255. 10.12775/CLR.2019.008
Norris A (2013) Legal issues relating to unmanned maritime systems monograph (unpublished manuscript). Available via: https://www.hsdl.org/?view&did=731705. Accessed on 27 July 2022
O Brian C (2018) Key advantages and disadvantages of ship autonomy. https://safety4sea.com/key-advantages-and-disadvantages-of-ship-autonomy/. Accessed on 27 July 2022

O'Rourke R (Updated February 17, 2022) Navy large unmanned surface and undersea vehicles: background and issues for congress, congressional research service report. https://crsreports.congress.gov/product/pdf/R/R45757/47. Accessed on 27 July 2022
Parker J (2021) The challenges posed by the advent of maritime autonomous surface ships for international maritime law. Aust N Z Marit Law J 35:31–42
Pedrozo RP (2020) US employment of marine unmanned vehicles in the South China Sea. In: Buszynski L, Hai DT (eds) The South China Sea from a regional maritime dispute to geo-strategic competition. Routledge, Oxford, pp 217–229
Perper R (2018) Drone boats filled with explosives are the new weapon in global terrorism. https://www.businessinsider.com/drone-boats-filled-with-explosives-houthis-saudi-arabia-2018-10?r=US&IR=T. Accessed on 27 July 2022
Petrig A (2020) The commission of maritime crimes with unmanned systems: an interpretive challenge for the United Nations Convention on the Law of the Sea. In: Evans MD, Galani S (eds) Maritime security and the law of the sea: help or hindrance? Edgar, Cheltenham, pp 104–131
Prasetya J (2020) The operation of unmanned vessel in light of Article 94 of the Law of the Sea Convention: seamanning requirement. Indones J Int Law 18:105–124
Pritchett PW (2015) Ghost ships: why the law should embrace unmanned vessel technology. Tulane Marit Law J 40:197–226
Richardson M (2004) A time bomb for global trade: maritime-related terrorism in an age of weapons of mass destruction. ISEAS Publications, Singapore
Ringbom H (2019) Regulating autonomous ships-concepts, challenges and precedents. Ocean Dev Int Law 50:141–169. https://doi.org/10.1080/00908320.2019.1582593
Schmitt MN, Goddard DS (2016) International law and the military use of unmanned maritime systems. Int Rev Red Cross 98:567–592
Sekimizu K (2014) The United Nations Convention on the Law of the Sea and the International Maritime Organization. Available via: https://www.imo.org/en/MediaCentre/SecretaryGeneral/Pages/itlos.aspx. Accessed 27 July 2022
Showalter S, Manley J (2009) Legal and engineering challenges to widespread adoption of unmanned maritime vehicles. Oceans 2009:1–5. https://doi.org/10.23919/OCEANS.2009.5422108
Talley WK (2008) Introduction. In: Talley WK (ed) Maritime safety, security and piracy. Informa, London, pp 1–8
The Guidelines on Cyber Security on board Ships (version 4) issued by ICS, IUMI, BIMCO, OCIMF, INTERTANKO, INTERCARGO, InterManager, WSC and SYBAss. Available via: https://wwwcdn.imo.org/localresources/en/OurWork/Security/Documents/ANNEX%20Guidelines%20on%20Cyber%20Security%20Onboard%20Ships%20v.4.pdf. Accessed on 27 July 2022
The Lloyd's Register Unmanned Marine Systems Code, available via: https://www.lr.org/en/unmanned-code/. Accessed on 27 July 2022
Turak N (2019) Saudi stock market dives, crude futures to jump after drone attack on oil plants. https://www.cnbc.com/2019/09/15/saudi-stock-market-dives-crude-to-jump-after-attack-on-oil-plants.html. Accessed on 27 July 2022
United Nations Conference on Trade and Development (UNCTAD) (2018), Review of Maritime Transport 2018. Available via: https://unctad.org/webflyer/review-maritime-transport-2018. Accessed on 27 July 2022
United Nations Conference on Trade and Development (UNCTAD), Review of Maritime Transport 2021. Available via: https://unctad.org/system/files/official-document/rmt2021_en_0.pdf. Accessed on 27 July 2022
United Nations General Assembly, Report of the Secretary-General, Oceans and the Law of the Sea (10 March 2008) UN Doc A/63/63 (UN A/63/63). Available via: http://daccess-ods.un.org/access.nsf/Get?Open&DS=A/63/63&Lang=E. Accessed on 27 July 2022

United Nations Office on Drugs and Crime (UNODC), Legislative Guide for the Implementation of the Protocol Against the Smuggling of Migrants by Land, Sea and Air, Supplementing the United Nations Convention Against Transnational Organized Crime. Available via: https://www.unodc.org/pdf/crime/legislative_guides/04%20Legislative%20guide_Smuggling%20of%20Migrants%20Protocol.pdf. Accessed on 27 July 2022

United Nations Office on Drugs and Crime (UNODC), UNODC provides a technological edge to address drug trafficking on the Mekong River. https://www.unodc.org/roseap/2021/09/drug-trafficking-mekong-river/story.html. Accessed on 27 July 2022

US Department of Defense, Unmanned Systems Integrated Roadmap FY2013–2038 (2014) Available via: https://www.hsdl.org/?abstract&did=747559. Accessed on 27 July 2022

US Department of the Navy, Navy Unmanned Undersea Vehicle (UUV) Master Plan (9 November 2004). Available via: https://www.hsdl.org/?abstract&did=708654. Accessed on 27 July 2022

Vallejo D (2015) Electric currents: programming legal status into autonomous electric currents: programming legal status into autonomous unmanned maritime vehicles unmanned maritime vehicles. Case West Reserve J Int Law 47:405–428

Vavasseur X (2021) Meet Singapore's New Maritime Security Unmanned Surface Vessels. https://www.navalnews.com/naval-news/2021/03/meet-singapores-new-maritime-security-unmanned-surface-vessels/. Accessed on 27 July 2022

Veal R, Tsimplis M, Serdy A (2019) The legal status and operation of unmanned maritime vehicles. Ocean Dev Int Law 50(1):23–48. https://doi.org/10.1080/00908320.2018.1502500

Xing W, Zhu L (2021) A functional approach to reassessing the legal status and navigational rights of ships and ship-shaped structures. Transp Policy 106:120–130. https://doi.org/10.1016/j.tranpol.2021.03.025

Zhu L, Xing RWW (2021) Developments of unmanned and autonomous merchant ships: legality under the UNCLOS. In: Zou K, Telesetsky A (eds) Marine scientific research, new marine technologies and the law of the sea. Brill Nijhoff, Leiden, pp 127–152

Salvage and Autonomous Maritime Navigation

Cecilia Severoni

Abstract The increasing use of Maritime Autonomous Surface Ships (MASS) in salvage operations at sea requires a reflection on the adequacy of the current applicable legal framework, while also considering that the ship providing assistance, or the ship to be assisted, may no longer have a master or a crew on board. It means, on the one hand, that the salvage operation presents a lower degree of risk for the salvor; on the other hand, we could say that if the shore-based operator is far from the place where the salvage operations are to be carried out, he may not realize the exact extent of the danger for property and the people. Furthermore, all obligations of conduct of the salvor should be reconsidered, in the light of a new element of evaluation such as the use of artificial intelligence in the management of the ship, also to determine a suitable insurance coverage. This chapter analyzes the problems that may arise from the application of the current conventional law to MASS; it deals with the problem of the salvage reward if the activity is carried out by a MASS; finally, it presents its own concluding remarks.

Keywords Unmanned aerial vehicles · Unmanned aircraft systems · Autonomous maritime navigation · Maritime autonomous surface ships · Salvage

1 Introductory Remark

The process of automation of the various human activities has had an impressive, recent acceleration.

The transport sector particularly shows a growing trend towards automation, in some areas more evident and already partially regulated, such as air navigation,[1] in

[1] On the EU legal framework on unmanned aerial vehicles, we can report the Regulation (EU) 2018/1139, which has provided a preliminary regulatory framework for drones within the European Union. The framework is further implemented by the delegated Regulation (EU) 2019/945 on

C. Severoni (✉)
Department of Legal Sciences, University of Udine, Udine, Italy
e-mail: cecilia.severoni@uniud.it

K. Noussia, M. Channon (eds.), *The Regulation of Automated and Autonomous Transport*, https://doi.org/10.1007/978-3-031-32356-0_6

others at the stage of first analysis of compatibility with the current international and internal regulations, as in the maritime field.

Both totally automated and remote piloted vehicles are included by the European Union[2] in the notion of autonomous vehicle, related to a phenomenon that now involves both air and water navigation and ground transport. In this perspective, the unmanned vessels are potentially supposed to be used in dangerous, dull, and dirty scenarios, in which it may be hazardous to send a ship with a crew on board. Among these are traditionally included many salvage hypotheses, in which there may be a risk also for the salvor, or repetitiveness and length of the operations themselves. The study focuses on the general duty to provide assistance to people in danger with autonomous vessels, but an examination of the type of operations that can be carried out with unmanned ships cannot even ignore the assessment of the cyber security aspects in the event that there are unauthorized accesses to the computer systems with the intention of stealing data for illegal purposes or taking control of the system. The chapter aims to address the problem if there is an obligation to assist people in distress only in the case of drone ships that are capable of providing salvage at sea, or if the salvage obligation exists even in the absence of technical equipment that allows to carry out the salvage operation. The chapter also aims to examine the extensibility of the obligation to the shore-based operator that operates the remote piloting of the ship.

It further analyzes the possible implications in terms of responsibility for the use of the unmanned vessel that is now capable to learn from experience and make autonomous decisions in carrying out salvage operations. In this regard, we are now aware that there are artificial intelligence systems for piloting the ship which makes learning from the surrounding environment possible, and that allows the ship to change its operating behavior by interacting with the elements, the circumstances, and the places in which it will operate. This circumstance may have important repercussions on the liability regime currently focused on the owner's role.

2 Salvage and Unmanned Ships

The salvage of property and people at sea represents one of the first uses of unmanned means of transport: in these hypotheses, the circumstances highlight a risk not only for the people or things to be rescued, but also often for the personnel who must operate, and that would be replaced by a remotely controlled ship. In this

unmanned aircraft systems and on third-country operators of unmanned aircraft systems, and by the implementing Regulation (EU) 2019/947 on the rules and procedures for the operation of unmanned aircraft.

[2] As indicated by the European Parliament resolution of 16 February 2017 with recommendations to the Commission on Civil Law Rules on Robotics (2015/2103(INL)) "autonomous transport covers all forms of remotely piloted, automated, connected and autonomous ways of road, rail, waterborne and air transport."

sense, since ancient times we have been assisting to the man's attempt to create a machine that is both intended for transport and other possible employment at sea, and that has no crew on board.[3] In 2017, a European Parliament resolution containing civil regulations on robotics with recommendations to the European Commission,[4] mentioned the autonomous means of transport, as the complex of "forms of remotely piloted, automated, connected and autonomous ways of road, rail, waterborne and air transport, including vehicles, trains, vessels, ferries, aircrafts, drones, as well as all future forms of developments and innovations in this sector." Thereby, all ships operated by an artificial intelligence are considered to be regulated by an AI system, i.e., a "software that is developed with one or more of the techniques and approaches listed in Annex I and can, for a given set of human-defined objectives, generate outputs such as content, predictions, recommendations, or decisions influencing the environments they interact with."[5]

[3] At the end of 1800 the famous inventor Nikola Tesla had already filed a patent request (no. 613,809, of 8 November 1898, application of 1 July 1898, serial number 684,934) entitled "Method of and apparatus for controlling mechanism of moving vessel or vehicles," in which he underlined the fundamental characteristic of his invention, given by the absence of cables or other mechanical connections: "I require no intermediate wires, cables, or other form of electrical or mechanical connection with the object save the natural media in space." With considerable clarity, the inventor listed all the possible applications of his invention, currently implemented in the various transport sectors: "The invention which I have described will prove useful in many ways. Vessels or vehicles of any kind may be used, as life, dispatch, or pilot boats or the like, or for carrying letters, packages, provisions, instruments, objects, or materials of any description, for establishing communication with inaccessible regions and exploring the conditions existing in the same, for killing or capturing whales or other animals of the sea, and for many other scientific, engineering, or commercial purposes; but the greatest value of my invention will result from its effect upon warfare and armaments, for by reason of its certain and unlimited destructiveness it will tend to bring about and maintain permanent peace among nations." On this subject, see the recent Soyer and Tettenborn (2021), pp. 63–80. Among the most dating back works we can mention Crisafulli Buscemi (1933), pp. 191–204. On the theme of the relationship between technological progress and international law of the sea, see also Craven (1985), pp. 1143–1159.

[4] See the European Parliament resolution of 16 February 2017 with recommendations to the Commission on Civil Law Rules on Robotics (2015/2103(INL)).

[5] On this matter, see the Art. 3 of the recent Proposal for a Regulation of the European Parliament and the Council laying down harmonized rules on Artificial Intelligence (Artificial Intelligence Act) and amending certain union legislative acts, 2021/0106 (COD), and the previous European Parliament resolution 20 October 2020, with recommendations to the Commission on a civil liability regime for artificial intelligence (2020/2014(INL)). The Proposal emphasizes that "Artificial intelligence is a fast evolving family of technologies that can contribute to a wide array of economic and societal benefits across the entire spectrum of industries and social activities. By improving prediction, optimising operations and resource allocation, and personalising digital solutions available for individuals and organisations, the use of artificial intelligence can provide key competitive advantages to companies and support socially and environmentally beneficial outcomes, for example, in healthcare, farming, education and training, infrastructure management, energy, transport and logistics, public services, security, justice, resource and energy efficiency, and climate change mitigation and adaptation." However, depending on the conditions involving its specific application and use, artificial intelligence may present risks and cause harm, both material and immaterial, to public interests and rights guaranteed by the Union law.

Likewise, the IMO Maritime Safety Committee 98[6] undertook a regulatory scoping exercise to determine how the safe, secure, and environmentally sound operation of Maritime Autonomous Surface Ships (hereafter generally indicated as MASS), also indicated as Unmanned Surface Vehicles (USV), could be introduced in IMO instruments, and the MSC 100 approved the framework and the methodology for the regulatory scoping exercise on MASS, identifying four degrees of autonomy.[7]

The analysis of the different degrees of autonomy shows a technological development of the ship, from a type with automated processes and seafarers on board to operate and control shipboard systems and functions (first degree of remotely operated ships), up to a ship with an operating system able to make decisions and determine actions by itself, without human intervention (fourth degree or autonomous ship). Particularly, the list of instruments considered in the Legal Committee's scoping exercise for MASS includes an examination of the London International Convention on Salvage, 1989 (hereafter, the Salvage Convention).

[6] Maritime Safety Committee (MSC), 98th session, 7–16 June 2017.

[7] Maritime Safety Committee (MSC), 100th session, 3–7 December 2018. The degrees of autonomy identified are: Degree one: Ship with automated processes and decision support: Seafarers are on board to operate and control shipboard systems and functions. Some operations may be automated and at times be unsupervised but with seafarers on board ready to take control. Degree two: Remotely controlled ship with seafarers on board: The ship is controlled and operated from another location. Seafarers are available on board to take control and to operate the shipboard systems and functions. Degree three: Remotely controlled ship without seafarers on board: The ship is controlled and operated from another location. There are no seafarers on board. Degree four: Fully autonomous ship: The operating system of the ship is able to make decisions and determine actions by itself.

The MASS is already having a certain development both in the military[8] and in the civil fields,[9] where it is used in the most different ways.

As for the first of the sectors identified, the development of USVs in the military field was very important for the evolution of the entire industry, and this progress has made it possible to arrive at the current situation in which they are employed, in their capacity as unmanned surface vehicles (USV), or autonomous surface vehicles (ASV) if they are totally autonomous, to provide additional warfighting capability with a tactical and strategic advantage for the military operation. Under a military perspective, indeed, the essence of the term "unmanned" is the physical removal or remote location of the human from the platform and the autonomy gives the ability to adapt interactively to the dynamic maritime environment, especially in scenarios

[8] On the use of USV for military operations, see Yan et al. (2010), p. 451, according to which "USVs have the potential, and in some cases the demonstrated ability, to reduce risk to manned forces, provide the necessary force multiplication to accomplish military missions, perform tasks which manned vehicles cannot, and do so in a way that is affordable for the navy." On the use of USVs by the US Navy, see the complete discussion in Savitz et al. (2013). In the study USVs are defined as "maritime vehicles uninhabited by personnel that maintain continuous, substantial contact with the surface" (xiii). For further considerations on the use of drones for military operations, see the most recent Department of the Navy (2021) Unmanned Campaign Framework: 1-37. As an example of military autonomous unmanned surface vehicle (USV) we can mention the Sea Hunter, launched in 2016 as part of the DARPA Anti-Submarine Warfare Continuous Trail Unmanned Vessel (ACTUV) program. See also O'Rourke (2022) Navy Large Unmanned Surface and Undersea Vehicles: Background and Issues for Congress. In fas.org. and further general considerations on the US Ghost Fleet Overlord Lagrone (2021). News.usni.org: "The Ghost Fleet Overlord program is currently in its second phase, which began in September 2019 and focuses on the integration of government-furnished command-and-control systems and payloads and more complex and challenging naval operations experimentation," SCO said in a statement. "Phase II is being conducted with the same vessels and industry teams that took part in Phase I and will conclude in early 2022, at which point both Ghost Fleet Overlord vessels will transition to the Navy for further experimentation. The ships will work as part of the Surface Development Squadron based out of San Diego, Calif. Surfdevron. The unit operates two Sea Hunter USVs, the Zumwalt-class destroyers, early Littoral Combat Ships and smaller unmanned aerial and sea vehicles."

[9] See, for example, the Yara Birkeland project, which is nearing completion and it is the result of a collaboration between Yara and Kongsberg. It is one of the first examples of autonomous and zero-emission container ship, totally electric. The ship is a 120 TEU (Twenty-foot Equivalent Units) open top container ship. It will be a fully battery powered solution, prepared for autonomous and unmanned operation. The vessel will reduce NOx and CO2 emissions by reducing diesel-powered truck transport by around 40,000 journeys per year. In addition, short sea shipping connections with autonomous vessels are currently being studied in Europe and the European Commission has founded a research project related to unmanned vessels, called Maritime Unmanned Navigation through Intelligence in Networks (MUNIN). More recently, the Norwegian maritime authority has identified in the Trondheim fjord the suitable place to test a commercial maritime connection with autonomous vehicles that has been implemented since 2018. At the beginning of 2019, the autonomous boat SEA-KIT Maxlimer successfully made the first sea crossing between the United Kingdom and Belgium: see Autonomous ships: Test areas and research centers making headlines, September 10, 2019, https://safety4sea.com/cm-autonomous-ships-test-areas-and-research-centers-making-headlines. On this topic, see Veal (2018), pp. 1–4; Ringbom (2019), pp. 141–169. Veal et al. (2019), pp. 23–48.

where special adaptation is required.[10] It is, therefore, possible to affirm that different missions call for different amounts of human-machine interdependence.

In the more recent past, the use of MASS for peacekeeping missions or commercial activities has also increased.

In the perspective indicated, the MASS can in fact be used in peacekeeping operations, or in the function of monitoring and control of individual areas, for example, to combat piracy, drugs, and dangerous or prohibited materials trafficking.[11]

In the second perspective, MASS is used in the transport of goods or for other civil uses (surveillance, monitoring of the degree of water pollution, use in agriculture, etc.).

It is further considered that one of the most interesting uses for remotely piloted or autonomous vessels concerns short sea shipping (SSS) and the European waterways. This would greatly reduce road transport, as well as pollution in general, given that electric or hybrid propulsion vehicles would be used.

Unlike the transport of passengers on short-distance ferries, or traffic in submarine tunnels or in general in highway ferries, for which it is considered that part of the crew should still remain on the ship for the purpose of assisting the passengers themselves for the entire duration of the journey, the use of unmanned ships is conceivable in SSS or particularly in last mile shipping, for the transport of goods from the hubs to more remote destinations that are not served by large feeder vessels. In this regard, recent studies underline the feasibility of a possible connection with unmanned ships in short-sea transport—so far considered less convenient and generally more expensive than road transport.[12]

This would not totally eliminate the use of trucks or trains, since it will be necessary to transport the goods to the final destinations, generally distant from the coast and ports, but it would still significantly reduce road traffic.

Some studies conclude that soon the transport of goods, and then later of people, could be entrusted to totally autonomous ships and led by an artificial intelligence.[13]

The benefit in terms of environmental protection is clear: an unmanned ship can use alternative sources of energy supply to power itself and, in any case, it produces

[10] Department of the Navy (2021) Unmanned Campaign Framework: 7.

[11] Department of Navy (2007) USV Masterplan: 5.

[12] On this, see the study conducted by Maritime Unmanned Navigation through Intelligence in Networks (MUNIN) (2015) Funding Scheme: SST.2012.5.2-5: E-guided vessels: the 'autonomous' ship - D10.1: Impact on Short Sea Shipping: 3. For MUNIN "Short sea shipping is less complex than deep sea in that an unmanned ship calls at ports more often and Salvage is much easier. Thus, some of the maintenance problems can be reduced or solved at lower cost. Also, close to shore operation is performed in emission control areas where also manned ships need to use cleaner fuels or exhaust cleaning systems. Thus, fuel costs are also less of a problem. Finally, coastal shipping will normally have access to much higher and lower cost communication infrastructure which also reduces cost of operation." On the MUNIN project, see further considerations in Burmeister et al. (2014), p. 1.

[13] As evidenced by Weiger and Pribyl (2017).

a lower fuel consumption; it also does not have to provide for the disposal of wastewater or the management of man-made waste.

Furthermore, the use of unmanned ships could be advantageous to the extent that it is possible to use a reduced crew, or even to operate unmanned or with a simple shore-based controller, if the ship is able to carry out dockside operations autonomously. Then, the unmanned vessel can interface with the automated cranes for cargo handling until loading onto the truck for the final destination.

Regarding the safety profile in maritime navigation, we cannot deny the relevance of the human error in most maritime accidents: some recent studies have estimated that 75% to 96% of marine accidents involve some level of human error, while human error is a primary factor in 75% of the value of all claims analyzed.[14] It often involves careless behavior on the part of the crew, over-reliance on other forms of technology, fatigue or a failure of organizational culture and behavior: in any case the human error is a key safety issue and a factor at the basis of many claims.

It must also be considered that one of the main cost items is represented by the salaries of the crew, as well as the costs of compensation for damages resulting from accidents involving them.

However, the use of unmanned ships can also present critical issues that must be considered.

Some difficulties in the use of the USVs for short sea shipping purposes can be identified in the port costs, which plausibly should be higher for the loading and unloading of goods with highly automated technologies.[15] It is clear that, from this point of view, the use of unmanned ships will receive the attention of the main shipping companies to the extent that there may be cost savings compared to the manned transport of the same goods by road to their final destination. Furthermore, traffic with unmanned ships must become competitive in operations in the various ports, compared to what currently occurs with ships with crew on board.

A further cause for concern is the traffic, often heavy, in which the unmanned ship would find itself operating in the restricted port areas, where not only commercial ships, but also pleasure craft, pass through. Unmanned vessels must be well equipped to detect obstacles to navigation and carry out safe maneuvers, but specific

[14] As reported in Allianz (2019) Shipping safety - Human error comes in many forms. Agcs.allianz.com. Similarly, on the human error in the road traffic incidents, see the Report of the eSafety Working Group (2002), reported by the Commission to the European Parliament and the Council (2016) Saving Lives: Boosting Car Safety in the EU - Reporting on the monitoring and assessment of advanced vehicle safety features, their cost effectiveness and feasibility for the review of the regulations on general vehicle safety and on the protection of pedestrians and other vulnerable road users {SWD(2016) 431 final}: 4.

[15] As reported in Maritime Unmanned Navigation through Intelligence in Networks (MUNIN) (2015) Funding Scheme: SST.2012.5.2-5: E-guided vessels: the 'autonomous' ship - D10.1: Impact on Short Sea Shipping: 10. On this matter, see further considerations in Ghaderi (2019), p. 152 ss.

protocols for the mooring of these vessels should also be required, as well as specific site for them.[16]

An additional consideration can be made on the risk of computer piracy in the use of MASS, which increases exponentially in the AI systems, resulting in the need for appropriate insurance cover, although the absence of a crew makes the ship less vulnerable against the request for a ransom for people taken hostage, a circumstance that could have a positive effect on insurance costs.

More generally, we know that the cybernetic system can transform the ship into a complex of interconnected systems, which increase the overall efficiency and safety of the means of transport.[17]

It is a consolidated opinion that "Network and information systems and electronic communications networks and services play a vital role in society and have become the backbone of economic growth. Information and communications technology (ICT) underpins the complex systems which support everyday societal activities, keep our economies running in key sectors such as (. . .) transport[18]"; however, the system requires constant control to protect it against any form of computer piracy, and it is difficult to predict the economic impact of a continuous control of the ship's management software against the risk of cybercrime.

In general terms, a cybercrime is intended as an "unauthorized access to computer systems and networks with the intention of accessing a computer system to steal, intercept or manipulate data to use for nefarious purposes; taking control of a system or assets; using another computer system to illegally access legitimate confidential

---

[16] Maritime Unmanned Navigation through Intelligence in Networks (MUNIN) (2015) Funding Scheme: SST.2012.5.2-5: E-guided vessels: the 'autonomous' ship - D10.1: Impact on Short Sea Shipping: 11.

[17] As evidenced by Lloyd's Register (2017) Cyber-enabled ships, ShipRight procedure assignment for cyber descriptive notes for autonomous & remote access ships - A Lloyd's Register guidance document: 1. Lloyd's adds that there are multiple reasons to advocate an increase in the ship's cyber management system: "the potential for better business performance (for example, better fuel economy); the need to comply with environmental and safety legislation; the increased bandwidth provided by modern satellite communications; a shortfall in crew competence (particularly for engineering roles, as complexity and demand for performance increase); the ability to capture and analyse a wide range of data, including operational, service, monitoring, regulation and off-ship storage data; the provision of operational support and guidance; the ability to conduct periodic inspection to enable preventative maintenance; the ability to easily update products based on software (these and other modern 'cyber' implementation techniques provide an easy path to product evolution); the fact that a single technology can present multiple design choices, including target, language, development tools, application programme interfaces (APIs) and protocols; the ability to future proof ships (by having system components designed to be adapted and extended in function, for example, through software changes); the ability to integrate, flexibility control and optimise systems; the potential for better communication both on and off ship (for example, for data sharing and performing updates and maintenance); the desire for a similar level of robustness as exists in systems based on shore."

[18] As indicated by Regulation (EU) 2019/881 of the European Parliament and of the Council (2019) ENISA (the European Union Agency for Cybersecurity) and on information and communications technology cybersecurity certification and repealing Regulation (EU) No 526/2013 (Cybersecurity Act): 1.

information (Banks, Governments, Companies, Individuals),"[19] while cyber security is defined as "a process used to control and protect an organizations computer systems, networks, and data from, and reduce the risk of, Cyberattack."[20]

Transport operators have also long been engaged in providing guidance on security against cybercrime aboard ships increasingly coordinated with ground operations through digital communication,[21] and which rely increasingly on digital solutions for the completion of everyday tasks. They are aware that the rapid developments in the information technology present all players in the maritime industry with increased possibilities for operational optimization, cost savings, safety improvements and a more sustainable business, but this should also increase the potential cyber vulnerabilities and risks.[22] In the above mentioned guidelines the

---

[19] Maritime UK (2020) Maritime Autonomous Ship Systems (MASS) UK Industry Conduct Principles and Code of Practice - A Voluntary Code Version: 33. The authors underline that "On the 30th September 2020, the United Nations International Maritime Organisation (IMO) suffered a Cyberattack which disrupted many of its systems. On the same day the French maritime transport and logistics giant CMA CGM S.A. revealed it was also the victim of a malware attack, on 28th September 2020, that affected some servers on its network. This follows similar attacks on Maersk (2017), COSCO (2018) and MSC (April 2020) raising fears that the maritime industry, which accounts for the transportation of 90% of global trade, is regarded as a highly valued target for Cybercriminals."

[20] Maritime UK (2020) Maritime Autonomous Ship Systems (MASS) UK Industry Conduct Principles and Code of Practice - A Voluntary Code Version: 36. As evidenced in the study "IT/OT convergence is the integration of information technology (IT) systems with operational technology (OT) systems. IT systems are used for data-centric computing; OT systems monitor events, processes and devices, and make adjustments in enterprise and industrial operations."

[21] On this matter, see the The Guidelines on Cyber Security Onboard Ships - Version 4, produced and supported by BIMCO, Chamber of Shipping of America, Digital Containership Association, International Association of Dry Cargo Shipowners (INTERCARGO), InterManager, International Association of Independent Tanker Owners (INTERTANKO), International Chamber of Shipping (ICS), International Union of Marine Insurance (IUMI), Oil Companies International Marine Forum (OCIMF), Superyacht Builders Association (Sybass) and World Shipping Council (WSC); it is also possible to analyze the Cyber Security Considerations in Maritime UK (2020) Maritime Autonomous Ship Systems (MASS) UK Industry Conduct Principles and Code of Practice - A Voluntary Code Version: 32 ss.

[22] Very extensive is the list of guidelines and regulatory sources relating to cyber risk management indicated by the Guidelines on Cyber Security Onboard Ships - Version 4, cit.: 1: "In 2017, the International Maritime Organization (IMO) adopted resolution MSC.428(98) on Maritime Cyber Risk Management in Safety Management System (SMS). The resolution stated that an approved SMS should consider cyber risk management in accordance with the objectives and functional requirements of the (International Safety Management) ISM Code. It further encourages administrations to ensure that cyber risks are appropriately addressed in SMS no later than the first annual verification of the company's Document of Compliance (DoC) after 1 January 2021. The same year, IMO developed guidelines that provide high-level recommendations on maritime cyber risk management to safeguard shipping from current and emerging cyber threats and vulnerabilities. As also highlighted in the IMO guidelines, effective cyber risk management should start at the senior management level. Senior management should embed a culture of cyber risk management into all levels and departments of an organisation and ensure a holistic and flexible cyber risk governance regime, which is in continuous operation and constantly evaluated through effective feedback mechanisms. In addition to the IMO resolution, the U.S. National Institute of Standards

main cyber incidents were identified in "a cyber security incident, which affects the availability and integrity of OT, for example corruption of chart data held in an Electronic Chart Display and Information System (ECDIS); an unintended system failure occurring during software maintenance and patching, for example through the use of an infected USB drive to complete the maintenance; loss of, or manipulation of external sensor data, critical for the operation of a ship. This includes but is not limited to Global Navigation Satellite Systems (GNSS), of which the Global Positioning System (GPS) is the most frequently used; failure of a system due to software crashes and/or 'bugs'; crew interaction with phishing attempts, which is the most common attack vector by threat actors, which could lead to the loss of sensitive data and the introduction of malware to shipboard systems."[23] In the perspective of a

---

and Technology (NIST) Cybersecurity Framework Version 1.1 (April 2018) has also been taken into account in the development of these guidelines. The NIST Cybersecurity Framework assists companies with their approach to risk assessments by helping them understand an effective approach to manage potential cyber risks both internally and externally. As a result of applying the Framework, a 'profile' is developed, which can help to identify and prioritise actions for reducing cyber risks. The profile can also be used as a tool for aligning policy, business and technological decisions to manage the risks. Sample framework profiles are publicly available for maritime bulk liquid transfer, offshore, and passenger ship operations. These profiles were created by the United States Coast Guard and NIST's National Cybersecurity Center of Excellence with input from industry stakeholders. The NIST's profiles can be used together with these guidelines to assist industry in assessing, prioritizing, and mitigating their cyber risks. Guidelines are also available from other associations, such as the Digital Container Shipping Association's (DCSA) 'DCSA Implementation Guide for Cyber Security on Vessels v1.0.' The DCSA's guidelines are based on an analysis of version 3 of these guidelines and the NIST framework. While the target audience for DCSA's guidelines is the container industry, other segments of shipping may also find them worthwhile to read. The International Association for Classification Societies (IACS) has issued a 'Recommendation on Cyber Resilience (No. 166).' This recommendation consolidates IACS' previous 12 recommendations related to cyber resilience (Nos. 153 to 164) and applies to the use of computer-based systems, which provide control, alarm, monitoring, safety or internal communication functions that are subject to the requirements of a classification society. The IACS recommendation applies to newbuild ships only but can also serve as guidance for existing ships. In due course, IACS is expected to develop Unified Requirements, which will also apply to newbuilds only. This publication is not intended to provide a basis for, and should not be interpreted as, calling for external auditing or vetting the individual company's and ship's approach to cyber risk management."

[23] The Guidelines on Cyber Security Onboard Ships—Version 4, cit.: 3. According to the Guidelines "the maritime industry presents a range of characteristics that affect its vulnerability to cyber incidents, such as involvement of multiple stakeholders in the operation and chartering of a ship potentially resulting in lack of accountability for the IT and OT system infrastructure and ship's networks; use of legacy IT and OT systems that are no longer supported and/or that rely on obsolete operating systems; use of OT systems that cannot be patched or run anti-virus due to type approval issues; ships that interface online with shoreside parties and other parts of the global supply chain; ship equipment that is remotely monitored and accessed, e.g. by the manufacturers or support providers; the sharing of business critical, data sensitive and commercially sensitive information with shore- based service providers, including marine terminals and stevedores and also, where applicable, public authorities; the availability and use of computer controlled critical systems, which may not have the latest patches installed or be properly secured, for the ship's safety and for environmental protection; a cyber risk management culture that still has potential for improvement,

more automated and networked vessel, cybercrime could take the most extreme forms of "hijacking of vessels and cargoes for personal use/financial gain; holding vessels and cargoes for ransom; environmental terrorism, (threatening to release chemical cargoes or running vessel aground); terrorism (taking control of vessels to damage other vessels, assets or reputations."[24]

In this regard, it is believed that already in the design phase the ship could be projected in such a way as to limit any possible access to the internal systems; protection *by design* can also include specific software for protection against computer intrusions aimed at sabotaging the management program.[25]

The MASS UK Code of Practice draws the attention on the circumstance that "The traditional approach to cyber security focuses on the protection of data and controlling access to the IT systems components. Advanced technology such as 'smart' equipment and the Internet of Things (IoT) have changed how IT is integrated into systems and is deployed and operated. These new systems, which are complicated, connected and can generate large amounts of data now demand a much more comprehensive cyber security approach. It is no longer sufficient for a cyber security system to consider just the IT system; it must also take into account the operational technology (OT) system. This is generally referred to as IT/OT

---

eg through more formalised training, exercises and clarified roles and responsibilities; frequently the automation system comprises of multiple sub-systems from numerous vendors that are integrated by shipyards with minimal regard to cyber issues."

[24] Maritime UK (2020) Maritime Autonomous Ship Systems (MASS) UK Industry Conduct Principles and Code of Practice - A Voluntary Code Version: 34.

[25] In the Guidelines on Cyber Security Onboard Ships - Version 4, cit.: 32, the case of a worm incident on maritime IT and OT is reported, where the ship was built recently, but this system was not connected to the Internet by design, and it was equipped with a power management system that could be connected to the Internet for software updates and patching, remote diagnostics, data collection, and remote operation: "The company's IT department made the decision to visit the ship and perform vulnerability scans to determine if the system had evidence of infection and to determine if it was safe to connect. The team discovered a dormant worm that could have activated itself once the system was connected to the internet and this would have had severe consequences. The incident emphasizes that even air gapped systems can be compromised and underlines the value of proactive cyber risk management. The shipowner advised the manufacturer about the discovery and requested procedures on how to erase the worm. The shipowner stated that before the discovery, a service technician had been aboard the ship. It was believed that the infection could potentially have been caused by the technician. The worm spread via USB devices into a running process, which executes a programme into the memory. This programme was designed to communicate with its command and control server to receive its next set of instructions. It could even create files and folders. The company asked cyber security professionals to conduct forensic analysis and remediation. It was determined that all servers associated with the equipment were infected and that the virus had been in the system undiscovered for 875 days. Scanning tools removed the virus. An analysis proved that the service provider was indeed the source and that the worm had introduced the malware into the ship's system via a USB flash drive during a software installation. Analysis also proved that this worm operated in the system memory and actively called out to the internet from the server. Since the worm was loaded into memory, it could affect the performance of the server and systems connected to the internet."

convergence."[26] More generally, the shipowner should implement technical and procedural measures to protect itself from a cyber incident, promptly detect incidents and ensure the continuity of operations, but in this new perspective, only OT cyber security system—understood as the "technologies, processes, and practices designed to prevent the intended or unintended use of a cyber technology system to do damage to the cyber technology (networks, computers, programs, data), and vessel or harm to people, and environment"—protects the complete system (vessel, people and environment). This may be achieved both by ensuring correct, safe, efficient, and reliable operation through software quality engineering, and by preventing malicious and non-malicious threats through the cyber security system. In this regard, a culture of security and the protection of ship systems from cyber risk should also be implemented at various levels, including senior managers on shore and shipboard personnel.

In addition, the use of unmanned ships for civil purposes presents further problems: they are equipped with a large technological apparatus, often expensive, which can hardly be inserted in the ship during the overhaul, but more reasonably it is related to newly built and newly designed ships. In fact, it is undeniable that the initial introduction costs of these types of ships in commercial traffic are not competitive with respect to existing fleets, for which shipowners have often already taken steps to optimize the use of the crew and equipment on board.

Based on this reasoning, it can be noted that if on the one hand the transport with tankers or with bulk carriers could technically be carried out with unmanned ships, it would, in any case, be necessary to evaluate the cost of the technological renewal of the fleet compared to the current costs, as well as the impact, in public opinion, of transport by oil tankers or tankers carrying polluting chemicals without a crew on board. Similarly, also for container ships, the use of unmanned vehicles should be analyzed from an economic point of view and under the perspective of the costs of renewing the fleet, compared to the current means of transport, widely tested and functional to the use for which they have been built.[27]

[26]Maritime UK (2020) Maritime Autonomous Ship Systems (MASS) UK Industry Conduct Principles and Code of Practice - A Voluntary Code Version: 36.

[27]In line with this opinion, we can find the statement by Moller-Maersk CEO Soren Skou, that surmises the unlikelihood of unmanned container ships operating at sea for the duration of his life: "Even if the technology advances, I don't expect we will be allowed to sail around with 400-meter long container ships, weighing 200,000 tonnes without any human beings on board, I don't think it will be a driver of efficiency, not in my time" (Bloomberg, February 16, 2018).

3 The Concept of Salvage Operation in the Perspective of the Use of MASS

In the aforementioned perspective, the use of unmanned ships can be extremely beneficial in the event of a maritime salvage.[28]

As with the employment of the Unmanned Aerial Vehicles (UAVs),[29] even MASS can be used in scenarios deemed *dangerous, dull, and dirty*, where sending a ship with crew can be extremely hazardous. Among these, we often find included many salvage hypotheses, which present a profile of danger for the salvor as well, or of repetitiveness and length of the operations themselves.

Consider, for example, the scenario in which the accident causes a spill of hydrocarbons from the ship: in this situation the intervention of small boats can involve a high level of danger for the salvors, who are directly exposed to combustible agents and pollutants. The operations may also be monotonous and repetitive, and they may involve an increase in costs for the whole salvage.

In this regard, some salvage projects, already in an advanced stage of development, concern ships that can assist a vessel without needing a crew on board, and which can therefore be carried out in unfavorable or hazardous environments, unsuitable for the human presence.[30]

The Salvage Convention defines the salvage operations as "any act or activity[31] undertaken to assist a vessel or any other property in danger in navigable waters or in any other waters whatsoever" (Art. 1(a)), while the concept of Vessel is defined as "any ship or craft, or any structure capable of navigation"[32] (Art. 1 (b)). The rule

[28] The concept of salvage operation, and the related legal framework in the Italian legal system, can be explored in Severoni (2005a, b) La remunerazione del soccorso tra interesse pubblico e interessi privati, vols. I e II.

[29] On this matter, see Calantropio (2019), p. 64; Kas and Johnson (2020).

[30] As indicated by Kurowsky et al. (2012), pp. 141–148. See also Ardito et al. (2013), related to a new concept of salvage operations of distressed ships at sea, based on the development of robotized unmanned marine platforms able to (semi-)automatically execute the high-risk operation of linking the emergency towing system of distressed ships to towing vessels, and, more recently, Yoo et al. (2020).

[31] See the Annex 2 to the CMI IWG Submission to MSC 99th Session, according to which "it is considered that under the present wording also the electronic restoration of a system from land could be a Salvage operation as long as the vessel was in danger and in navigable waters or any other waters of navigation. Unless this is further specified, this suggests that an entirely shore-based IT-expert who helps to reestablish communication and command of an unmanned ship entitled to a Salvage award."

[32] Severoni (2018), pp. 67–85, which also takes as an example the definition of ship provided by the Italian Navigation Code (Art. 136), as "any construction intended for transport by water, including for the purpose of towing, fishing, pleasure, or for other purposes." The broad notion that derives from it primarily relies on the element of construction, understood as *res connexa* and therefore as a set of heterogeneous elements brought together by the work of men, in an element juridically considered in a unitary sense. The construction must have the aptitude to float, a necessary element for it to also possess the aptitude to navigate, or to move by water regardless of the means of

does not expressly provide for the hypothesis of a remotely operated or autonomous ship, but it is believed that the definition is sufficiently generic to also include the salvage carried out with MASS.

Nonetheless, it is evident that the aforementioned convention, and in general any maritime regulation, is not designed to intervene on the phenomenon of remotely operated, or totally autonomous, navigation. Indeed, it appears that the international legislation does not contain references to the hypothesis of salvage rendered with a MASS, or even to a MASS itself; and yet, there are no elements that indicate the absolute incompatibility of the current regulatory system with the presence of autonomous means of transport, although in some cases it will be necessary to clarify or even modify the current discipline to also include MASS.

Consequently, any legislative provision in this regard will have to be harmonized with the new proposed scenarios of a salvage by any means of remotely operated, or even completely autonomous, ships.

An analysis of the above-mentioned phenomenon should consider two different kinds of salvage if a MASS is involved: it can be a hypothesis of a salvage to a MASS, or there can be a case of a salvage brought with a MASS.

In both cases, we must consider that the master of the MASS, except in the first degrees of autonomy, is not on board the ship.

In any case, the application of principles foreseen for salvage with manned ships is not always simple.

For instance, if the salvage is rendered to a manned ship "notwithstanding the express and reasonable prohibition of the owner or master of the vessel or the owner of any other property in danger which is not and has not been on board the vessel," there is no salvage reward under the Salavage Convention (Art. 19). This rule could also apply to MASS, but, in the absence of a master on board the ship, a subject must be identified who has the authority to prohibit the execution of the salvage operation, such as, for example, the shore-based operator/commander who controls and manages the ship remotely.[33] The problem arises, in this regard, for fully automated ships

propulsion used. It follows that the ship, today as in the times of imperial Rome, is identified with the function that the floating construction is intended to perform, that being navigation—"*navis etenim ad hoc paratur ut naviget*"—understood in the broad sense of destination for transport as a movement in the water of a manufact used to carry out any activity and regardless of the means of propulsion used. The notion, on the other hand, does not contain any reference to the ship's crew as a necessary component for its configuration. Soyer and Tettenborn (2021), pp. 63–89.

[33] On the role of the shore control center operators and the shore-based operators in general, see Saha (2021). According to the author in the Shore Control Center (SCC) concept "command, control, and/or monitoring of ships will take place from the shore." Among the definitions reported by the Maritime UK (2020) Maritime Autonomous Ship Systems (MASS) UK Industry Conduct Principles and Code of Practice - A Voluntary Code Version: 21, we can highlight those of Remote Control, i.e., an "Operational control of some or all ship operations or functions, at a point remote from the ship" and of Remote Control Centre (RCC), that being a site off the ship from which control of an autonomous vessel can be executed. The RCC may be located either ashore or afloat and may exercise varying degrees of control as defined under "Levels of Control." An RCC may consist of more than one Control Station or Room. A further definition reported by the Code of

without shore-based support, because in this case there is no master on board or shore-based operator, and it should be clarified whether only the owner of the ship or cargo can provide the necessary consent or denial of the operation, or someone delegated by him to carry out the aforementioned activity.

In this regard, the recent MASS UK Code of Practice distinguished the figure of the MASS Watch Officer, intended as "the individual who has responsibility for the MASS when it is operational"; the Operator, i.e., "An entity (e.g. a company) that discharges the responsibilities necessary to maintain the MASS in a seaworthy condition and compliant with all relevant IMO Instruments and national legislation. The operator is also responsible for ensuring that all staff concerned with the control of MASS hold appropriate qualifications as required by IMO instruments and national legislation"; the Owner, namely, "The title holder of the MASS." If the "Company" is not the Owner, then the Owner shall report the name and details of the Operating Company to the Maritime Administration; the Shipowner, that is "One who owns or operates a ship, whether a person, a corporation or other legal entity, and any person acting on behalf of the Owner or Operator" and the Ship Security Officer, that "means a person accountable to the master, designated by the Company as responsible for the security of the ship, including implementation and maintenance of the ship security plan and for liaison with the company security officer and port facility security officers."[34]

4 The Authority of the Master to Conclude Contracts for Salvage Operations

Similarly, a clarification should be provided under the Salvage Convention, regarding the authority of the master of the ship to conclude salvage contracts, when a MASS is involved.

The salvage reward is often agreed upon by the master of the ship to be rescued based on the indications and in the interest of the ship owner.

The legitimacy of the master of the ship in distress to also enter into salvage contracts on behalf of the owner of the property on board the vessel is another important aspect of the issues relating to the assignment of the salvage reward that emerges from the convention.

In this regard, the Salvage Convention expressly provides that the master is entitled to enter into salvage contracts on behalf of the owner of the vessel and that the master or the owner of the vessel may conclude contracts on behalf of the owner of the goods on board the ship, under Art. 6.2: "The master shall have the

Practice is that of Remote Monitoring, i.e., "Monitoring some or all ship operations or functions at a point remote from the ship."

[34]Maritime UK (2020) Maritime Autonomous Ship Systems (MASS) UK Industry Conduct Principles and Code of Practice - A Voluntary Code Version: 21.

authority to conclude contracts for Salvage operations on behalf of the owner of the vessel. The master or the owner of the vessel shall have the authority to conclude such contracts on behalf of the owner of the property on board the vessel."

The provisions of Art. 6.2 represent the legislative response to the difficulty often encountered in carrying out an effective salvage, due to the impossibility of quickly contacting all those interested in the shipment to be salved, to obtain their respective consent to the operations. Major problems arise, in this regard, for ships with autonomy of the second degree, namely, remotely controlled ships with seafarers on board: in this case it is necessary to clarify whether a shore-based operator can be considered as "master" and if so, whether such authority to conclude contracts is reserved only to him or it can be attributed to a seafarer present onboard, or to other persons besides the owner himself.[35] For vessels with autonomy of third degree, i.e., remotely controlled ships without the crew on board, no seafarers are on board, and it is necessary to clarify in the text of the Salvage Convention whether only the owner of the vessel can conclude the salvage contract, or if someone other than him has the authority to enter into contracts for salvage operations: in this case Art. 6 must be amended unless the term "master" can be interpreted as including such persons, or it should be clarified if the shore-based operator of a MASS can constitute the "master" for this purpose. The master of the ship is therefore formally recognized as representing the interests not only of the owner of the ship in danger, but also of the owner of the properties on board the ship. For ships with fourth-degree autonomy there is no shore-based support: and it should be clarified if only the owner of the vessel can enter into the salvage contract or if other individuals are provided with the authority to conclude contracts for salvage operations, with a consequent amendment to the Art. 6.[36] In the case of fully automated ships without shore-based support there is no master on board, and somebody else should be delegated by the shipowner to conclude the aforementioned contracts: for example, it can be assumed that in this case the authority to conclude salvage contracts can be assigned to the shipping agent, in addition to the shipowner.

Far more complex is the analysis of the limits on the master's activity in signing the salvage contract. On this point, it is believed, in accordance with what happens in practice, that he has the right to include the *no cure no pay clause* in the salvage contract.

A further problem emerges from the identification of a new interest by the Salvage Convention, namely, the interest of the owner of property "which is not and has not been on board the vessel" (Art. 19). In this case, these are properties which have no contact with the ship, but which may have been endangered by the same event that involved the ship.

[35] See the Work Conducted by the CMI International Working Group on Unmanned Ships. Consolidated analysis of legal Conventions, Annex II, sub-Art. 6 of the Salvage Convention.

[36] See the Work Conducted by the CMI International Working Group on Unmanned Ships. Consolidated analysis of legal Conventions, Annex II, sub-Art. 6 of the Salvage Convention.

In fact, the question arises on whether the master of the ship to be salved, and consequently the shore-based operator, can equally represent the interests of the owner of goods that were not, at the time of the salvage, on board the ship, thus carrying out acts that fall not only on the goods entrusted to him but also on goods that have no connection with these, but which are also affected by the same event (for example, in the case of a fire that involves not only the ship but also a floating dock that was nearby).

On this point, Art. 6.2 remains silent, so it can be considered that in this case no representation of such interests is conceivable.

5 The Duty to Assist Persons in Distress at Sea

In this legal framework, one of the most relevant problems related to the provision of salvage with MASS is represented by the general obligation to render assistance to people in distress.[37]

This obligation was introduced and progressively specified in various international legislative texts starting from the beginning of the twentieth century,[38] also due to the need to adapt the regulatory framework to the historical context, and progressively to the technological developments.

In the current regulation, to understand the exact content of the obligation to assist distressed individuals at sea and define the related scope in the salvage with MASS, we must analyze the content of the International Convention for the Safety of Life at Sea (SOLAS) (rule 33, Chapter V), of the Art. 98 of Montego Bay Convention, as well as Art. 10 of the aforementioned Salvage Convention.

Under rule 33 of the SOLAS Convention "The master of a ship at sea which is in a position to be able to provide assistance, on receiving a signal from any source that persons are in distress at sea, is bound to proceed with all speed to their assistance," thereby giving the master the possibility of evaluation of the most appropriate conduct to follow; with a slightly different legal provision, the State shall require the master of a ship flying its flag to render assistance to any person found in danger at sea, if there is no serious danger for the ship, the crew and passengers, or to proceed with all possible speed to the rescue of persons in distress in so far as such action may reasonably be expected of him (Art. 98 of the Montego Bay Convention). In this second article, the master has the duty to proceed to the rescue of individuals, to the extent that such behavior could reasonably be assumed from him ("in so far as such action may reasonably be expected of him"); in the first article, the master is asked to proceed to the assistance of people in danger if in a position to do so ("in a position to be able to provide assistance"), so that, if, in particular circumstances, the master does not consider himself "able," or, according to a subjective assessment,

[37] Severoni (2018), pp. 67–85.

[38] In this sense Papanicolopulu (2018), p. 187.

the circumstances are not "reasonable," he can abstain from conducting assistance operations.

Under the Salvage Convention "Every master is bound, so far as he can do so without serious danger to his vessel and persons thereon, to render assistance to any person in danger of being lost at sea" (Art. 10, par. 1). This provision allows us to state that the master's obligation to provide assistance to people in distress at sea is expressly recognized in the conventional law, as well as, according to another perspective, the master's duty to provide salvage at sea can be considered as a general principle of international law which informs all the conventional rules on the subject.[39]

The legislative framework here reported does not contain exceptions for the commercial transport. It follows that the obligations of the shipowner are fundamentally the same in the case he operates container vessels, or instead bulk carrier, tankers, cruise liners of fishing vessels. The duty to render assistance is implemented in practice, based on the circumstances in which the danger occurs: the salvor can save passengers from a fire, can tow the ship with passengers on board to a safe port, can assist them on the ship by providing food or other supplies, but we cannot deny the costs and danger of using commercial ships for large-scale salvages. Salvage of people cand lead to increased direct costs, such as extra fuel consumed, port charged assessed during disembarkation of salved people, additional wages or repairing or cleaning the vessel. But we can also hypothesize indirect costs, the most important of which is the delay, if the vessel is obliged to deviate from its intended voyage, to embark rescued persons, up to the final destination to a safe port. To limit the danger for the crew employed in the salvage operation, and a part of the costs of the salvage activity, we can imagine a possible use of a MASS. It is believed that in principle such activities are not precluded to a salvage with MASS.[40]

However, the absence of a master on board the ship should be compensated by other professional figures, related to the ship providing assistance and possibly identified by the legislation, who take on the task of assisting people in distress, even if not present aboard the rescue ship.

Therefore, on the applicability of this principle to salvage with MASS, it will be necessary to further ascertain the compatibility of its use with respect to the salvage obligation on a case-by-case basis.[41] The duty exists as long as there is no serious

[39] On this matter, see Kilpatrick (2010), p. 407; Starita (2019), p. 9. Italian navigation code (Art. 489) contains a basic regulation that imposes the obligation to provide assistance to people in danger in the water, in compliance with the spirit of solidarity that informs the maritime world and the degree of civilization of the people of sea, without a right to a reward for the salvage od people: "Absit, o Jupiter, ut lucrum captemus tale ex hominum infortunio" (DIONE, Praes. Orat., VII, quoted by Piantanida L (1806) Della giurisprudenza marittima-commerciale antica e moderna: 13).

[40] As evidenced for example by Kenney and Tasikas (2003), p. 151.

[41] Veal et al. (2019), p. 23. Some authors (Barnes (2017), p. 185; Papanicolopulu I (2016) The Duty to Rescue at Sea. Peacetime and in War: A General Overview. War and Security at Sea: 495) also apply this duty to the masters of warships or public ships, although Art. 4.1 of the Salvage Convention excludes it: "Without prejudice to article 5, this Convention shall not apply to warships

danger for the salvor: in the case of a salvage with MASS, if the ship is autonomous and there is no crew on board, it does not expose the salvor to risks of personal injury.

It is further important to clarify if the duty to assist people in distress at sea remains for a MASS not equipped to carry out salvage, as in the case of a ship commonly used in the commercial traffic. In this case, MASS is not designed to host people on board and its capability to render assistance is limited. Surely, the shore-based operator can request assistance from ships that are in the vicinity of the one to be assisted, can send out a distress signal, or warn the competent authorities that a ship is in danger. More uncertain is that there is an obligation to provide suitable equipment to provide help to the ship in distress, based on the duty to render assistance at sea.

For the duty of salvage to be respected, it may be sufficient for the salvors to provide assistance to the extent possible. In this sense, some authors[42] have hypothesized that flag States can impose precise ship construction standards to guarantee assistance to people in distress. However, this solution may be excessively burdensome for shipowners, unless rewarding measures are envisaged for those who want to provide specific salvage equipment in the ship. Moreover, the crew is often outnumbered by the rescued subjects, and neither physically nor psychologically trained to carry out this type of operation.[43]

This imposition of an obligation could serve to make the implementation of the salvage obligation less uncertain in the event that the ship operates remotely, or it is totally autonomous. In these cases, in fact, the master of a private ship could be asked to make a deviation of the route to assist people in distress, which in itself involves a considerable economic burden, in a situation in which he could find himself not on the ship,[44] but in the vicinity of the owner who should bear the related costs.

However, many of these costs do not derive directly from the assistance operations, but instead concern the deviation of the route and the delays in the disembarkation phase of the rescued people. As recently evidenced,[45] P&I Insurance covers some rescue-related delays: particularly, The Gard P&I Club Rules, applicable, in the example given in note, to the oil tanker Maersk Etienne, contain a provision that

---

or other non-commercial vessels owned or operated by a State and entitled, at the time of Salvage operations, to sovereign immunity under generally recognized principles of international law unless that State decides otherwise."

[42] Mandrioli (2020), p. 91.

[43] Attard and Kilpatrick (2020).

[44] On these considerations, see Starita (2019), p. 6; Davies (2003), p. 109; Attard and Kilpatrick (2020). In the case of the Maersk Etienne reported by the two authors: it is a Danish-flagged oil tanker, which was instructed by the Maltese rescue coordination center to attend to a small fishing vessel in distress off the coast of Tunisia in the Gulf of Gabes on 4 August 2020. The ship was diverted off its commercial route, and it rescued 27 migrants. After arriving in Malta, the ship was denied entry into the port, and it remained anchored outside the island's contiguous zone. Only on 13 September, the migrants were finally disembarked in Pozzallo.

[45] Attard and Kilpatrick (2020).

allows cover for the extra costs of fuel, port charges, wages and provisions for the crew, “attributable to a diversion. . .for the purpose of saving persons at sea” (rule 31), while rule 32 explicitly identifies cover for “costs and expenses directly and reasonably incurred in the consequence of the Ship having. . .refugees or persons saved at sea on board. . .” The same rule, however, also states that “cover does not include consequential loss of profit or depreciation.”

It is further highlighted that charterparties, i.e., the contracts between shipowners and charterers which regulate the interests between the parties in the transport of goods, are often standardized (they use forms drawn up by organizations such as INTERTANKO or BIMCO[46]) and reveal that there is generally no reference to the party who assumes the risk of rescue-related delays, or if the charterer must continue to pay the agreed rate of hire in the time charter, or of freight in the voyage charter, for the entire delay period.[47] In this regard, an author suggested the use of a “refugee clause” in commercial shipping contracts,[48] in the belief that either the shipowner or the charterer would benefit from clear language precisely allocating rescue risk, while the void of contractual silence would entail an increase in the costs of legal disputes.

To identify the juridical figure to which to attribute the salvage activity, especially in the case of an autonomous ship, we should consider the role of the shipowner, who has the management and the responsibility of the ship, according to a criterion which, generally, makes him responsible for the events concerning the crew and for the obligations contracted by the master of the ship, as regards the ship and the shipment.[49] Considering the due differences, we can therefore assume that the obligation to render assistance can be attributed to the shipowner himself, who took over the management of a MASS.[50]

[46] Charterparties are contracts that define the rights and the obligations of shipowner and charterer who generally employ the shipowner’s vessel (depending on the specific provisions) for a fixed period or for a specific voyage. On the uncertainty of who bears the risk of rescue-related costs, see in particular Kilpatrick (2010), p. 412 ss. The author adds that the international Conventions also address the issue of deviation as it relates to the rights and liabilities allocated between carriers and shippers under a bill of lading. In this regard, the International Convention for the Unification of Certain Rules of Law Relating to Bills of Lading (Hague Rules), Article IV (4) reads: “Any deviation in saving or attempting to save life or property at sea or any reasonable deviation shall not be deemed to be an infringement or breach of this Convention or of the contract of carriage, and the carrier shall not be liable for any loss or damage resulting therefrom.” Similar provisions are contained in the more recent Hague-Visby Rules, Hamburg Rules, and Rotterdam Rules, exculpating the carrier for losses caused by efforts to save lives at sea.

[47] Attard and Kilpatrick (2020).

[48] Kilpatrick (2010), p. 403, 433 ss. The author highlights that “While these acts of heroism have been lauded as compliant with entrenched moral and legal obligations, it is often overlooked that they have also come at great expense to shipping industry participants” and the direct and indirect costs arising out of these operations could be staggering.

[49] In the Italian legal system under Art. 274 of the navigation code.

[50] In compliance with the principle of safeguarding lives at sea, it can be considered that an obligation in this sense derives from Art. 94, paragraphs 3 and 4, letter c of the Montego Bay Convention, which considers the duty of the State to take measures for ships flying its flag to

Another issue is the one indicated by Art. 10, third paragraph, of the Salvage Convention, which denies any liability of the shipowner, for the master's failure to provide assistance to people in difficulty: "10.3. The owner of the vessel shall incur no liability for a breach of the duty of the master under paragraph 1." This provision will also have to consider that a master, if any, is present in a remote station, and that the absence of a crew could make it more difficult to assist people in difficulty, at sea or on a ship in danger.

As previously said, Art. 10 of the Salvage Convention establishes the duty of the master, who is physically on the ship, to render assistance to people in danger, unless this exposes the ship itself or people on board to serious danger.[51]

The absence of a crew on board on the one hand, means that the salvage operation presents a lower degree of risk for the salvor; on the other hand we could say that, if the shore-based operator is closer to the owner and less close to the place where the salvage operations are to be carried out, he/she could be influenced by his/her decisions, diminishing the command function that he/she has, with respect to values such as the obligation to assist people in danger.[52]

Certainly, such an obligation exists if the MASS is equipped to bring assistance to people in distress; it is, therefore, necessary to understand whether the obligation to provide, for example, automated equipment for salvage operations is also configurable for a MASS not equipped for salvage,[53] or whether an autonomous ship may be obliged to provide *by design* the on-board equipment or instruments suitable for salving people in danger "in navigable waters or in any other waters whatsoever," given the duty to provide assistance to people in distress: the solution will greatly depend on the costs that such equipment requires and on the economic incentives for the shipping industry that could be provided by the governments to encourage such virtuous behaviors, also considering that the Salvage Convention

"ensure safety at sea," among which that to ensure "that the master, officers and, to the extent appropriate, the crew are fully conversant with and required to observe the applicable international regulations concerning the safety of life at sea, the prevention of collisions, the prevention, reduction and control of marine pollution, and the maintenance of communications by radio." It is further intended that "In taking the measures called for in paragraphs 3 and 4 each State is required to conform to generally accepted international regulations, procedures and practices and to take any steps which may be necessary to secure their observance" (para. 5).

[51] In addition to the danger for the salvor, it is also necessary to consider the pressure sometimes exerted by the shipowners or by the charterers, who indefinitely bear the economic costs deriving from the deviations of the route and the delays imposed by the salvage operation: Davies (2003), p. 109.

[52] As indicated by Ancis (2019), p. 460.

[53] As mentioned in the previous notes, the CART (Cooperative Autonomous Robotic Towing System) project, has developed a system of unmanned robotic marine ships, capable of semi-automatically performing the high-risk operation of connecting the emergency towing system to ships in distress. The innovation aims to reduce the risk to human lives and increase the protection of the environment, helping, for example, to prevent oil pollution at sea during rescue operations. The key idea is to collect a floating object coming from the vessel to be rescued, such as a floating buoy, by tying a knot around it with a floating rope by an unmanned robotic ship that pulls the floating rope, which is connected to the tugboat ring.

provides for the express obligation for the States to adopt "the measures necessary to enforce the duty" to render assistance to any person in danger of being lost at sea (Art. 10, para. 1 and 2).

In conclusion, it is necessary to understand whether the master of the remotely operated ship, influenced on the one hand by the distance from the rescue site and, on the other, by proximity to the shipowner, can refrain from providing assistance to people in distress, since this conduct could lead to a failure in assistance.

In fact, even if the notion of support may be uncertain, the master cannot evade the duty to assist people in distress simply because the ship is not fit to physically perform the rescue in the absence of a crew. By the other hand, CMI clearly stated that "the assistance that MASS (n.d.r.) can render will necessarily be more limited than in the case of manned ships. Depending on the scope of the term 'master' for ships with no seafarers on board, there may be a need to extend this provision to make sure that such ships are not exempted from this obligation. Expected assistance by MASS ships could be specifically clarified."[54]

6 The Provision of a Reward for the Salvage with MASS

In general terms, the Salvage Convention provides that "Salvage operations which have had a useful result give right to a reward" (Art. 12, para.1); likewise, "Except as otherwise provided, no payment is due under this Convention if the Salvage operations have had no useful result" (Art. 12, para. 2).

The determination of reward does not respond to strict rules, but to a principle that can be defined "liberal," i.e., not closely related to the evaluation of the actual costs, nor to a sum that represents the simple consideration (*quantum meruit*) of the salvage services rendered,[55] but it is added a component of prize, peculiar element of the salvage reward, which responds to incentive reasons for the assistance rendered.

The Salvage Convention highlights the right of the salvor to obtain a salvage reward, except in the case of a salvage of persons, where "No remuneration is due from persons whose lives are saved" (Art. 16, para. 1). Nevertheless, "A salvor of human life, who has taken part in the services rendered on the occasion of the accident giving rise to Salvage, is entitled to a fair share of the payment awarded to the salvor for salving the vessel or other property or preventing or minimizing damage to the environment" (Art. 16, para. 2).

[54] See the Work Conducted by the CMI International Working Group on Unmanned Ships. Consolidated analysis of legal Conventions, Annex II, sub-Art. 10 of the Salvage Convention.

[55] *The Nagasaki Spirit*, Court of appeal, 4, 5, 6 e 21 dec. 1995. In: Lloyd's Law Reports, 1996 (I): 459. For the Court "The need to encourage salvors to undertake unusual risks in the general public interest, combined with recognition of the fact that unsuccessful services or ones where no property was saved resulted in payment of any kind, meant that the rewards for success were generous. The jurisdiction was equitable, and it took account of these factors which were extraneous to the individual case."

The aforementioned rules have been foreseen for the hypothesis of a salvage rendered by manned ships; also considering the fact that "the reward shall be fixed with a view to encouraging Salvage operations" (Art. 13), we can hypothesize that the salvor should have the right to the reward even if the vessel is operated remotely or if it is totally autonomous.

The criteria for determining the remuneration set by Art. 13 should also apply in the case of an autonomous salvage ship, even if some of them are provided for a traditional salvage operation, such as indicated by the criterion of "the skill and efforts of the salvors in salving the vessel, other property or life" (letter e), or by the reference to the risk of liability or other risks run by salvors (letter g).

The perspective of a salvage with autonomous ships must also consider the further duty of the salvor to operate with due diligence in assisting a ship in danger, even in the specific case of environmental salvage (Art. 8.1 (b) of the Salvage Convention), also seeking assistance and accepting the contribution of other salvors reasonably requested by the shipowner or by the owner of other property at risk (Art. 8.1 (c) and (d)). The master of the vessel, together with the owner of the vessel and the owner of other property in danger, has a duty to the salvor to also co-operate fully with him/her during the course of the salvage operations; in doing so, to exercise due care to prevent or minimize damage to the environment; and when the vessel or other property has been brought to a place of safety, to accept redelivery when reasonably requested by the salvor to do so (Art. 8.2).

In general terms, Art. 8 of the Salvage Convention provides a detailed list of the main duties of the salvor, which the heading of the article also extends to the ship's master, among which undoubtedly the obligation to operate "with due care" is highlighted: "The salvor shall owe a duty to the owner of the vessel or other property in danger: (a) to carry out the Salvage operations with due care; (b) in performing the duty specified in subparagraph (a), to exercise due care to prevent or minimize damage to the environment; (c) whenever circumstances reasonably require, to seek assistance from other salvors; and (d) to accept the intervention of other salvors when reasonably requested to do so by the owner or master of the vessel or other property in danger; provided however that the amount of his reward shall not be prejudiced should it be found that such a request was unreasonable." It is therefore necessary to clarify whether an obligation of this kind can also be envisaged in the hypothesis of salvage carried out by a MASS, and whether it is attributable to the shore-based operator and to the shipowner. Instead, it should remain the duty to seek, or if circumstances require it, to accept the help of third parties who are near the ship to be salved, as indicated in letters c and d.

In any case, it can be expected that the duty to carry out the salvage operations with the due care will be applied to a ship specifically equipped to provide assistance, where the shipyard is supposed to take greater care in preparing the remote viewing software of the situation adjacent to the ship to assist and specific equipment for the salvage operations, also for the benefit of the environment.

Likewise, also "the owner and master of the vessel or the owner of other property in danger shall owe a duty to the salvor: (a) to co-operate fully with him during the course of the Salvage operations; (b) in so doing, to exercise due care to prevent or minimize damage to the environment" (Art. 8, para. 2). In the assistance to a MASS, it can be assumed that such obligations are attributable to the shore-based operator or/and to the shipowner of the ship in danger.

Having ascertained that in a MASS there is no crew in the traditional sense of the term, we can however imagine that the duty to act with the due diligence can be attributable to the shipowner, in the sense indicated above. Namely, as aforementioned, for ships with second degree of autonomy it is necessary to clarify, in the Salvage Convention, whether a shore-based operator can be considered as master, or if the role can be attributed to somebody else on board the ship. For ships with a third degree autonomy, in which there are no crew on board, it should be clarified if the shore-based operator can be considered as master, while in totally autonomous ships, with an autonomy of fourth degree, it is necessary to further clarify, in the absence of personnel on board and of a remote control, whether the role of salvor can be assumed by other figures such as the programmer of the piloting and management software or the ship or system builder.[56]

It can be assumed that these criteria are applicable to the reward for a salvage carried by MASS, if we consider that the skills and efforts are no longer those of the master and crew of the salvor, but those of the shore-based operator, or of the shipowner who provides for autonomous management software and adequate equipment on the ship, in the case of totally autonomous vessels. Furthermore, if the MASS is professionally equipped to provide salvage, it can increase remuneration under Art. 13 (i) and (j), which sets the salvage compensation on "i) the availability and use of vessels or other equipment intended for Salvage operations; j) the state of readiness and efficiency of the salvor's equipment and the value thereof." This item can include the investments and economic efforts made by professional salvors in the purchase of equipment and software of a MASS, which are supposed to reach high costs.

7 Special Compensation in the Case of Salvage Operations with MASS

The Salvage Convention provides that the salvage operation which has had a useful result give right to a reward (Art. 12), which is furthermore fixed considering, inter alia, the skill and efforts of the salvors in preventing or minimizing damage to the environment (Art. 13.1 (b)).

If although "the salvor has carried out Salvage operations in respect of a vessel which by itself or its cargo threatened damage to the environment and has failed to earn a reward under Article 13 at least equivalent to the special compensation assessable in accordance with this article, he shall be entitled to special

[56] See the Work Conducted by the CMI International Working Group on Unmanned Ships. Consolidated analysis of legal Conventions, Annex II, sub-Art. 8 of the Salvage Convention.

compensation from the owner of that vessel equivalent to his expenses" (Art. 14.1). If, in the above-mentioned circumstances, the salvor by his salvage operations has prevented or minimized damage to the environment, the special compensation payable by the owner to the salvor may be increased up to a maximum of 30% of the expenses incurred by the salvor. However, the tribunal, if it deems it fair and just to do so and bearing in mind the relevant criteria set out in Art. 13, paragraph 1, may increase such special compensation further, but in no event shall the total increase be more than 100% of the expenses incurred by the salvor (Art. 14.2). This rule could be applied also in the case of a salvage operation rendered with MASS, but it is therefore necessary to clarify to whom "skills and efforts" refer, given that in a MASS (degree three and four of autonomy) neither the master nor the crew are on board the ship. If the ship is remotely operated, it can be assumed that they are attributed to the shore-based operator who directs the operations; if the ship is autonomous, they may, instead, be referred to the shipowner.

But also in this case, it must be considered that the subject is not in the place where the danger occurs and therefore, he/she may not have the exact perception of the event of danger in all its complexity, nor regarding the ship or the persons in danger, nor, as specified in letter b, with regard to the environment in which the salvage operations are carried out.

The Salvage Convention further states that salvor's expenses mean the out-of-pocket expenses reasonably incurred by the salvor in the salvage operation and a fair rate for equipment and personnel actually and reasonably used in the salvage operation (Art. 14.3). If the salvage operation is carried out by a MASS it should be clarified, in the Salvage Convention, whether "equipment and personnel" in this paragraph includes the equipment and personnel onshore. It should be specified, in this regard, if the cost of labor of a shore-based operator can be compensated under this provision.[57]

If the vessel is totally autonomous, there is no personnel, neither on board nor operating remotely: in this case it would be necessary to clarify, in the Salvage Convention, whether the reference to the concept of fair rate for personnel contained in Art. 14.3 is still valid.

Furthermore, the salvor must behave diligently: this requirement is highlighted specifically in the provision of a special compensation for the environmental salvage, where it is emphasized, in negative terms, that the negligence in the salvage operation, or the failure in the prevention of a damage to the environment, can deprive the salvor in whole or in part of the compensation (Art. 14, para. 5). Also in this case, the reference to the salvor's negligence does not fit easily with fully automated operations and should be clarified.

The centrality of the role of a diligent behavior in the definition of the salvage activity itself and in the determination of the remuneration due, requires a more in-depth reflection on the imputation of the criterion itself in the salvage with MASS.

---

[57] See the Work Conducted by the CMI International Working Group on Unmanned Ships. Consolidated analysis of legal Conventions, Annex II, sub-Art. 14 of the Salvage Convention.

Article 18 of the Salvage Convention states that "a salvor may be deprived of the whole or part of the payment due under this Convention to the extent that Salvage operations have become necessary or more difficult because of fault or neglect on his part or if the salvor has been guilty of fraud or other dishonest conduct." If the salvage operation is carried out by a remotely controlled ship with personnel on board, it is necessary to ascertain whose act can be attributed to a salvor, even more if the ship is totally autonomous (fourth degree). And if the ship is totally autonomous, it is also necessary to examine the conduct of other professional figures, such as programmers or manufacturers.

The provision refers to salvor's negligence, fraud, and dishonesty, which are behaviors or attitudes that are difficult to attribute to a management software.

8 Conclusions

The most recent studies point out that AI ecosystem will bring the benefits of technology to the European economy, particularly for the citizens to reap new benefits for safer and cleaner transport systems, for the business development: for example, a new generation of products and services in areas such as transport and cyber security; and for services of public interest: for example, by reducing the costs of providing services in transport, by improving the sustainability of products and by equipping law enforcement authorities with appropriate tools to ensure the security of citizens, with proper safeguards to respect their rights and freedoms.[58]

The bright descriptive picture referred to so far, on the use of MASS in salvage operations, does not hide the shadow areas that arise in this specific and very particular matter.

On the one hand, in fact, it is not clear what evolution the human work will have, since the use of artificial intelligence will lead to the need for increasingly specific professions in this sector. The conclusions of the first studies underline that the machines will replace the man in the heavier, repetitive, and dangerous jobs and that the human contribution will increasingly be an activity of conception, design, and construction.[59] However, it is not yet clear how this will affect the many skills currently employed in shipping, at all levels of activity.

On the other hand, we cannot overlook the fact that the artificial intelligence intended for the management of the vessel will be increasingly based on the characteristics indicated by the European Parliament of the Autonomy, obtained thanks to sensors and through the exchange of data with the environment in which it operates as well as on Self-learning from the experience that derives from the

[58] On this matter, see the European Commission (2020) White Paper On Artificial Intelligence - A European approach to excellence and trust. Brussels, 19.2.2020 COM(2020) 65 final: 2.

[59] European Commission (2020) White Paper On Artificial Intelligence - A European approach to excellence and trust. Brussels, 19.2.2020 COM(2020) 65 final: 6.

interaction with the environment and on the adaptation of its behavior and actions to the environment.

Thus, we can imagine possible risks deriving from the use of AI for salvage purposes: these risks may be related to a material harm, including loss of life or damage to property, or to an immaterial damage, such as loss of privacy or limitations of human rights. For example, as result of a flaw in the object recognition technology, a MASS could wrongly identify the subject to assist, producing a further personal injury. These damages can be caused by flaws in the design of the AI technology or be related to problems with the availability of data or to other problems stemming from machine learning.[60]

Indeed, an autonomous ship will be able to learn from the surrounding environment and change its operational conduct by interacting with the elements, circumstances, and places in which it will operate. A recent Resolution of the European Parliament provided detailed recommendations for drawing up a European Parliament and Council Regulation on liability for the operation of an Artificial Intelligence system, intended as "system that is either software-based or embedded in hardware devices, and that displays behaviour simulating intelligence by, inter alia, collecting and processing data, analysing and interpreting its environment, and by taking action, with some degree of autonomy, to achieve specific goals," which is autonomous, since it "operates by interpreting certain input and by using a set of pre-determined instructions, without being limited to such instructions, despite the system's behaviour being constrained by, and targeted at, fulfilling the goal it was given and other relevant design choices made by its developer."[61]

This new perspective, indeed, can have important repercussions on the liability regime currently centered on the shipowner. One wonders to what extent the same, called to be responsible for the facts of the crew and the obligations contracted by the master, will be liable for the damages produced by an artificial intelligence controlling a ship managed by the same shipowner. In fact, we must consider that the artificial intelligence resets its own behavior calibrating it on the circumstances and it is capable of self-learning, to modify its conduct with respect to the initial parameters set by the builder, or by the owner himself.

In this regard, the European Parliament notes that "all physical or virtual activities, devices or processes that are driven by AI-systems may technically be the direct

[60] As indicated by European Commission (2020) White Paper On Artificial Intelligence - A European approach to excellence and trust. Brussels, 19.2.2020 COM(2020) 65 final:13, "Under the Product Liability Directive, a manufacturer is liable for damage caused by a defective product. However, in the case of an AI based system such as autonomous cars, it may be difficult to prove that there is a defect in the product, the damage that has occurred and the causal link between the two. In addition, there is some uncertainty about how and to what extent the Product Liability Directive applies in the case of certain types of defects, for example if these result from weaknesses in the cybersecurity of the product."

[61] European Parliament resolution of 20 October 2020 with recommendations to the Commission on a civil liability regime for artificial intelligence (2020/2014(INL)), sub-Art. 3.

or indirect cause of harm or damage, yet are nearly always the result of someone building, deploying or interfering with the systems; notes in this respect that it is not necessary to give legal personality to AI-systems; is of the opinion that the opacity, connectivity and autonomy of AI-systems could make it in practice very difficult or even impossible to trace back specific harmful actions of AI-systems to specific human input or to decisions in the design; recalls that, in accordance with widely accepted liability concepts, one is nevertheless able to circumvent this obstacle by making the different persons in the whole value chain who create, maintain or control the risk associated with the AI-system liable."[62]

The European Parliament focused on the liability of the operator intended both as 'frontend operator', i.e., "any natural or legal person who exercises a degree of control over a risk connected with the operation and functioning of the AI-system and benefits from its operation," and backend operator, as "any natural or legal person who, on a continuous basis, defines the features of the technology and provides data and an essential backend support service and therefore also exercises a degree of control over the risk connected with the operation and functioning of the AI-system."

The operator, under the European Regulation, exercises a degree of control over a risk connected with the operation and functioning of an AI-system, which is comparable to that of an owner of a car.

The control is intended as "any action of an operator that influences the operation of an AI-system and thus the extent to which the operator exposes third parties to the potential risks associated with the operation and functioning of the AI-system; such actions can impact the operation at any stage by determining the input, output or results, or can change specific functions or processes within the AI-system; the degree to which those aspects of the operation of the AI-system are determined by the action depends on the level of influence the operator has over the risk connected with the operation and functioning of the AI-system." Although the existing fault-based liability law of the Member States offers in most cases a sufficient level of protection for persons that suffer harm or damage caused by an interfering third person—as that interference regularly constitutes a fault-based action, where the third-party uses the AI system to cause harm—it is necessary to distinguish the operator of a high-risk AI-system, who will be strictly liable for any harm or damage that was caused by a physical or virtual activity, device or process driven by that AI-system (Art. 4.1 of the Regulation).[63] He would not be able to exonerate himself from liability by arguing that he acted with due diligence, or that the harm or damage was caused by an autonomous activity, device or process driven by their AI-system, except in the event of harm or damage caused by force majeure.

---

[62] European Parliament resolution of 20 October 2020 with recommendations to the Commission on a civil liability regime for artificial intelligence (2020/2014(INL)), sub nr. 7.

[63] Under Art. 4.1 of the European Parliament resolution of 20 October 2020 with recommendations to the Commission on a civil liability regime for artificial intelligence (2020/2014(INL)), "The operator of a high-risk AI-system shall be strictly liable for any harm or damage that was caused by a physical or virtual activity, device or process driven by that AI-system."

Because of this approach, the European Parliament believes that all operators of high-risk AI-systems should hold liability insurance, based on the significant potential to cause harm or damage and by taking Directive 2009/103/EC of the European Parliament and of the Council of 16 September 2009 relating to insurance against civil liability in respect of the use of motor vehicles.[64]

On the contrary, the operator of an AI-system that does not constitute a high-risk AI-system will be "subject to fault-based liability for any harm or damage that was caused by a physical or virtual activity, device or process driven by the AI-system." He will not be liable if he can prove that the harm or damage was caused without his fault, relying on either of the fact that "the AI-system was activated without his knowledge while all reasonable and necessary measures to avoid such activation outside of the operator's control were taken, or due diligence was observed by performing all the following actions: selecting a suitable AI-system for the right task and skills, putting the AI-system duly into operation, monitoring the activities and maintaining the operational reliability by regularly installing all available updates."

Finally, the operator will not be able to escape liability by arguing that the harm or damage was caused by an autonomous activity, device or process driven by his or her AI-system, except if the harm or damage was caused by force majeure (Art. 8 of the Regulation).[65]

---

[64] European Parliament resolution of 20 October 2020 with recommendations to the Commission on a civil liability regime for artificial intelligence (2020/2014(INL)), sub nr. 24, according to which the European Parliament "Is of the opinion that, based on the significant potential to cause harm or damage and by taking Directive 2009/103/EC of the European Parliament and of the Council of 16 September 2009 relating to insurance against civil liability in respect of the use of motor vehicles, and the enforcement of the obligation to insure against such liability into account, all operators of high-risk AI-systems listed in the Annex to the proposed Regulation should hold liability insurance; considers that such a mandatory insurance regime for high-risk AI-systems should cover the amounts and the extent of compensation laid down by the proposed Regulation; is mindful of the fact that such technology is currently still very rare, since it presupposes a high degree of autonomous decision making and that, as a result, the current discussions are mostly future-oriented; believes, nevertheless, that uncertainty regarding risks should not make insurance premiums prohibitively high and thereby an obstacle to research and innovation." Under Art. 4.4 of the aforementioned resolution "The frontend operator of a high-risk AI-system shall ensure that operations of that AI-system are covered by liability insurance that is adequate in relation to the amounts and extent of compensation provided for in Articles 5 and 6 of this Regulation. The backend operator shall ensure that its services are covered by business liability or product liability insurance that is adequate in relation to the amounts and extent of compensation provided for in Article 5 and 6 of this Regulation. If compulsory insurance regimes of the frontend or backend operator already in force pursuant to other Union or national law or existing voluntary corporate insurance funds are considered to cover the operation of the AI-system or the provided service, the obligation to take out insurance for the AI-system or the provided service pursuant to this Regulation shall be deemed fulfilled, as long as the relevant existing compulsory insurance or the voluntary corporate insurance funds cover the amounts and the extent of compensation provided for in Articles 5 and 6 of this Regulation."

[65] Under Art. 8.2 of the European Parliament resolution of 20 October 2020 with recommendations to the Commission on a civil liability regime for artificial intelligence (2020/2014(INL)), "The

Considering these new regulatory perspectives, the relationships between the shipyard, the supplier of the management software, and the operator who takes over the control will also have to be rethought, without wanting to indulge in easy fantasy suggestions,[66] but in the awareness that a new element has been definitively introduced into the legal sphere that can be assessed independently with respect to the subjects considered so far for the purpose of defining liability. Where the activity of the MASS falls within those envisaged as a high-risk AI system, we will therefore have to assume that a strict liability regime will apply, which will concern the operator, including both the frontend and the backend operator, who have control of the MASS in the sense indicated above.

Concepts such as the due diligence (due care: Art. 8, first paragraph, letter a, of the Salvage Convention) in salvage operations and the negligent behavior of the salvor who "has thereby failed to prevent or minimize damage to the environment" (Art. 14.1 of the Salvage Convention) will then have to be reconsidered, in the light of a new element of evaluation such as the artificial intelligence or the management software, which certainly cannot be neglected in the provision of a suitable insurance coverage for damage caused by autonomous ships.

Regulatory Guidance and Decisions

Allianz (2019) Shipping safety - Human error comes in many forms. Agcs.allianz.com, available at https://www.google.com/url?sa=t&rct=j&q=&esrc=s&source=web&cd=&ved=2ahUKEwjLiobBjvD1AhWPhv0HHfOlDsEQFnoECAIQAQ&url=https%3A%2F%2Fwww.agcs.allianz.com%2Fnews-and-insights%2Fexpert-risk-articles%2Fhuman-error-shipping-safety.html&usg=AOvVaw25blg001DlNbjgVUWKMK-9

operator shall not be liable if he or she can prove that the harm or damage was caused without his or her fault, relying on either of the following grounds: (a) the AI-system was activated without his or her knowledge while all reasonable and necessary measures to avoid such activation outside of the operator's control were taken, or (b) due diligence was observed by performing all the following actions: selecting a suitable AI-system for the right task and skills, putting the AI-system duly into operation, monitoring the activities and maintaining the operational reliability by regularly installing all available updates. The operator shall not be able to escape liability by arguing that the harm or damage was caused by an autonomous activity, device or process driven by his or her AI-system. The operator shall not be liable if the harm or damage was caused by force majeure."

[66] As stated by Nevejans (2016), p. 5, "Once a new legal and ethical sector surfaces, a general approach to the big theoretical questions needs to be found in the first instance, so as to eliminate any misunderstanding or misconceptions about robotics and artificial intelligence. When we consider civil liability in robotics, we come up against fanciful visions about robots. Here we must resist calls to establish a legal personality based on science fiction. This will become all the more crucial once the liability law solutions adopted in respect of autonomous robots determine whether this new market booms or busts."

Autonomous ships: Test areas and research centers making headlines, September 10, 2019, https://safety4sea.com/cm-autonomous-ships-test-areas-and-research-centers-making-headlines

Commission Delegated Regulation (EU) 2019/945 of 12 March 2019 on unmanned aircraft systems and on third-country operators of unmanned aircraft systems

Commission Implementing Regulation (EU) 2019/947 of 24 May 2019 on the rules and procedures for the operation of unmanned aircraft

Commission Regulation (EU) 2018/1139, which has provided a preliminary regulatory framework for drones within the European Union

European Commission (2020) White Paper On Artificial Intelligence - A European approach to excellence and trust. Brussels, 19.2.2020 COM(2020) 65 final: 2.

European Parliament resolution 20 October 2020, with recommendations to the Commission on a civil liability regime for artificial intelligence (2020/2014(INL)

European Parliament Resolution of 16 February 2017 with recommendations to the Commission on Civil Law Rules on Robotics (2015/2103(INL))

Lloyd's Register (2017) Cyber-enabled ships, ShipRight procedure assignment for cyber descriptive notes for autonomous & remote access ships - A Lloyd's Register guidance document: 1, available at https://www.google.com/url?sa=t&rct=j&q=&esrc=s&source=web&cd=&ved=2ahUKEwiPpMblj_D1AhWE_7sIHWAiBNEQFnoECBcQAQ&url=http%3A%2F%2Finfo.lr.org%2Fl%2F12702%2F2016-07-07%2F32rrbk&usg=AOvVaw063BYOxU6jdOeG6_wayoa_

Maritime Safety Committee (MSC), 100th session, 3-7 December 2018

Maritime Safety Committee (MSC), 98th session, 7-16 June 2017

Maritime Unmanned Navigation through Intelligence in Networks (MUNIN) (2015), available at https://www.google.com/url?sa=t&rct=j&q=&esrc=s&source=web&cd=&ved=2ahUKEwjA6uOFivD1AhXqgv0HHQBbD7gQFnoECAgQAQ&url=https%3A%2F%2Fcordis.europa.eu%2Fproject%2Fid%2F314286%2Fit&usg=AOvVaw1IxFOmgpL6NCZ0yzvtTwFa

Maritime UK (2020) Maritime Autonomous Ship Systems (MASS) UK Industry Conduct Principles and Code of Practice - A Voluntary Code Version, available at https://www.google.com/url?sa=t&rct=j&q=&esrc=s&source=web&cd=&ved=2ahUKEwjFsZyxkPD1AhUIlP0HHXx6CCAQFnoECA0QAQ&url=https%3A%2F%2Fwww.maritimeuk.org%2Fmedia-centre%2Fpublications%2Fmaritime-autonomous-surface-ships-industry-conduct-principles-code-practice-v4%2F&usg=AOvVaw3k2IFOm7GWO5Jj435Xj83g

Proposal for a Regulation of the European Parliament and the Council laying down harmonized rules on Artificial Intelligence (Artificial Intelligence Act) and amending certain union legislative acts, 2021/0106 (COD)

Regulation (EU) 2019/881 of the European Parliament and of the Council (2019) ENISA (the European Union Agency for Cybersecurity) and on information and communications technology cybersecurity certification and repealing Regulation (EU) No. 526/2013 (Cybersecurity Act)

The Guidelines on Cyber Security Onboard Ships - Version 4, produced and supported by BIMCO, Chamber of Shipping of America, Digital Containership Association, International Association of Dry Cargo Shipowners

(INTERCARGO), InterManager, International Association of Independent Tanker Owners (INTERTANKO), International Chamber of Shipping (ICS), International Union of Marine Insurance (IUMI), Oil Companies International Marine Forum (OCIMF), Superyacht Builders Association (Sybass) and World Shipping Council (WSC); it is also possible to analyze the Cyber Security Considerations in Maritime UK (2020) Maritime Autonomous Ship Systems (MASS) UK Industry Conduct Principles and Code of Practice - A Voluntary Code Version, available at https://www.google.com/url?sa=t&rct=j&q=&esrc=s&source=web&cd=&ved=2ahUKEwj-kIXSkPD1AhXb_rsIHTHFCmUQFnoECAYQAQ&url=https%3A%2F%2Fwww.bimco.org%2Fabout-us-and-our-members%2Fpublications%2Fthe-guidelines-on-cyber-security-onboard-ships&usg=AOvVaw3S5qLCJ9O2dwRV61O1Z9GC

The US Navy Unmanned Surface Vehicle (USV) Masterplan (2007), available at https://www.google.com/url?sa=t&rct=j&q=&esrc=s&source=web&cd=&ved=2ahUKEwjVqKK9jPD1AhXm_7sIHf3yBKUQFnoECAUQAQ&url=https%3A%2F%2Fwww.hsdl.org%2F%3Fview%26did%3D479083&usg=AOvVaw0DRx5SQguUu0vfySlwvY0v

US Department of the Navy (2021) Unmanned Campaign Framework: 1-37, available at https://www.google.com/url?sa=t&rct=j&q=&esrc=s&source=web&cd=&ved=2ahUKEwjJv-qSjPD1AhVMhv0HHXT8CEUQFnoECAYQAQ&url=https%3A%2F%2Fwww.navy.mil%2FPortals%2F1%2FStrategic%2F20210315%2520Unmanned%2520Campaign_Final_LowRes.pdf%3Fver%3DLtCZ-BPlWki6vCBTdgtDMA%253D%253D&usg=AOvVaw3WuwwF5Y7rhIVXxoBVxbcd

Report of the eSafety Working Group (2002), reported by the Commission to the European Parliament and the Council (2016) Saving Lives: Boosting Car Safety in the EU - Reporting on the monitoring and assessment of advanced vehicle safety features, their cost effectiveness and feasibility for the review of the regulations on general vehicle safety and on the protection of pedestrians and other vulnerable road users {SWD(2016) 431 final}, available at https://www.google.com/url?sa=t&rct=j&q=&esrc=s&source=web&cd=&ved=2ahUKEwiIhfLfjvD1AhXog_0HHd3JAFkQFnoECBMQAQ&url=https%3A%2F%2Feur-lex.europa.eu%2Flegal-content%2FEN%2FTXT%2F%3Furi%3DCELEX%253A52016DC0787&usg=AOvVaw1Tkk6o_YnFdxHIGd49AvHY

The Nagasaki Spirit, Court of appeal, 4, 5, 6 e 21 dec. 1995. In: Lloyd's Law Reports, 1996 (I): 459

Work Conducted by the CMI International Working Group on Unmanned Ships. Consolidated analysis of legal Conventions, Annex II, available at https://www.google.com/url?sa=t&rct=j&q=&esrc=s&source=web&cd=&ved=2ahUKEwjQ1cvVqvD1AhWWuKQKHSbQCCoQFnoECAgQAQ&url=https%3A%2F%2Fcomitemaritime.org%2Fwork%2Fmass%2F&usg=AOvVaw3_UZcKM6qTQAwH8yV4XqIp

References

Ancis L (2019) Navi pilotate da remoto e profili di sicurezza della navigazione nel trasporto di passeggeri. Dir. trasp.: 460.

Ardito S, Lazarevs D, Vasiliniuc B, Vukic ZK, Masabayashi ZK, Caccia M (2013) Cooperative Autonomous Robotic Towing system: definition of requirements and operating scenarios. https://doi.org/10.3182/20120919-3-IT-2046.00045

Attard FG, Kilpatrick RL (2020) Reflections on the Maersk Etienne Standoff and its Ramifications for the Duty to Render Assistance at Sea. EJIL: Talk!, available at https://www.google.com/url?sa=t&rct=j&q=&esrc=s&source=web&cd=&ved=2ahUKEwjVytDVqfD1AhWCHewKHTaWB10QFnoECAkQAQ&url=https%3A%2F%2Fwww.ejiltalk.org%2Freflections-on-the-maersk-etienne-standoff-and-its-ramifications-for-the-duty-to-render-assistance-at-sea%2F&usg=AOvVaw2w5fE0Yf78AApRJplayarP. Accessed 27 July 2022

Barnes RA (2017) Article 18 United Nations Convention on the Law of the Sea. A Commentary, edited by A. Proelß: 185

Burmeister HC, Bruhn WCØ, Rodseth J, Porathe T (2014) Can unmanned ships improve navigational safety?. Transport Research Arena, available at https://www.google.com/url?sa=t&rct=j&q=&esrc=s&source=web&cd=&ved=2ahUKEwiQ69WgjfD1AhXahP0HHbK7CxIQFnoECAMQAQ&url=https%3A%2F%2Fciteseerx.ist.psu.edu%2Fviewdoc%2Fdownload%3Fdoi%3D10.1.1.846.5385%26rep%3Drep1%26type%3Dpdf&usg=AOvVaw1ufP8NsBO_uxVg6_EaZ3cf. Accessed 27 July 2022

Calantropio A (2019) The Use of UAVs for Performing Safety-Related Tasks at Post-Disaster and Non-Critical Construction Sites. MDPI 5, 64; doi:10.3390/safety5040064, available at https://www.google.com/url?sa=t&rct=j&q=&esrc=s&source=web&cd=&ved=2ahUKEwjgnYawkfD1AhWx57sIHXBTAd4QFnoECAkQAQ&url=https%3A%2F%2Fwww.mdpi.com%2F2313-576X%2F5%2F4%2F64&usg=AOvVaw0PRz9PW5Chx_MTvNM5wO4p. Accessed 27 July 2022

Craven JP (1985) Technology and the law of the sea: the effect of prediction and misprediction. Louisiana Law Rev 1143–1159, Available at https://www.google.com/url?sa=t&rct=j&q=&esrc=s&source=web&cd=&ved=2ahUKEwibzOfIiPD1AhVL8bsIHSnWAV4QFnoECAYQAQ&url=https%3A%2F%2Fdigitalcommons.law.lsu.edu%2Fcgi%2Fviewcontent.cgi%3Farticle%3D4928%26context%3Dlalrev&usg=AOvVaw2OcoqY7SgyRrNvui70Fnm4. Accessed 27 July 2022

Crisafulli Buscemi S (1933) Alcune considerazioni sulla situazione giuridica della nave manovrabile da lontano. Studi in onore di Francesco Berlingieri 191–204

Davies M (2003) Obligations and implications for ships encountering persons in need of assistance at sea. Pac Rim Law Policy J 109, Available at https://www.google.com/url?sa=t&rct=j&q=&esrc=s&source=web&cd=&ved=2ahUKEwiY2IqCqvD1AhWZLOwKHWxOCVsQFnoECAcQAQ&url=https%3A%2F%2Fdigitalcommons.law.uw.edu%2Fcgi%2Fviewcontent.cgi%3Farticle%3D1305%26context%3Dwilj&usg=AOvVaw2HTgDGzA-l2Q8kaE0DDAqF. Accessed 27 July 2022

Ghaderi H (2019) Autonomous technologies in short sea shipping: trends, feasibility and implications. Transport Rev 39:152 ss., Available at https://www.google.com/url?sa=t&rct=j&q=&esrc=s&source=web&cd=&ved=2ahUKEwjUucSkj_D1AhVIh_0HHcHLDV4QFnoECAMQAQ&url=https%3A%2F%2Fwww.tandfonline.com%2Fdoi%2Fabs%2F10.1080%2F01441647.2018.1502834&usg=AOvVaw1EAnZBSiw8KeNEgsWlwFUo. Accessed 27 July 2022

Kas KA, Johnson GK (2020) Using unmanned aerial vehicles and robotics in hazardous locations safely. Process Saf Progr 39(1): https://doi.org/10.1002/prs.12096., Available at https://www.google.com/url?sa=t&rct=j&q=&esrc=s&source=web&cd=&ved=2ahUKEwi6j_DLkfD1AhUgi_0HHWtvAuYQFnoECAUQAQ&url=https%3A%2F%2Faiche.onlinelibrary.wiley.com%2Fdoi%2Fabs%2F10.1002%2Fprs.12066&usg=AOvVaw3cyT0f-6Ul8kx06FwqJdmj. Accessed 27 July 2022

Kenney FJ Jr, Tasikas V (2003) The Tampa incident: IMO perspectives and responses on the treatment of persons rescued at sea. Pac Rim Law Policy J 12:151, available at https://www.google.com/url?sa=t&rct=j&q=&esrc=s&source=web&cd=&ved=2ahUKEwi19-SEp_D1AhUQ8LsIHbimDfAQFnoECAQQAQ&url=https%3A%2F%2Fcore.ac.uk%2Fdownload%2Fpdf%2F267981593.pdf&usg=AOvVaw0c40I0sGMyS53B6QMJgqXt. Accessed 27 July 2022

Kilpatrick RL (2010) The "Refugee Clause" for commercial shipping contracts: why allocation of rescue costs is critical during periods of mass migration at sea. Georgia J Int Comp, Available at https://www.google.com/url?sa=t&rct=j&q=&esrc=s&source=web&cd=&ved=2ahUKEwif_a6XpvD1AhXz7rsIHdvSC8MQFnoECAMQAQ&url=https%3A%2F%2Fdigitalcommons.law.uga.edu%2Fcgi%2Fviewcontent.cgi%3Farticle%3D2409%26context%3Dgjicl&usg=AOvVaw2vpcepvu5bo87U5GY-ptoE. Accessed 27 July 2022

Kurowsky M, Korte H, Lampe BP (2012) Search-and-Rescue-Operation with an Autonomously Acting Rescue Boat. Autonomous and intelligent Systems: third international Conference, in Lecture Notes in Artificial Intelligence, AIS, Available at https://www.google.com/url?sa=t&rct=j&q=&esrc=s&source=web&cd=&ved=2ahUKEwjcjOX2kfD1AhUjlP0HHZKACi0QFnoECAkQAQ&url=https%3A%2F%2Fwww.springerprofessional.de%2Fen%2Fsearch-and-rescue-operation-with-an-autonomously-acting-rescue-b%2F3917330&usg=AOvVaw1fWW56YOBXbSaCHRfxsZHL. Accessed 27 July 2022

Lagrone S (2021) Ghost Fleet Ship 'Nomad' Arrives in California After 4,421 Nautical-Mile, '98 Percent' Autonomous Trip, Available at https://www.google.com/url?sa=t&rct=j&q=&esrc=s&source=web&cd=&ved=2ahUKEwjUgaLGifD1AhU057sIHUckCTEQFnoECBMQAQ&url=https%3A%2F%2Fnews.usni.org%2F2021%2F06%2F07%2Fghost-fleet-ship-nomad-arrives-in-california-after-4421-nautical-mile-98-percent-autonomous-trip&usg=AOvVaw26bqwk03BgxkZnpIGB_I0C. Accessed 27 July 2022

Mandrioli D (2020) The international duty to assist people in distress at sea in the era of unmanned navigation: no place for people on board. Revista multidisciplinar humanidades e tecnologia: 91, Available at https://boa.unimib.it/retrieve/handle/10281/280886/411970/Humanidades%20e%20Tecnologia%202020.pdf. Accessed 27 July 2022

Nevejans N (2016) European Civil Law Rules on Robotics, Available at https://www.google.com/url?sa=t&rct=j&q=&esrc=s&source=web&cd=&ved=2ahUKEwi36Zinq_D1AhVR2aQKHf-hDD0QFnoECA8QAQ&url=https%3A%2F%2Fwww.europarl.europa.eu%2FRegData%2Fetudes%2FSTUD%2F2016%2F571379%2FIPOL_STU(2016)571379_EN.pdf&usg=AOvVaw3LIK4V6oNcjaZdSf3K4Mf6. Accessed 27 July 2022

O'Rourke R (2022) Navy Large Unmanned Surface and Undersea Vehicles: Background and Issues for Congress, Available at https://www.google.com/url?sa=t&rct=j&q=&esrc=s&source=web&cd=&ved=2ahUKEwjw6eCPh_D1AhX97rsIHSSIArIQFnoECAcQAQ&url=https%3A%2F%2Fsgp.fas.org%2Fcrs%2Fweapons%2FR45757.pdf&usg=AOvVaw1T-onV-ik2Cc7PvTDA2RkL. Accessed 27 July 2022

Papanicolopulu I (2018) International law and the protection of people at sea. Oxford: 187, https://doi.org/10.1093/oso/9780198789390.001.0001

Ringbom H (2019) Regulating autonomous ships—concepts, challenges and precedents. Ocean Dev Int Law 50(2–3):141–169, Available at https://www.google.com/url?sa=t&rct=j&q=&esrc=s&source=web&cd=&ved=2ahUKEwiN-LPAi_D1AhXJif0HHXReAuwQFnoECAoQAQ&url=https%3A%2F%2Fwww.tandfonline.com%2Fdoi%2Fabs%2F10.1080%2F00908320.2019.1582593&usg=AOvVaw19SYB63AuwLxw-A6t8Ty6s. Accessed 27 July 2022

Saha R (2021) Mapping competence requirements for future shore control center operators, Maritime Policy Manage 10.1080/03088839.2021.1930224, also available at https://www.google.com/url?sa=t&rct=j&q=&esrc=s&source=web&cd=&ved=2ahUKEwjwlMOypfD1AhXYk_0HHXmKBdUQFnoECAcQAQ&url=https%3A%2F%2Fwww.tandfonline.com%2Fdoi%2Ffull%2F10.1080%2F03088839.2021.1930224&usg=AOvVaw32iRV1aZnPJXNaPuNm3V1G. Accessed 27 July 2022

Savitz SI, Blickstein P, Buryk RW, Button P, Deluca J, Dryden J, Mastbaum J, Osburg P, Padilla A, Potter CC, Price L, Thrall SK, Woodward RJ, Yardley J, Yuchak (2013) U.S. Navy Employment Options for Unmanned Surface Vehicles (USVs). Rand – National Defense Research Institute

Severoni C (2005a) La remunerazione del soccorso tra interesse pubblico ed interessi privati – Premesse storico-dogmatiche, vol, I, available (resctricted access, last access 26.07.2022) at https://air.uniud.it/retrieve/handle/11390/876922/225554/Severoni%20soccorso%20I%28M%29%20ISBN%209788814121418%20%20COD%20%200518875.pdf

Severoni C (2005b) La remunerazione del soccorso tra interesse pubblico ed interessi privati – Profili sistematici e lineamenti evolutivi, vol, II, Available (restricted access, last access 26.07.2022) at https://air.uniud.it/retrieve/handle/11390/876910/225562/Severoni%20soccorso%20II%20%28M%29%20ISBN%209788814121418%20COD%200518876.pdf

Severoni C (2018) Soccorso e mezzi di trasporto autonomi. Dir Trasp 31(1):67–85, Available (restricted access, last access 26.07.2022) at https://air.uniud.it/retrieve/handle/11390/1131368/247271/Severoni%20Soccorso.pdf

Soyer B, Tettenborn A (2021) Artificial intelligence and autonomous shipping: developing the international legal framework. Bloomsbury Publishing:63–89

Starita M (2019) Il dovere di soccorso in mare e il diritto di obbedire al diritto (internazionale) del comandante della nave privata. Diritti um dir Intern https://doi.org/10.12829/93310

Veal R (2018) Maritime surface ships: autonomy, manning and the IMO. Lloyd's Ship Trade Law 18(5):1–4, Available at https://www.google.com/url?sa=t&rct=j&q=&esrc=s&source=web&cd=&ved=2ahUKEwiM8aSQi_D1AhUL8rsIHRBrAGcQFnoECAMQAQ&url=https%3A%2F%2Feprints.soton.ac.uk%2F435581%2F&usg=AOvVaw3EAXSlObUZQbMHeDRT6Zts. Accessed 27 July 2022

Veal R, Tsimplis M, Serdy A (2019) The legal status and operation of unmanned maritime vehicles. Ocean Dev Int Law 50(1):23–48, Available at https://www.google.com/url?sa=t&rct=j&q=&esrc=s&source=web&cd=&ved=2ahUKEwjAkNDti_D1AhXi_bsIHa43BicQFnoECAMQAQ&url=https%3A%2F%2Fwww.tandfonline.com%2Fdoi%2Fabs%2F10.1080%2F00908320.2018.1502500&usg=AOvVaw378XlkZ5N1XfDpUh7cRaVk. Accessed 27 July 2022

Weiger AM, Pribyl ST (2017) The future is now: unmanned and autonomous surface vessels and their impact on the maritime industry. Blank Maritime Bulletin

Yan R-J, Pang S, Sun H-B, Pang YJ (2010) Development and missions of unmanned surface vehicles. J Mar Sci Appl 451, Available at https://www.google.com/url?sa=t&rct=j&q=&esrc=s&source=web&cd=&ved=2ahUKEwiK98n0iPD1AhWmlf0HHdbUAy8QFnoECAQQAQ&url=https%3A%2F%2Fui.adsabs.harvard.edu%2Fabs%2F2010JMSA....9..451Y%2Fabstract&usg=AOvVaw1_mGDvlIrwwNKLEVpywCGZ. Accessed 27 July 2022

Yoo J, Goerlandt F, Chircop A (2020) Unmanned remotely operated search and rescue ships in the Canadian arctic: exploring the opportunities, risk dimensions and governance implications. In: Chircop A, Goerlandt F, Aporta C, Pelot R (eds) Governance of arctic shipping. Springer Polar Sciences. Springer. https://doi.org/10.1007/978-3-030-44975-9_5, or available at https://www.google.com/url?sa=t&rct=j&q=&esrc=s&source=web&cd=&ved=2ahUKEwjRnb2ypPD1AhUug_0HHY1MDgkQFnoECAMQAQ&url=https%3A%2F%2Flink.springer.com%2Fchapter%2F10.1007%2F978-3-030-44975-9_5&usg=AOvVaw0GPhqeJWQMpJo1UaPcGmEJ. Accessed 27 July 2022

Uncharted Legal Waters: "The Applicability of the Law on Seaworthiness & Good Seamanship to Autonomous Vessels."

Shanice N. Trowers

Abstract This chapter explores the concepts of seaworthiness and good seamanship and their applicability to autonomous vessels. The author provides a comprehensive overview of what is an autonomous vessel and explores some of the key advantages and disadvantages of autonomous vessels. The author then dissects the concept of seaworthiness in its traditional sense and does a critical examination of whether an autonomous vessel could apply to the traditional concept of seaworthiness. In this respect, the author examines key international legal frameworks such as the Hague/ Hague Visby Rules and the International Convention for the Safety of Life at Sea (SOLAS). The final part of this chapter explores the concept of good seamanship and critically evaluates whether an autonomous vessel could satisfy the requirements of good seamanship under the Convention on the International Regulations for Preventing Collisions at Sea, 1972 (COLREGs). The author ends with a comprehensive discussion and conclusion on both concepts' applicability to autonomous vessels and proposes amendments that can be made to the existing legal frameworks.

Keywords Seaworthiness · Good seamanship · Autonomous vessels · Unmanned vessels · Autonomous shipping

1 Introduction

In every industry across the globe, technology has continuously evolved over the years to develop products and machineries that are foreign to societies which existed in previous centuries. In the transport industry, there has been an influx of automated vehicles which uses smart technology and Artificial Intelligence to operate on their own. In the motor vehicle industry, there are driverless cars such as Tesla and Waymo that are currently maneuvering our roads. Similarly, in the aeroplane industry, there are many small, automated flying objects which are called drones

S. N. Trowers (✉)
Faculty of Law, University of Technology, Jamaica, Kingston, Jamaica
e-mail: shanice.trowers@utech.edu.jm

K. Noussia, M. Channon (eds.), *The Regulation of Automated and Autonomous Transport*, https://doi.org/10.1007/978-3-031-32356-0_7

in operation. The shipping industry has also not been immune to technology's evolution in the form of automation. In fact, the use of automation in the shipping industry is not novel, as vessels have been relying on automation to varying degrees for several decades. It is commonplace to find that a traditional merchant vessel has automated sensors to support navigation, communication, to manage cargo loading and discharge and to generally operate safely. However, what is novel is and what is currently being developed are vessels that are fully autonomous. Autonomous technology is currently being developed and adapted for use in every type of vessel including but not limited to ferries, tankers, container ships, tugboats, and war ships. Autonomous vessels are the future of the maritime industry, as autonomous technology will radically change the design and operation of vessels across the board. It will also radically change the design and infrastructure of ports, as they will have to adapt to this new technology. Currently, there are several developers of autonomous vessels. One of the most popular developers, Rolls Royce, recently announced a partnership in September 2021 with a company called Sea Machines Robotics. Both companies aspire to deliver comprehensive and intelligent crew support systems to the shipping industry. They intend to develop and sell fully and semi-autonomous vessel control systems to the maritime market. Michael Johnson, CEO, and founder of Sea Machines stated in a recent press release that:

> Our autonomous vessel control products and advanced perception systems are pioneering the revolutionary shift of conventional and manual vessel control effort from human to intelligent technology. Autonomous systems take over routine efforts, reduce stress for crews and thus increase operational predictability and safety. The technology makes shipping more productive, economical and contributes to more sustainable operations. We partner with those best-in-class and Rolls-Royce stands out as being most trusted. We look forward to serving the market together and furthering the technology that provides solutions for customers.[1]

Similarly, a Norwegian company has developed the world's first crewless, zero emission container ship, which is set to make its first journey between two Norwegian towns by the end of the year 2021. This vessel is being developed by Yara International and is called the Yara Birkeland. According to the International Maritime Organization (IMO), in its fourth IMO Greenhouse gas study, there has been a significant increase in the greenhouse gas (GHG) emissions produced by the shipping industry including carbon dioxide (CO2), methane (CH4), and nitrous oxide (N2O), expressed in CO2e. The IMO stated further in the said report that the total CO2e has increased from 977 million tons in 2012 to 1,076 million tons in 2018 which represents a 9.6% increase.[2] As such, the Yara Birkeland was designed to reduce toxic fumes and green house cases which typically emit from traditional

[1]Online document "Rolls-Royce and Sea Machines agree autonomous vessel tech deal" https://smartmaritimenetwork.com/2021/09/22/rolls-royce-and-sea-machines-agree-autonomous-vessel-tech-deal/ Accessed on August 10, 2022.

[2]Online document "Fourth IMO Greenhouse gas study 2020" https://wwwcdn.imo.org/localresources/en/OurWork/Environment/Documents/Fourth%20IMO%20GHG%20Study%202020%20-%20Full%20report%20and%20annexes.pdf Accessed on August 10, 2022.

vessels. The present design of the Yara Birkeland will see the loading and unloading of the ship being done in the initial trial stage. However, it is expected that all loading, discharging and mooring activities will be done autonomously eventually on the Yara Birkeland.

These autonomous vessels, such as those by Yara International and Rolls Royce, are being developed at a relatively rapid state but the laws to regulate them are not being developed as quickly. As such, it is very likely that there will be a legal vacuum when autonomous vessels start to sail the seas soon. We are indeed navigating uncharted legal waters and as such, it is very important that we consider whether existing maritime law and conventions can accommodate an autonomous vessel. The selected focus of this chapter will be on two important aspects of maritime law which are seaworthiness and good seamanship and their applicability to an autonomous vessel.

1.1 Definition of Autonomous Vessels

An autonomous vessel has been defined as one that can navigate without human interaction to a varying degree. The Lloyd's Register has outlined different levels of autonomy which range from AL0 TO AL6. AL0 represents the traditionally manned vessel, where all decision-making and actions are performed by a human located on board. AL1 and AL2 are referred to as Low and Partial Automation, where all actions are taken by human, but these decisions may be supported or influenced by tools. AL3 is referred to as Conditional Automation where decisions and actions are performed autonomously with human supervision and authorization. AL4 is referred to as high autonomation where decisions and actions are performed autonomously with human supervision. High impact decisions give human operators the opportunity to intervene and override. AL5 is referred to as Full Automation where the operation is rarely supervised, and decisions are made and actioned by the system entirely. AL6 is referred to as Unmanned which means that operations are unsupervised, and decisions are made entirely and actioned by the system during the mission.[3]

The focus of this chapter is on vessels who fall within AL4–AL6, as the current lex maritima can cover AL1–AL3 as they have humans on board/direct supervision.[4] Vessels which fall under AL4–AL6 are often referred to as remote-controlled unmanned vessels and fully autonomous vessels. A remote-controlled unmanned vessel is one that operates without seafarers on board but have shore-based operators

[3]Online document "LR defines 'autonomy levels' for ship design and operation." https://www.lr.org/en/latest-news/lr-defines-autonomy-levels-for-ship-design-and-operation/ Accessed on August 7, 2022.

[4]Online document "The Good, the Bad and the Ugly: Unmanned Ships" https://www.hellenicshippingnews.com/the-good-the-bad-and-the-ugly-unmanned-ships/ Accessed on August 5, 2022.

who oversee its operation from a remote hub, whereas a fully autonomous vessel (AL6) fully controls its own operations.

The author will collectively throughout this chapter refer to both remote-controlled unmanned vessels and fully autonomous vessels as "autonomous vessels." Further, although there are several types of autonomous vessels being developed, the focus of this chapter will be solely on autonomous cargo vessels.

1.2 Advantages and Disadvantages of Autonomous Vessels

The debate on whether autonomous vessels are necessary usually sees opposing parties examining the pros and cons of these vessels. It is the authors' view that it is imperative that a discussion be had around the potential advantages and disadvantages of autonomous/unmanned vessels to give the reader a wholistic overview of the area.

1.2.1 Advantages

Cutting the Cost of Manning It has been said that one of the biggest impacts that autonomous vessels will have is that it will cut the operating costs related to salaries and living expenses for an on-board crew. It has been reported that eliminating a crew on an autonomous vessel will save shipping companies an approximately US $1.5 million dollars of on-board crew expenses.[5] This is particularly attractive having regard to the fact that the COVID-19 pandemic has reportedly resulted in a 10–15% rise in manning costs in the year 2020.[6] This is because companies were keen to keep seafarers employed. There were also difficulties in undertaking crew changes during the periods of lockdowns during the pandemic, which resulted in seafarers having to serve far beyond their contracted period onboard and shipping companies had to make additional payments to compensate. It was also reported that for every seafarer who was stranded at sea and held over beyond the end of their contracted period, there were also another set of crew that was stuck at home, who were unable to start their new contracts and some companies were actually retaining and paying these said seafarers.[7] Shipping companies also faced further incidental crew costs including paying for repatriation flights and booking hotels for sea farers

[5] Online document "Autonomous Shipping and Safety" http://www.crewtoo.com/crewtoo-blogs/autonomous-shipping-and-safety/ Accessed on August 10, 2022.

[6] Online document "Crew costs to rise 10–15% this year due to Covid-19" https://www.seatrade-maritime.com/ship-operations/crew-costs-rise-10-15-year-due-covid-19-drewry Accessed on July 20, 2022.

[7] Online document "Crew costs to rise 10–15% this year due to Covid-19" https://www.seatrade-maritime.com/ship-operations/crew-costs-rise-10-15-year-due-covid-19-drewry Accessed on July 20, 2022.

to quarantine, as well as the cost of COVID tests.[8] It is the authors' view that if autonomous vessels were fully in operation during the pandemic, then these incidental additional crew costs and increase would not have been incurred by shipping companies.

Greater Revenue Generally, the greater the quantity of cargo that a cargo ship can hold, is the greater the revenue. Traditional cargo ships are legally required however to make adequate provisions for crew accommodation on board vessels. The International Labour Organization (ILO) through the Maritime Labour Convention, 2006, mandates member states to require vessels flying its flag to comply with the requirements of the said convention. Regulation 3.1 of the Maritime Labour Convention, 2006 states that:

> Each Member shall ensure that ships that fly its flag provide and maintain decent accommodations and recreational facilities for seafarers working or living on board, or both, consistent with promoting the seafarers' health and well-being.

Regulation 3.1 (9) (f) of the said Maritime Labour Convention, 2006 states that when sleeping accommodation on board ships is required, a single berth seafarers sleeping room shall not be less than:

(i) 4.5 square metres in ships of less than 3000 gross tonnage;
(ii) 5.5 square metres in ships of 3000 gross tonnage or over but less than 10,000 gross tonnage;
(iii) 7 square metres in ships of 10,000 gross tonnage or over.

There are other regulations in the Maritime Labour Convention which mandate that the applicable vessels provide inter alia furniture for each officer, sanitary facilities including showers, toilets and wash basins, table or desk, comfortable seating, and hospital facilities.[9] The regulations also require that there be furnished laundry facilities on board as well as an open deck for seafarers to have access to when they are off duty. It is the author's view that without a crew on board, the spaces that were once used for crew accommodation could be transformed into more cargo hold. As such, it is possible for a ship owner to carry more cargo on an autonomous vessel and thus increasing his revenue.

Lower Voyage Expenses Fuel accounts for a large part of a vessels operating expenses. The larger and bulkier a vessel, is the more fuel it consumes. It is estimated that a large vessel might consume up to 250 tons of fuel per day.[10] That is equivalent to approximately 80,000 gallons of fuel a day. A regular-sized vessel however may use up to 150 tons of fuel per day and as such, as a vessel size shrinks, so does its fuel

[8] Online document "Crew costs to rise 10–15% this year due to Covid-19" https://www.seatrade-maritime.com/ship-operations/crew-costs-rise-10-15-year-due-covid-19-drewry Accessed on July 20, 2022.

[9] See Regulations 2, 3, and 4 of the Maritime Labour Convention, 2006.

[10] Online document https://www.colorado.edu/mechanical/2016/07/25/how-much-fuel-does-cruise-ship-use Accessed on July 15, 2022.

consumption.[11] Autonomous vessels are being designed to be at least 5% lighter than traditional manned vessels. It is expected that they will thus use approximately 15% less fuel than traditional manned vessels and as such, this will lower a shipowner's overall voyage expenses if commerce is done by an autonomous vessel.[12]

Human Error Eliminated Studies have found that human error accounts for almost 96% of maritime accidents.[13] Human error can occur because of insufficient communication, inexperience, inadequate knowledge, fatigue, poor maintenance and faulty standards and procedures being followed amongst other things.[14] Human error can occur in many forms including small fires to loud explosions, to large scale collisions.[15] Developers of autonomous vessels theorize that eliminating humans on board will see a significant reduction in maritime accidents.[16]

1.2.2 Disadvantages

Regulatory Framework There are currently no international regulations that were developed with autonomous vessels in mind. It is debatable whether these international rules and regulations are currently able to accommodate autonomous vessels. It has been reported that the IMO is working assiduously to develop new rules and regulations that are specific to autonomous vessels.[17] However, it has been difficult as all the regulations and conventions that previous exist contemplated humans on-board and as such, autonomous vessel requires major overhauling of existing legislations. As such, if autonomous vessels begin sailing soon, they will most likely be operating in a legal vacuum.

Safety A major potential drawback of autonomous vessels is that they may not be safe. Questions have been raised around what would happen if the autonomous vessel began to malfunction and humans cannot override it.[18] Although human error

[11] Online document https://www.windstarcruises.com/blog/how-much-fuel-cruise-ship-uses/ Accessed on July 15, 2022.

[12] Online document https://www.windstarcruises.com/blog/how-much-fuel-cruise-ship-uses/ Accessed on July 15, 2022.

[13] Online document https://www.marineinsight.com/marine-safety/the-relation-between-human-error-and-marine-industry/ Accessed on August 10, 2022.

[14] Online document https://www.marineinsight.com/marine-safety/the-relation-between-human-error-and-marine-industry/ Accessed on August 10, 2022

[15] Online document https://www.marineinsight.com/marine-safety/the-relation-between-human-error-and-marine-industry/ Accessed on August 10, 2022

[16] See Rolls Royce Project and Yara Birkeland Project.

[17] Online document https://www.imo.org/en/MediaCentre/PressBriefings/pages/MASSRSE2021.aspx Accessed on June 10, 2022.

[18] Online document https://www.equaltimes.org/will-autonomous-ships-be-the#.YY3Q72DMJyw Accessed on June 10, 2022.

is eliminated, another type of risk is created. Additionally, there are issues as it relates to cyber security. Safety from this aspect is also quite important as it may be very possible for hackers to access an autonomous vessels system and block signals with the shipping companies' hub that is tracking the autonomous vessel. It is thus possible that a cyberattack like that which affected Maersk in 2017 could happen to an autonomous vessel. That attack ended up costing the company 255 million euros (US$298 million).[19]

Liability and Insurance Marine Insurance is the oldest form of insurance. It protects shipowners and cargo owners against the loss of a ship/and or cargo.[20] It helps to manage risks in the event of accidents, damage to environment and property and/or loss of life. One of the arguments against autonomous vessels is that there are no insurance products currently available to protect shipowners and cargo owners who choose to conduct commerce via autonomous vessels.[21] There are also discussions that the risks associated with an autonomous vessel may be too risky to underwrite and that insurers may be unwilling to insure an autonomous vessel when they begin to sail.[22]

2 Seaworthiness and Autonomous Vessels

Seaworthiness is a very important concept in the context of maritime law and carriage of goods by sea generally. A vessel being unseaworthy can lead to many devastating consequences, most serious one being a collision. Whether a vessel is seaworthy also impacts potential claims for personal injuries and loss of goods where an incident happens at sea.

The concept of seaworthiness has its roots as far back as in Rhodian Sea Law, which is reportedly the first consolidated maritime law in history. Rhodian Sea Law, oftentimes called *Lex Rhodia*, was the body of regulations governing commercial trade and navigation in the Byzantine Empire which dates back to the seventh century.[23] These laws and regulations focused mainly on the liability for the cost of lost or damaged cargo.[24] Cargo loss was greatest during storms, when part or all of

[19] Shackelford (2020), p. 88.

[20] Rose (2013) para 30.

[21] Online document http://www.crewtoo.com/crewtoo-blogs/autonomous-shipping-and-safety/ Accessed on August 10, 2022.

[22] Online document http://www.crewtoo.com/crewtoo-blogs/autonomous-shipping-and-safety/ Accessed on August 10, 2022.

[23] Online document https://www.britannica.com/event/Rhodian-Sea-Law Accessed on July 12, 2022.

[24] Online document https://www.britannica.com/event/Rhodian-Sea-Law Accessed on July 12, 2022.

it had to be thrown overboard to save the ship.[25] On careful examination of these laws, it is observed that there was no obligation for seaworthiness in ordinary voyages, but Rhodian Sea Law required that a vessel be seaworthy when it was carrying valuable cargo.[26] Rhodian Sea Law persisted, through the 12th century. However, by the 13th and 14th centuries eventually the law became obsolete.[27]

The concept of seaworthiness continued to evolve over the 15th century to the present-day realities, where it is legally required by the IMO and member states that a vessel be deemed seaworthy.[28] Article 94 of the United Nations Convention on the Law of the Sea (UNCLOS) on Duties of the flag States stipulates that the flag State shall:

> "effectively exercise its jurisdiction and control in administrative, technical and social matters over ships flying its flag" and take "such measures for ships flying its flag as are necessary to ensure safety at sea with regard, inter alia, to the construction, equipment and seaworthiness of ships. . .," including "those necessary to ensure that each ship, before registration and thereafter at appropriate intervals, is surveyed by a qualified surveyor of ships. . ."

Seaworthiness is also not static and as such it can vary depending on the weather as well as the conditions of the voyage. In Hong Kong Fir Shipping Ltd v Kisen Kaisha Ltd,[29] Diplock LJ opined that seaworthiness:

> can be broken by the presence of trivial defects easily and rapidly remediable as well as by defects that must inevitably result in a total loss of the vessel.

Despite its importance, there is no universal definition of seaworthiness in the context of maritime law and as such, we must examine the concept as it is understood in each international legal instrument.[30]

Prof. Tetley, however, provides a comprehensive definition of what seaworthiness means. He states that seaworthiness is:

> the state of the vessel in such a condition, with such equipment, and manned by such a master and crew, that normally the cargo will be loaded, carried, cared for and discharged properly and safely on the contemplated voyage.[31]

From this definition, it is evident that seaworthiness is not just related to the physical state of the vessel but also includes voyage worthiness and cargo

---

[25] Online document https://www.britannica.com/event/Rhodian-Sea-Law Accessed on July 12, 2022.

[26] Online document (Kampantais, 2016) "Seaworthiness in Autonomous Unmanned Cargo Ships" p. 6. https://www.researchgate.net/publication/311452364_Seaworthiness_in_autonomous_unmanned_cargo_ships Accessed on May 31, 2022.

[27] Online document (Kampantais, 2016) "Seaworthiness in Autonomous Unmanned Cargo Ships" p. 6. https://www.researchgate.net/publication/311452364_Seaworthiness_in_autonomous_unmanned_cargo_ships Accessed on May 31, 2022.

[28] Online document https://www.imo.org/en/OurWork/IIIS/Pages/Survey-Verification-Certification.aspx Accessed on May 19, 2022.

[29] (1962) EWCA Civ 7.

[30] Soyer (2006), p. 50.

[31] Tetley (1978), p. 877.

worthiness. To properly analyze the concept of seaworthiness and its applicability to autonomous vessels, the author is of the view that it is important to dissect these different elements of seaworthiness.

2.1 Physical Seaworthiness

The physical aspect of seaworthiness refers to the actual physical state of the vessel. This usually means that the vessel's equipment, hull, machinery, bunkers, and stores must be able to withstand the ordinary perils of the voyage it is about to undertake.[32] Support for this position can be found in the case of Kopitoff v Wilson,[33] where the court stated that a vessel is seaworthy if it is:

> fit to meet and undergo the perils of the sea and other incidental risks to which of necessity she must be exposed in the course of a voyage.

Other factors including the time of voyage, state of voyage, the type of cargo, and the age and model of the vessel are also considered when determining whether a vessel is physically seaworthy.[34] However, the law has established that there is no duty that a ship owner must provide a perfect ship *"which might withstand all conceivable hazards,"*[35] but they must provide a ship that is reasonably suitable for its intended voyage.[36] In the case of The Rover, the court stated that as it relates to seaworthiness that *"perfection is unattainable."*[37] An important case to consider when discussing seaworthiness is the case of Burges v Wickham,[38] where Blackburn J stated that *"the standard of seaworthiness must rise with the improved knowledge of shipbuilding and navigation."* The author ponders whether this judgment could mean that traditional vessels would need to eventually automate fully in the future when autonomous vessels become commonplace. In considering this point, the author examined the 1962 US case of The President of India v West Coast Steamship Company[39] *"The Portland Trader."* In that case, The President of the Republic of India, acting by and through the Director of the India Supply Mission in the United States, brought a case against West Coast Steamship Company, as owner of the *S.S.*

---

[32]Online document (Kirchner) "Rise of the Machines – A Legal analysis of Seaworthiness in the context of autonomous shipping" https://lup.lub.lu.se/luur/download?func=downloadFile&recordOId=8977227&fileOId=8984495 Accessed on August 11, 2022.

[33](1876) 1 QBD 377, 380.

[34]Online document (Kirchner) "Rise of the Machines – A Legal analysis of Seaworthiness in the context of autonomous shipping" https://lup.lub.lu.se/luur/download?func=downloadFile&recordOId=8977227&fileOId=8984495 Accessed on August 11, 2022.

[35]President of India v West Coast Steamship Co, [1963] 2 Lloyd's Rep. 278, Killenny. J.

[36]President of India v West Coast Steamship Co, [1963] 2 Lloyd's Rep. 278, Killenny. J.

[37]The Rover (D.C.) 33 Fed. 515, p. 521.

[38](1836) B&S 669, p. 693.

[39]President of India v West Coast Steamship Co, [1963] 2 Lloyd's Rep. 278, Killenny. J.

Portland Trader, for damage to cargo which occurred when the vessel ran aground on Tubbataha Reef in the Sulu Sea, Republic of the Philippines, January 5, 1961. Their primary contention was that the vessel was unseaworthy by reason of its not being equipped with either a radar or loran. District Judge John F. Kilkenny found that the vessel was seaworthy and that the sole cause of the accident was negligent navigation by the master.[40] The court further found that although the radar was a valuable aid for navigation and that it could have avoided the accident, there was no established practice worldwide to use radar at that time.

Legal scholars have also argued that there is also no obligation for the owner of a ship to constantly update his vessel and its equipment if his vessel is able to meet the demands of its intended voyage.[41]

2.2 *Seaworthiness in the Context of Master and Crew*

As enunciated above, the concept of seaworthiness is not restricted to the physical state of the vessel, but it also considers the master and crew. In fact, it has been said that a seaworthy vessel is one that has been properly manned.[42] A properly manned vessel is one that has been staffed with a crew that has the requisite competence, experience, and training to handle the specific tasks for that specific vessel.[43] A vessel will also be rendered unseaworthy if the shipowner does not provide the master and crew with the requisite instructions or supervision in relation to the particular vessel and voyage. If a crew member is generally competent but is not specifically competent for a particular type of vessel, then the vessel may still be considered unseaworthy.[44] See the case of Hong Kong Fir Shipping Ltd v Kisen Kaisha Ltd[45] where the court found that because of the old age of the engines of the vessel, the type of crew that should have been hired by the shipowner specifically for the engine room was one of exceptional ability, experience and dependability. The court found that the engine staff aboard this ship was not competent and was not sufficient in terms of size and as such, the vessel was rendered unseaworthy as a result. The case of Papera Traders Co Ltd v Hyundai Merchant Marine Co Ltd (The Eurasian Dream)[46] is also instructive on this point. In that case, the court found that

[40] President of India v West Coast Steamship Co, [1963] 2 Lloyd's Rep. 278, Killenny. J.

[41] Online document (Kirchner) "Rise of the Machines – A Legal analysis of Seaworthiness in the context of autonomous shipping" https://lup.lub.lu.se/luur/download?func=downloadFile&recordOId=8977227&fileOId=8984495 Accessed on August 11, 2022.

[42] Güner-Özbek (2008), p. 121.

[43] Güner-Özbek (2008), p. 121.

[44] Online document (Kirchner) "Rise of the Machines – A Legal analysis of Seaworthiness in the context of autonomous shipping" https://lup.lub.lu.se/luur/download?func=downloadFile&recordOId=8977227&fileOId=8984495 Accessed on August 11, 2022.

[45] (1962) EWCA Civ 7.

[46] (No.1) [2002] EWHC 118 (Comm).

the only way within which a shipowner could absolve himself from liability where a vessel is deemed unseaworthy due to the incompetence or inefficiency of the master and/or crew, is if the said shipowner can show that he has exercised proper care when he appointed this master and/or crew. As illustrated earlier in this chapter, human error is a major cause of maritime accidents and collisions and as such, it is expected that autonomous vessels will reduce the frequency of accidents and incidents that occur at sea.

Having outlined the importance of having a competent master and crew, it is now imperative to discuss whether an autonomous vessel can be considered as seaworthy since it will no longer be travelling with a master and crew in the traditional sense. It is the authors' opinion that an autonomous vessel could still be considered as seaworthy if the courts and regulators are willing to exercise its discretion and view that the elimination of the master and crew on board leads to less accidents and as such increases the seaworthiness overall. Alternatively, the courts could also view the various algorithms as fictitious masters and crews and once these algorithms and data base devices are in good condition, software licenses are intact, it could be argued that the vessel is seaworthy although it does not have a master and crew. Similarly for remote controlled autonomous vessels, the shore-based operators could be seen as equivalent to the traditional master and crew and as such, it is possible to argue that a remote-controlled autonomous vessel would still be considered as seaworthy although it is not manned in the traditional way.

2.3 Cargo Worthiness

Cargo worthiness refers to the overall condition of the vessel and the stowage of cargo on board a vessel.[47] In essence, it considers the capability of the vessel to deliver its cargo safely to its final destination.[48] As it relates to the overall condition of the vessel, the shipowner has to show that the vessel's hatches and holds are cleaned and fumigated before calling at a port or before the cargo is loaded.[49] The pumps on the vessel must also be fully capable of draining water from the cargo and there should be no leakage of pipes and hatches.[50] *The Aconcagua*[51] is instructive on the stowage of cargo. The court in that case held that *"A vessel may be unseaworthy if there is no system in operation to deal with the ordinary incidents of a voyage,*

[47] Online document (Kassem) "The Legal Aspects of Seaworthiness: Current Law and Development" https://discovery.ucl.ac.uk/6988/1/6988.pdf Accessed on May 13, 2022.

[48] Elder, Dempster and Company, Limited, and Others v Paterson, Zochonis and Company, Limited and Others, [1924] A.C. 522, Viscount Cave at p. 530.

[49] Mediterranean Freight Services Ltd. v BP Oil International Ltd., (The Fiona), [1994] 2 Lloyd's Rep. 506.

[50] Rathbone Brothers & Co. v D. Maciver, Sons & Co [1903] 2 K.B. 378.

[51] Compania Sud Americana De Vapores SA v Sinochem Tianjin Limited "The Aconcagua" [2010] EWCA Civ 1403.

including the need for the cargo to be stowed in a way that does not endanger the ship."

However, it is important to note that poor stowage does not automatically make the vessel unseaworthy as in the case of The Aconcagua.[52] It is only where the improper stowage endangers the safety of the vessel by making it unstable, then the vessel would be seen as unseaworthy. If it simply jeopardizes the safety of the cargo alone then that would not make the ship unseaworthy.[53]

2.4 Legal Seaworthiness v Contractual Seaworthiness

The regulation of seaworthiness can be divided into two main categories:

1. Legal seaworthiness; and
2. Contractual seaworthiness.

The author will now discuss these two categories of seaworthiness and critically explore whether both contractual seaworthiness and legal seaworthiness in their current state can accommodate autonomous vessels.

2.4.1 Legal Seaworthiness

Common Law

In practice, the contract of freight or bill of lading will almost always require the carrier to provide a seaworthy vessel.[54] If seaworthiness is not contractually provided for and international conventions are not applicable, then a claimant must turn to the common law for guidance as it relates to the concept of seaworthiness. The common law has an implied obligation in almost all contracts for carriage of goods by sea that the vessel must be seaworthy. The common law definition of seaworthiness can be found in the case of McFadden v Blue Star Line where it was said that *seaworthiness is:*

> ...that degree of fitness which an ordinary careful and prudent owner would require his vessel to have at **the commencement of her voyage** having regard to all the probable circumstances of it.[55]

[52] Compania Sud Americana De Vapores SA v Sinochem Tianjin Limited "The Aconcagua" [2010] EWCA Civ 1403.

[53] The Elder Demspter & Co. v Paterson Zochonis & Co. and The Thorsa.

[54] Mitchelhill (1990), p. 7.

[55] (1905) 1 KB 697.

Under common law, the shipowner has an absolute duty to provide a seaworthy vessel.[56] The carrier is expected to provide a vessel:

> that is fit to meet and undergo the perils of the sea and other incidental risks which of necessity she must be exposed in the course of the voyage.[57]

Thus, under common law, it does not matter if the carrier can show that he took all reasonable precautions to provide a seaworthy vessel, he will still be liable because of the absolute nature of the duty. It is the author's view that an autonomous vessel can be viewed as seaworthy at common law, as no reference has been made to there being a master and crew but simply that the vessel must be fit to undergo the perils of the sea at the commencement of the voyage.

Hague and Hague-Visby Rules

The Hague/Hague-Visby Rules are international conventions that assist with regulating the international carriage of goods by sea. Over eighty (80) countries have ratified these rules and as such they are widely applicable to many international contracts of affreightment.[58] Given its wide reach, it is quite imperative to examine what the Hague/Hague-Visby Rules state about seaworthiness and to examine if in its current form, it would be applicable to autonomous vessels.

Article 3.1 of the Hague/Hague-Visby Rules states that:

> The carrier shall be bound before and at the beginning of the voyage to exercise due diligence to make the ship seaworthy, **properly man,** equip, and supply the ship, make the holds, refrigerating and cool chambers, and all other parts of the ship in which goods are carried, fit and safe for their reception, carriage and preservation.

The case of Papera Traders Co. Ltd. and Others v Hyundai Merchant Marine Co. Ltd. and Another—The "Eurasian Dream"[59] is instructive on the interpretation of the Hague/Hague-Visby Rules and how the rules address seaworthiness. On July 23, 1998, a fire started on deck 4 of the pure car carrier Eurasian Dream while in port at Sharjah, a port located in the United Arab Emirates (UAE). The fire, which was not contained or extinguished by the master and crew, eventually destroyed, or damaged the vessel's cargo of new and second-hand vehicles and rendered the vessel itself a constructive total loss. The relevant cargo interests commenced proceedings in London against the carrier. The court held that:

> Seaworthiness is not an absolute concept; it is relative to the nature of the ship, to the particular voyage and even to the particular stage of the voyage on which the ship is engaged

---

[56] Kopitoff v Wilson (1876) 1 Q.B.D 377.

[57] Kopitoff v Wilson (1876) 1 Q.B.D 377 at 380 per Field J, giving the Judgment of the Divisional Court of the Queen's Bench.

[58] Djadjev (2017), p. 36.

[59] [2002] 1 Lloyd's Rep. 719.

> and must be judged by the standards and practices of the industry at the relevant time, at least so long as those standards and practices are reasonable.
>
> The duty of "due diligence" is an "inescapable personal obligation": it is non-delegable. The carrier will therefore be responsible for negligence of those to whom it delegates due diligence.
>
> The question is whether unseaworthiness is due to any lack of diligence in those who have been implicated by the carrier in the work of keeping or making the vessel seaworthy.
>
> Such persons are the carriers' agents whose diligence or lack of it is attributable to the carrier. This principle is relevant in two respects:
>
> (1) the carrier under the bills of lading is liable for the want of due diligence by the owners or managers;
> (2) the carrier is liable for the want of due diligence of the master insofar as the carrier or the owners or managers have delegated to him their duties as to seaworthiness.[60]

It is quite evident from this case that the Hague/Hague-Visby Rules consider the different aspects of seaworthiness discussed above including physical seaworthiness, human seaworthiness, and cargo worthiness. It is arguable that an autonomous or unmanned vessel would be capable of satisfying the requirement of physical and cargo worthiness under the Hague/Hague-Visby Rules based on the technological advancement in the area. Where it becomes difficult to satisfy is the human seaworthiness element.

As highlighted above, Article 3.1 of the Hague/Hague-Visby Rules has stated that the vessel must be properly manned. Thus, it is quite arguable that an autonomous vessel would prima facie be deemed as unseaworthy because it is not manned. However, especially in the case of shore-based remote controlled autonomous vessels, it is arguable that the remote-controlled operators could be deemed as being virtually "on board," since they are required to control the vessel remotely. The human element is thus still arguably present, and the argument can be confidently made that an autonomous remote-controlled vessel is still seaworthy under the Hague/Hague Visby Rules. However, it is equally arguable that a fully autonomous vessel without any human element would not be able to satisfy this requirement under the Hague/Hague-Visby Rules.

The United Nations Convention on the Carriage of Goods by Sea 1978 "Hamburg Rules"

The Hamburg Rules were adopted by the United Nations at a conference on the 31st day of March 1978 in Hamburg, Germany.[61] The Rules were devised and adopted with the intention that they should supersede the 1924 Hague Rules together with the

[60] [2002] 1 Lloyd's Rep. 719.

[61] Online document https://www.euro-marine.eu/hague-visby-rules Accessed on May 25, 2022.

1968 Hague Visby amendments.[62] Nevertheless, both rules are still in existence and the Hamburg Rules have been ratified in thirty-five (35) countries as of 2021.[63] After five years of ratification, it is expected that ratifiers will denounce other conventions, specifically the Hague/Hague Visby Rules.[64]

As it relates to seaworthiness, Article 5.1 of the Hamburg Rules states that:

> The carrier is liable for loss ... unless the carrier proves that he, his servants or agents took all measures that could reasonable be required to avoid the occurrence and its consequences.

It is important to note that the Hamburg rules do not explicitly speak to manning of the vessel nor does it explicitly mention seaworthiness in the traditional way that it is mentioned under the Hague/Hague-Visby Rules. The Hamburg Rules also apply do not have any due diligence requirement as is done under the Hague/Hague-Visby Rules and arguably the obligation is a continuing obligation because it speaks to liability up to the point of delivery, whereas under the Hague/Hague-Visby Rules, the liability is simply at the commencement of the voyage.

It is arguable that the Hamburg Rules it is more accommodating of an autonomous vessel than the Hague/Hague-Visby Rules, since there is no explicit mention of manning of vessels in the traditional way. As such, the author suggests that an autonomous vessel shipowner might initially find it easier to conduct business in one of the thirty-five (35) countries that have ratified the Hamburg Rules, as it would be arguably very easy for that vessel to be viewed as seaworthy under these said rules.

Rotterdam Rules

The United Nations Convention on Contracts for the International Carriage of Goods Wholly or Partly by Sea, 2009, known as the "Rotterdam Rules," was signed by sixteen countries.[65] However, at the time of the writing of this chapter, it was not yet in force because it requires ratification by twenty (20) countries, but it currently only is ratified by five.[66] It is the author's view that it is still imperative to discuss these rules nonetheless because it is very possible that at the point when autonomous vessels begin sailing frequently, that these rules may very well be in force.

Article 14 of the Rotterdam Rules states that:

> The carrier is bound before, at the beginning of, and during the voyage by sea to exercise due diligence to: a) make and keep the ship seaworthy b) properly crew, equip and supply the ship and keep the ship so crewed, equipped and supplied throughout the voyage; [..]

---

[62] Online document https://www.euro-marine.eu/hague-visby-rules Accessed on May 25, 2022.

[63] Online document https://unis.unvienna.org/unis/en/pressrels/2021/unisl312.html Accessed on May 25, 2022.

[64] Lüddeke and Johnson (1995), p. 5.

[65] Güner-Özbek (2008), p. 65.

[66] Online document https://uncitral.un.org/en/texts/transportgoods/conventions/rotterdam_rules/status Accessed on May 25, 2022.

Upon careful examination of this Rule, it is evident that it has the said requirement of due diligence as can be found in Hague and Hague-Visby Rules. However, it also has similarities with the Hamburg Rules in that it requires seaworthiness throughout the entire duration of the voyage. It is distinct from the Hamburg Rules but similar to the Hague/Hague-Visby Rules because it requires that the vessel be properly crewed. As such, as it relates to fully autonomous vessels, it is arguable that the Rotterdam Rules would view them as unseaworthy.

However, just as was discussed above as it relates to the Hague/Hague-Visby Rules, it is arguable that the shore-based operators of a remote based unmanned vessel can be seen as being virtually crewed and as such, it is arguable that those types of autonomous vessels will be within the realm of the Rotterdam Rules if the rules do become effective.

The author however suggests that an amendment could be made to the Rotterdam Rules before they come into force, to remove the requirement of being properly crewed so that autonomous vessels are not seen as unseaworthy according to these rules in the future.

The International Convention for the Safety of Life at Sea (SOLAS)

The SOLAS Convention was adopted in 1974 to specify minimum standards for the construction, equipment, and operation of ships, compatible with their safety.[67] Similarly, the IMO introduced the International Safety Management Code in November 1993, and it was formally adopted and included as a part of the SOLAS Convention in the year 1994.[68] The aim of the code is to provide a uniformed platform for marine safety for vessels of all nationalities. The ISM Code requires all shipowners, managers, and bareboat charterers to develop a Safety Management System (SMS) for every vessel.[69] The SMS ensures that every vessel comply with the mandatory safety rules and regulations, and follow the codes, guidelines, and standards recommended by the IMO and other affiliated maritime organizations. Under Chapter IX of the SOLAS Convention, a vessel that is compliant with the ISM Code is issued a Document of Compliance.[70] This author carefully perused the latest revision of the ISM Code (2015) to see whether an autonomous vessel would be able to comply with the requirements of the ISM Code. Article 6.2 states that:

> The Company should ensure that each ship is:
>
> 1. manned with qualified, certificated and medically fit seafarers in accordance with national and international requirements; and

---

[67] Liu (2009), p. 126.

[68] Liu (2009), p. 126.

[69] Branch and Robarts (2014), p. 36.

[70] Branch and Robarts (2014), p. 36.

2. appropriately manned in order to encompass all aspects of maintaining safe operation on board.

The author is of the view that this article is clearly contemplating a physically manned vessel and as such, not even shore-based operators of unmanned autonomous vessels could be seen as analogous to satisfy this article. Article 6.2 (2) specifically states that it must be appropriately manned to maintain safe operation on board. As such, the author is of the opinion that an autonomous or a remotely operated vessel would be viewed as noncompliant in the eyes of the ISM Code and the SOLAS Convention. Thus, an autonomous vessel would not be able to comply with the requirements of the ISM Code. The author is of the view that the capability of an autonomous vessel being able to satisfy the ISM Code is important when considering if an autonomous vessel will be viewed as legally seaworthy. A comparison can be made between the cases of Papera Traders Co Ltd v Hyundai Merchant Marine Co Ltd (The Eurasian Dream)[71] and "The Torpeo."[72]

In the Torpeo, the court referred to the ISM-record keeping facilities on board the ship in concluding that the vessel was seaworthy.[73] However, in the Eurasian Dream, where there was a fire on board a ferry, the court was of the view that the ISM emergency manual onboard did not provide guidance as to the best way to fight a fire to the master since it was *"too voluminous to be digestible"* by the master.[74]

It is thus the author's opinion that compliance with the ISM Code and a Document of Compliance on board an autonomous vessel will raise the presumption that an autonomous vessel is seaworthy under present maritime law. Further, it is the authors' opinion that when autonomous vessels begin operations, that the ISM Code will have to be further revised to make accommodation for autonomous vessels.

International Convention on Standards of Training, Certification and Watchkeeping for Seafarers (STCW) Convention

The STCW Convention was adopted in 1978 by the IMO. As of 2018, the convention has been signed by 164 countries, representing 99.2% of the world's shipping tonnage.[75] The intention behind the convention is to promote safety of life and property at sea and to protect the oceans and marine environment. The Convention outlines the minimum standards of training, certification and watchkeeping for seafarers.[76] The convention applies to all vessels flying the flag of a contracting

[71] (No.1) [2002] EWHC 118 (Comm); [2002] 1 Lloyd's Rep. 719.

[72] The Torepo [2002] 2 Lloyd's Rep 535.

[73] The Torepo [2002] 2 Lloyd's Rep 535.

[74] The Torepo [2002] 2 Lloyd's Rep 535.

[75] Online document https://www.maritimeskillsacademy.com/stcw-convention-explained/ Accessed on May 30, 2022.

[76] Online document https://www.maritimeskillsacademy.com/stcw-convention-explained/ Accessed on May 30, 2022.

state and as such, it is imperative to examine whether or not an autonomous vessel will be able to satisfy the STCW Convention.[77]

Paragraph 3 of Regulation 1/14 of the STCW 2010 outlines that owners of vessels must ensure that their *"ships are manned in compliance with the applicable safe manning requirements."*

Under paragraph 6 of the said Regulation, shipowners must ensure that the

> ship's complement can effectively coordinate their activities in an emergency situation in performing function vital to safety, security and to prevention of mitigation of pollution.

Additionally, paragraph 7 of the said Regulation states that shipowners must also ensure that

> at all times on board its ships there shall be effective oral communication.

It is clear from these paragraphs that an autonomous vessel will generally not be able to satisfy the requirements of the STCW Convention. The autonomous vessel contemplates no human element on board and as such there is no way to comply with paragraph 3 of the Regulation 1/14 of the STCW.

However, it is arguable that as for paragraphs 6 and 7, the technology being used on some of these proposed autonomous vessels should be able to effectively coordinate activities in the event of an emergency. Paragraph 7 is particularly easy to satisfy especially if Artificial Intelligence on board these autonomous vessels are employed. This is arguably best satisfied by remote controlled unmanned vessels that have shore-based operators, as they could effectively communicate by giving oral commands remotely to the unmanned autonomous vessel at sea. Compliance with the STCW impacts the seaworthiness of a vessel and a shipowner who does not comply with the requirements of the STCW will arguably have their vessel rendered as unseaworthy if a claim was to arise in marine insurance.[78]

2.4.2 Contractual Seaworthiness

A charterparty is a common contract in the marine industry. It is a contract by which the owner of a ship lets it to others for use in transporting a cargo. The shipowner generally controls the navigation and management of the vessel, but its carrying capacity is engaged by the charterer. It is quite common that these charterparties will have requirements which contractually impose seaworthiness.[79] Additionally, to call at most ports, it is a requirement that the vessel should be seaworthy. Given the importance of these charterparties, the author will now examine the concept of

[77] Online document https://www.maritimeskillsacademy.com/stcw-convention-explained/ Accessed on May 30, 2022.

[78] Online document (Girvin) "The Obligation of Seaworthiness Shipowner and Charterer" https://law.nus.edu.sg/wp-content/uploads/2020/04/019_2017_Stephen-Girvin-CML.pdf Accessed on June 5, 2022.

[79] Baatz (2014), p. 125.

seaworthiness under charterparties and critically discuss if autonomous vessels can be viewed as seaworthy under some common standard form charterparties.

There are three main types of charterparties. These are time charterparties, voyage charterparties and bareboat charterparties.[80] In this modern era, charterparties are almost always based in standard forms. The type of standard forms that exist differ based on the type of charterparty.

Seaworthiness Under Time Charterparties

A time charter is a time-bound agreement where the shipowner leases a vessel to a charterer for a fixed period and the charterer is free to transport any cargo subject to legal regulations.[81] Although the charterer controls the ship, the maintenance of the vessel still falls under the purview of the owner. They are responsible for ensuring that the vessel meets internationally accepted maritime standards, throughout the course of the agreement.[82] They regularly employ marine surveyors to prepare reports on the seaworthiness of the vessel and make repairs as and when required. The owner will face legal action in case the vessel is found to have some major problem and deemed unseaworthy.[83]

The most common and widely used time charter party is the New York Produce Exchange Form (NYPE) and the most widely used versions are the NYPE46 and 93. However, the most recent form of the NYPE is the NYPE 2015.

The NYPE46, clause 21, states that the:

> Vessel on her delivery to be ready to receive cargo with clean swept holds and tight, staunch, strong and in every way fitted for the service, having water ballast, winches and donkey boiler with sufficient steam power, or if not equipped with donkey boiler, then other power sufficient to run all the winches at one and the same time (and with full complement of officers, seamen, engineers and firemen for a vessel of her tonnage), to be employed, in carrying lawful merchandise...

Although this clause does not specifically refer to the word seaworthy, based on what is known about the requirements of seaworthiness, it is clear that this is the clause which speaks to seaworthiness on delivery of the vessel in this standard form charterparty. Upon analysis of this clause, the NYPE46 considers a vessel seaworthy that has been manned by competent officers, seamen, engineers, and firemen suited for a vessel of her tonnage. As such, it is the authors' view that without the use of any

[80] Todd (2016), p. 5.

[81] Todd (2016), p. 5.

[82] Online document https://www.marineinsight.com/maritime-law/voyage-charter-vs-time-charter/#:~:text=A%20time%20charter%20is%20a,cargo%2C%20subject%20to%20legal%20regulations Accessed on July 1, 2022.

[83] Online document https://www.marineinsight.com/maritime-law/voyage-charter-vs-time-charter/#:~:text=A%20time%20charter%20is%20a,cargo%2C%20subject%20to%20legal%20regulations Accessed on July 1, 2022.

rider clauses, a shipowner issuing its autonomous vessel out for hire using the NYPE 46 would be providing an unseaworthy vessel on delivery.

The NYPE93, clause 2, which is arguably the seaworthy clause, states that:

> The Vessel on her delivery shall be ready to receive cargo with clean-swept holds and tight, staunch, strong and in every way fitted for ordinary cargo service, having water ballast and with sufficient power to operate all cargo-handling gear simultaneously.

On careful examination of this clause, it is evident that the NYPE93 does not refer to the vessel being manned with competent master and crew on delivery. However, clause 6 of the said NYPE93 states that the ship owners are to provide a full complement of officers and crew. Throughout the NYPE93 as well, there is reference being made to crew and master and as such, it is the authors' opinion that the NYPE93 does not contemplate autonomous vessels. As such, like the NYPE46, if a shipowner of an autonomous vessel were to contract out his vessel to a charterer using the NYPE93 in its standard form, it would be viewed as unseaworthy.

The NYPE 2015 is the most recent version of the NYPE and as such an examination of seaworthiness in the context of autonomous vessel is quite relevant.

Clause 2 (b) of the NYPE 2015 states that:

> The Vessel on delivery shall be seaworthy and in every way fit to be employed for the intended service, having water ballast and with sufficient power to operate all cargo handling gear simultaneously, and, with full complement of Master, officers and ratings who meet the Standards for Training, Certification and Watchkeeping for Seafarers (STCW) requirements for a vessel of her tonnage.

On careful comparison between the NYPE46, NYPE93, and NYPE 2015, the author notes that the NYPE 2015 is the first version to include the word *"seaworthy."* Interestingly, in comparison to the NYPE 93, the 2015 version specifically makes mention of seaworthiness including a full complement of masters and officers who meet the STCW requirements. The author previously discussed the STCW and seaworthiness of autonomous vessels under that convention. The author finds it interesting that the 2015 version of the NYPE did not contemplate autonomous vessels especially since research and development for these vessels have been actively going on since 2012.[84] It is the authors' opinion that similarly to the NYPE 46 and NYPE 93, an autonomous vessel would not be able to utilize the NYPE 2015 in its standard form, as it would be viewed as unseaworthy.

Further, the author is of the opinion that the only way autonomous vessels would be viewed as seaworthy if trading was done via any version of the NYPE, is if rider clauses were introduced which replaced or amended the standard form clauses which relate to seaworthiness. However, if this approach is taken, every clause which mentions a Master and crew in the traditional form in the various versions of the NYPEs would need rider clauses. Further, it is the authors' view that the more amendments/rider clauses that are done to standard charterparties, is the more they

[84] Felski and Zwolak (2020), pp. 15, 41.

lose their effectiveness and credibility.[85] It is the authors' view that the language of standard charterparties should reflect the current practice in the shipping industry and as such, when autonomous vessels begin sailing frequently, then a different version of the NYPE would have to be drafted to accommodate autonomous vessels.

Another commonly used standard form time charter party is that of the Baltime Standard Time Charter 1939 Revised in 2001. It is imperative to examine the Baltime Standard Time Charter to determine if it can accommodate charters that take place on autonomous vessels. Clause 12 of the Baltime states that:

> The Owners shall be responsible for delay in delivery of the Vessel or for delay during the currency of the Charter and for loss or damage to goods onboard, if such delay or loss has been caused by want of due diligence on the part of the Owners or their Manager in making the Vessel seaworthy and fitted for the voyage or any other personal act or omission or default of the Owners or their Manager. The Owners shall not be responsible in any other case nor for damage or delay whatsoever and howsoever caused even if caused by the neglect or default of their servants. The Owners shall not be liable for loss or damage arising or resulting from strikes, lockouts or stoppage or restraint of labour (including the Master, officers or crew) whether partial or general. The Charterers shall be responsible for loss or damage caused to the Vessel or to the Owners by goods being loaded contrary to the terms of the Charter or by improper or careless bunkering or loading, stowing or discharging of goods or any other improper or negligent act on their part or that of their servants.

Upon careful perusal of the Baltime Standard Form the author noted that clause 12 is the only clause which mentions the word *"seaworthy."* However, the author notes that this clause specifically deals with responsibility of the Owner vis-a-vis the Charterer and does not specifically state what is a seaworthy vessel in the context of the Standard Form unlike the different version of the NYPEs. The argument could thus be made that an autonomous vessel could be viewed as seaworthy in the context of the Baltime Time Charter Standard Form.

However, an argument could also be advanced that this entire clause contemplates a vessel that is manned by a Master and has officers and crew. As such it would not be possible for an autonomous vessel to be considered as seaworthy in the context of the Baltime Standard Form. It is the authors' opinion that the Baltime 1939 (Revised 2001) would need to be revised to accommodate charters via autonomous vessels in the future.

In the interim, like the different versions of the NYPE, an autonomous vessel owner could introduce rider clauses to the Baltime 1939 (Revised 2001) if he wishes to trade using the Baltime 1939 (Revised 2001).

[85] Online document https://www.itic-insure.com/our-publications/intermediary/current-developments-in-charterparty-clauses-2863/ Accessed on June 15, 2022.

Voyage Charter

A voyage charter is a type of charter party in which the vessel is leased for a particular voyage.[86] This is different from time charters which, as explained above, are contracts for the use of the vessel for a specified period within agreed contractual limits, as directed by the charterer, in consideration for the payment of hire.[87] However, under a traditional voyage charter, the shipowners promise to make the vessel available to the charterers for the charterers to carry an agreed cargo on an agreed voyage, in exchange for the payment of freight. The ship owners usually remain responsible for the execution of the agreed voyage and as such, the shipowners bear most of the operational risks.[88] The voyage charterers promise to provide the cargo which would enable the ship owners to earn their freight. Voyage charters are not regulated by any compulsory international conventions and as such the parties have freedom to agree to whatever contractual terms they please.[89] Given the differences between a time charter and a voyage charter, the allocation of risks between the shipowners and charterers differ. However, some concepts are commonly found in both charters including that the vessel must be supplied with a master and a crew by the shipowners and the expectation that the vessel should be seaworthy. Similarly, to time charters, voyage charters also have standard forms, the most common one being the GENCON 94. It is therefore imperative that the author examine the GENCON 94 to determine if an autonomous vessel would be viewed as seaworthy under this standard charter party.

Clause 2 of the GENCON 1994 Uniform General Charter states that:

> The Owners are to be responsible for loss of or damage to the goods or for delay in delivery of the goods only in case the loss, damage or delay has been caused by personal want of due diligence on the part of the Owners or their Manager to make the Vessel in all respects seaworthy and to secure that she is properly manned, equipped and supplied, or by the personal act or default of the Owners or their Manager. And the Owners are not responsible for loss, damage or delay arising from any other cause whatsoever, even from the neglect or default of the Master or crew or some other person employed by the Owners on board or ashore for whose acts they would, but for this Clause be responsible, or from unseaworthiness of the Vessel on loading or commencement of the voyage or at any time whatsoever.

On close examination of this clause, it is evident that seaworthiness is viewed in the context of a properly manned vessel and as such, an autonomous vessel would not be viewed as seaworthy under the GENCON 94. As such, if a shipowner of an autonomous vessel is interested in undertaking a voyage charter using the GENCON 94, then it would have to utilize rider clauses.

[86] Force (2004), p. 25.

[87] Force (2004), p. 25.

[88] Online document https://www.bimco.org/-/media/bimco/training/elearning-contingency/bimco-training-curriculum-voyage-chartering.ashx Accessed on August 5, 2022.

[89] Online document https://www.bimco.org/-/media/bimco/training/elearning-contingency/bimco-training-curriculum-voyage-chartering.ashx Accessed on August 5, 2022.

It is imperative to note that a revised version of the GENCON 94 is underway and was expected to be finalized by the end of 2021.[90] It has been reported that the BIMCO drafting teams have specifically revised clause 2 entitled *"owners responsibility."* This said clause has colloquially been called the *"owners non-responsibility clause"* because it arguably provides the shipowners with a great degree of immunity.[91] It has been further reported that the new clause will offer a more balanced risk allocation between the parties and the owners will be required to provide a seaworthy vessel at the beginning of the voyage and a cargo worthy vessel at the beginning of loading.[92]

The 1994 clause 2 does not specify at what point of the voyage does the owner need to make the vessel seaworthy and as such, this new revision will provide more clarity to the shipowners' position.

The 1994 clause also mentions "personal act" which creates a scheme that is different from what exists under the Hague/Hague Visby Rules. As illustrated above, under the Hague Rules the carriers' due diligence obligation is non delegable. This does not mean that the shipowners are required to personally perform these obligations, but they will remain liable for any damage that occurs because of lack of their due diligence, irrespective of the persons at fault, whether they be employees, independent contractors, or employees. However, under clause 2 of the GENCON 94, the owner is only responsible for damage that resulted from their personal act or default. The case of Gesellschaft Buergerlichen Rechts v Stockholms Rederiaktiebolag Svea (The Brabant)[93] is instructive on the meaning of personal act or default in this context. In this case:

> The Brabant was charted on the Baltime form clause 13 of which was similar to clause 2 of the Gencon form. She was ordered to carry woodpulp and paper, but the crew failed to properly clean the holds of coal dust, which was deleterious to the prospective cargo. It was found that the failure to remove the coal dust was not due to any personal fault of the owners or managers, but only of the crew. The charterers argued that the decision in The Muncaster Castle should govern the case and that fault by the crew was fault for which the owners were liable under clause 13.
>
> McNair J. held that because the clause referred to "personal" fault, the fault of employees did not, without more, render the owners liable. However, the owners were held liable under a specific typed clause in the charter for which clause 13 provided no protection.

With this new proposed clause 2, the argument has been made that GENCON users will have a much clearer picture of the owner's responsibilities, and that the responsibilities will be slightly more limited than those that exist under the Hague/ Hague-Visby Rules.

---

[90] Online document https://www.bimco.org/news/priority-news/20200923-new-gencon Accessed on August 7, 2022.

[91] Online document https://www.hellenicshippingnews.com/owners-responsibilities-clarified-in-new-gencon-charter/ Accessed on August 7, 2022.

[92] Online document https://www.bimco.org/news/priority-news/20200923-new-gencon Accessed on August 7, 2022.

[93] [1965] 1 Lloyd's Rep. 546.

Another major change that has been proposed for the GENCON 2021 is a Force Majeure Clause which includes not only traditional types of events such as natural disasters, explosions, and strikes, but also interestingly includes more modern events such as cyberattacks and pandemics.[94]

However, there is no reports from BIMCO that the author could find that autonomous vessels were given any consideration under the proposed GENCON 2021 and as such, it is quite likely that the 2021 version of clause 2 which addresses the owners responsibility to provide a seaworthy vessel will still contemplate traditional vessels only. As such, the author is of the opinion that the GENCON would need further amendment to accommodate autonomous vessels beyond its proposed 2021 revision.

Bareboat Charter

A bareboat charter is often times called a lease of a vessel, as most of the customary liabilities of the owners are shifted to the charterer.[95] This therefore means that traditionally under a bareboat charterparty, the owner of the vessel delivers it up to the charterer for the agreed period without crew, stores, insurance, or any other provision, and the charterer is responsible for running the ship as if he were the owner.[96] The charterer is also usually responsible for any damages or injuries to third parties or to the vessel which arises from the vessel being unseaworthy.[97] Notwithstanding this, the courts have held that the owner is not completely exculpable of liability for delivering an unseaworthy vessel. It has been said that the bareboat charterer still has an implied duty to deliver a seaworthy vessel to the charterer. The extent to which an owner under a bareboat charter can be shielded from liability caused by unseaworthiness was discussed in the US case of Baker v Raymond International Inc,[98] where the court held that a bareboat charter would no longer shield owners from personal liability for third party injuries caused by the unseaworthiness of a vessel, although the owner had no control over the vessel, and regardless of whether it was the owner or charterer who created the unseaworthy condition.

It is the author's view that the owner of an autonomous vessel could be viewed as seaworthy under a bareboat charter, as there is no obligation for the owner to provide a vessel that is seaworthy in the context of a competent crew. If the owner of an autonomous vessel is desirous of conducting trade via a bareboat charter, the author recommends that this owner uses one of the internationally recognized standard

[94] Online document https://www.bimco.org/news/priority-news/20200923-new-gencon Accessed on August 7, 2022.

[95] McDowell and Gibbs (1954), p. 187.

[96] McDowell and Gibbs (1954), p. 187.

[97] McDowell and Gibbs (1954), p. 187.

[98] 656 F. 2d 173, 184 (5th Cir. 1981).

charter forms such as the BIMCO BARECON 2017. In this regard, it is important to examine some aspects of the BARECON 2017. Clause 3 (a) states that:

> The Owners shall deliver the Vessel in a seaworthy condition and in every respect ready for service under this Charter Party and in accordance with the particulars stated in Boxes 4 to 6.

Boxes 4 to 6 address the vessel class, last survey, and the validity of certifications for the vessel.

Clause 13 (d) of the BARECON 2017 also states that:

> The Charterers shall at their own expense crew, victual, navigate, operate, supply, fuel, maintain and repair the Vessel during the Charter Period and shall be responsible for all costs and expenses whatsoever relating to their use and operation of the Vessel, including any taxes and fees. The Crew shall be servants of the Charterers for all purposes whatsoever, even if for any reason appointed by the Owners.

In comparing clause 3 (a) and clause 13 (d), it is evident that the BARECON 2017 does not contemplate seaworthiness in the context of a master and crew and as such, the owner of an autonomous vessel could utilize the BARECON 2017 when conducting commerce. It is the authors' view that nevertheless clause 13 (d) of the BARECON 2017 could be amended to state that:

> The Charterers shall at their own expense crew **(if any),** victual, navigate, operate, supply, fuel, maintain and repair the Vessel during the Charter Period and shall be responsible for all costs and expenses whatsoever relating to their use and operation of the Vessel, including any taxes and fees. The Crew **(if any)** shall be servants of the Charterers for all purposes whatsoever, even if for any reason appointed by the Owners.

The author is of the view that these two small amendments would add clarity and prevent confusion when autonomous vessels start sailing.

The author is of the overall view that the best course of action is to create an international charter party specifically for autonomous vessels, as this will prevent the need to butcher essential and traditional clauses of standard charter parties. The author's view is in tandem with BIMCO's as they recently announced that they have started to develop a standard charter party for autonomous vessels.[99] BIMCO announced on November 6, 2020, that it is developing a specially adapted version of its widely used SHIPMAN 2009 agreement. It is to be called "Autoshipman."[100] BIMCO announced that the current development process has been hampered by the fact that there are no autonomous ships currently in operation that is conducting international trade consistently.[101] As seen above, most of these vessels are still in the development stages. As such, the success of "Autoshipman" will depend on its actual usage by autonomous vessels in the future. Nevertheless, the author is still of

[99] Online document https://www.bimco.org/news/contracts-and-clauses/20201106-first-ever-standard-contract-for-autonomous-ship-operation-underway Accessed on June 30, 2022.

[100] Online document https://www.bimco.org/news/contracts-and-clauses/20201106-first-ever-standard-contract-for-autonomous-ship-operation-underway Accessed on June 30, 2022.

[101] Online document https://www.bimco.org/news/contracts-and-clauses/20201106-first-ever-standard-contract-for-autonomous-ship-operation-underway Accessed on June 30, 2022.

the opinion that seaworthiness under "Autoshipman" cannot not be viewed in the same manner as it has been done under existing standard form charter parties.

3 Good Seamanship and Autonomous Vessels

Good seamanship is another critical concept of maritime law and as such the author found it necessary to explore this concept and to see if it would still be applicable to an autonomous vessel. Good seamanship refers to the skill and knowledge of a master and his crew in the work of navigating, maintaining, and operating a vessel.[102] The standard of good seamanship that is required internationally is that of the average good master and crew.[103] There is no requirement that a good seaman should have an extraordinary ability or unusually high degree of care and caution.[104] The concept of good seamanship underpins the need to prevent collisions at sea. According to Halsbury Laws of England, regulations outlining rules necessary for the prevention of collisions at sea has been developed as far back as the middle of the 19th century.[105] These collision regulations were and are still based on the important concept of good seamanship and ship management. The collision rules were first given statutory authority in the year 1846. If a vessel infringed these regulations in civil proceedings, between 1873 and 1911, there existed a statutory presumption of fault. The presumption could have been rebutted if it could be proven that the infringement did not contribute in any way to the collision.[106] The regulations have been amended several times since the 19th century. The Convention on the International Regulations for Preventing Collisions at Sea, 1972 (amended) (COLREGs) is the current version of the regulations in force. The present COLREGs set out the duties that are incumbent on the master and crew of a ship and breaches of the regulations are evidence of fault or negligence. Proof is still required that such breach caused or materially contributed to the collision and subsequent damage. The COLREGs must be adopted by each member country of the IMO that is signatory to the convention.[107] Each IMO member country must then establish an administration for implementation of the COLREGs so that it applies to ships flying its flag. The

[102] Online document (Suppiah) "Departure from COLREGS infringement or good seamanship Departure from COLREGS infringement or good seamanship" https://commons.wmu.se/cgi/viewcontent.cgi?article=1116&context=all_dissertations Accessed on June 10, 2022.

[103] Online document (Suppiah) "Departure from COLREGS infringement or good seamanship Departure from COLREGS infringement or good seamanship" https://commons.wmu.se/cgi/viewcontent.cgi?article=1116&context=all_dissertations Accessed on June 10, 2022.

[104] Online document (Suppiah) "Departure from COLREGS infringement or good seamanship Departure from COLREGS infringement or good seamanship" https://commons.wmu.se/cgi/viewcontent.cgi?article=1116&context=all_dissertations Accessed on June 10, 2022.

[105] Halsbury's Laws of England (2017), Shipping and Maritime Law, Volume 93.

[106] Halsbury's Laws of England (2017), Shipping and Maritime Law, Volume 93.

[107] See Articles I and II of the COLREGs.

national navigational rules must also conform to the COLREGs.[108] Additionally, it has been argued that the concept of good seamanship fills gaps in the COLREGs that the said regulations do not answer. Given its importance and impact, the author is of the view that it is pertinent to discuss whether the COLREGs in its current form would be relevant to autonomous vessels or whether they would be deemed as obsolete in the context of autonomous vessels.

To examine whether an autonomous vessel would be able to comply with the COLREGs, it is first important to examine how the COLREGs define vessel. Rule 3 (a) of the COLREGs states that the word "vessel" includes every description of watercraft, including non-displacement craft, WIG craft and seaplanes, used or capable of being used as a means of transportation on water. The author notes that the definition does not mention that the vessel must be manned or crewed and as such, it is possible to argue that an autonomous vessel would come within the ambits of the COLREGs. The author will now dissect some of the salient rules of the COLREGs and examine their applicability of autonomous vessels.

3.1 Responsibility

Rule 2 of the COLREGs is of paramount importance and is often misunderstood. It is unique from the other rules in the COLREGs because it does not specifically tell owners, masters and or crew what to do or when to do it.[109]

Rule 2 (a) states that:

> Nothing in these Rules shall exonerate any vessel or the owner, master or crew thereof, from the consequences of any neglect to comply with these Rules or of the neglect of any precaution, which may be required by the ordinary practice of seamen, or by the special circumstances of the case.

Rule 2(a) of the COLREGs specifically means that there is nothing in this rule that would allow an owner, master, or crew to escape liability by neglecting to follow the rules. Rule 2 requires that ship owners, masters and or crew should follow both the rules and ordinary practice of seamen. It is oftentimes viewed as confusing because it requires that one must carry out their duties according to the rules but also exercise precaution which required by ordinary seamen. Ordinary practice of seamen translates into what is commonly known as "good seamanship."[110] The concept of good seamanship in Rule 2 fills gaps in the COLREGs where there is not a clear answer on how a vessel should operate in a specific situation. It is the author's view

[108] See Articles I and II of the COLREGs.

[109] Online document https://naututor.com/rule-2-responsibility-explained/ Accessed on May 10, 2022.

[110] R (on the application of the Public and Commercial Services Union and another) v Secretary of State for the Home Department; R (on the application of Channel Rescue) v Secretary of State for the Home Department [2022] EWHC 517 (Admin).

that it is possible to program an autonomous vessel with some of the basic tenets of good seamanship.[111] It is the authors view that algorithms could be developed by using historical data and relying on the input of experienced captains to program the autonomous vessel. However, it is impossible to program every possible situation and as such, Rule 2 (a) would need reforming for a fully autonomous vessel to comply with it with precision. A remote operated unmanned vessel however could probably comply with Rule 2(a) if it yields to the discretion of shore-based operators to take actions necessary of an ordinary seaman in practice.

It is arguable however that Rule 2(b) is easier for an autonomous vessel to comply with.

Rule 2 (b) states that:

> In construing and complying with these Rules due regard shall be had to all dangers of navigation and collision and to any special circumstances including the limitations of the vessels involved which may make a departure from these Rules necessary to avoid immediate danger.

Rule 2(b) only allows departure from the rules if it is the only way to avoid immediate danger. Essentially, it means that if following the rules would make the situation worse, then it would be prudent for the person who is faced with the danger, to take an action that is different from the rules but would ensure the safety of the respective vessels.[112] This therefore requires technical analysis and the exercise of one's discretion in the dangerous situation. It is thus arguable that with an autonomous vessel an override function or an override button could be programmed in the vessel so that if it is that the programmed rules of the COLREGs cannot be followed, then the vessel could arguably override and operate in a manner that avoids the danger.

Thc author posits that it may be easier for a remote-controlled unmanned vessel to satisfy Rule 2 (b), since it would be monitored by shore-based operators who could accurately assess if the rules of the COLREGs should not be followed to avoid immediate danger.

3.2 Proper Look-Out

Rule 5 is one of the most relevant rules of the COLREGs. It states that:

> Every vessel shall at all times maintain a proper look-out by sight and hearing as well as by all available means appropriate in the prevailing circumstances and conditions so as to make a full appraisal of the situation and of the risk of collision.

[111] Online document https://knowledgeofsea.com/colreg-in-easy-language-rule-257-8/ Accessed on May 9, 2022.

[112] Online document https://naututor.com/rule-2-responsibility-explained/ Accessed on May 10, 2022.

A proper look-out is very important in the operation of traditional manned vessels and an improper look-out has been the cause of collisions between vessels. The case of "The Sea Star" is illustrative of this. In The Horta Barbosa v Sea Star ("The Sea Star")[113] a collision occurred in the Gulf of Oman between two vessels, Horta Barbosa and the Sea Star. It was reported that visibility was good and both vessels had their radar in use which disclosed the presence of the other at 14 to 16 miles. They could also see the masthead lights of each other at a substantial distance and would have had a warning of the other's course eight miles away either by eye or binoculars.

The second officer was on watch on Horta Barbosa and her automatic steering was on 324 degrees. Before the collision, the cadet on the lookout, who had also been on the bridge left, to call the relief. When Sea Star was about three to four miles away, the second officer thought she would pass safely at one mile distance. He, therefore, left the bridge for the charthouse to get a radar fix. The cadet on the lookout returned to the bridge and saw that Sea Star was on a crossing course. He then called the second officer, who altered the steering to manual and ordered the engines full astern. The collision took place at once. The Sea Star was destroyed by fire and eleven of the crewmembers lost their lives.

It was held by the learned Judge Brandon, J. that the Sea Star was seriously to blame for altering course to starboard at an improper time. This improper maneuver was caused by previous defective lookout or defective appreciation of the situation, or both. The court said it was negligent of her second officer to go to the chartroom and remain there for six to seven minutes before the collision. He should have remained in the wheelhouse or on the starboard wing of the bridge, watching the approach of Sea Star. Apportionment of blame was Sea Star 75% and Horta Barbosa 25%.

Similarly, the case of The Golden Mistral[114] is instructive on this point. In this case, the overriding cause of the collision was a bad lookout on both vessels and neither vessel saw the other until one minute or less before the collision. In that case, Sheen, J. had the following to say:

> If a good look-out had been maintained aboard Andhika Patra it would have become apparent that Golden Mistral was not taking appropriate action in compliance with the regulations long before the moment when Andhika Patra took action. In such circumstances it would have been correct for Andhika Patra to take action immediately. Andhika Patra was proceeding down the wrong side of the fairway towards a bend, in the vicinity of which the arc of vision for an inward-bound ship on its correct side of the fairway was restricted to some extent; that was unseamanlike.[115]

It is thus evident that the court views the concept of keeping a proper lookout as a function to be performed predominantly by humans. However, upon careful analysis of Rule 5 of the COLREGs, the author observed that the clause does not specify that

[113] (1976) 1 Lloyd's Rep 115. Appeal by Sea Star was dismissed, see (1976) 2 Lloyd's Rep. 477.
[114] [1986] 1 Lloyd's Rep. 407.
[115] [1986] 1 Lloyd's Rep. 407.

keeping a proper lookout by sight and hearing is dependent on a human. It is the authors' opinion that the COLREGs relate to the vessel as a fictional person carrying out these functions. As such, the author is of the view that a high-definition camera could function as the "eye" and a well-tuned microphone and/or speaker could function as the "ear" on an autonomous vessel.

3.3 *Safe Speed*

It is also pertinent to examine Rule 6 of the COLREGs when discussing the concept of good seamanship. Rule 6 states that:

> Every vessel shall at all times proceed at a safe speed so that she can take proper and effective action to avoid collision and be stopped within a distance appropriate to the prevailing circumstances and conditions.

In the case of "Roseline,"[116] the judge stated that in a well-run ship, the ship will be navigated at a safe speed. Similarly, in the recently decided case of Evergreen Marine (UK) Limited v Nautical Challenge Ltd,[117] the lower court judge found that one of the vessels involved in the collision "EVER SMART" contributed far more to the damage resulting from the collision than the very much lower (and safe) speed of the other vessel "ALEXANDRA 1." The judge also found that "EVER SMART" proceeded at an excessive speed which was a direct consequence of her failure to keep a good lookout.

It is important to note however that a safe speed does not necessarily mean a low speed. In fact, it has been said that if the speed of a vessel is very low and the water current is strong, she may drift on to any other vessel in the water way and cause a collision.[118] As such, *"safe speed"* means that the vessel sails at a speed so that if any emergency occurs, she is able to navigate her way out of this situation, without anybody getting injured.[119]

It is the authors' view that an autonomous vessel could be pre-programmed to consistently maintain a safe speed throughout its voyage, having regard to the potential presence of other vessels, current and environmental factors.

[116] (1981) 2 Lloyd's Report. 410.

[117] [2021] UKSC 6.

[118] Online document (Suppiah) "Departure from COLREGS infringement or good seamanship Departure from COLREGS infringement or good seamanship" https://commons.wmu.se/cgi/viewcontent.cgi?article=1116&context=all_dissertations Accessed on June 10, 2022.

[119] Online document https://cultofsea.com/colregs/part-b-steering-and-sailing-rules-4-19/rule-6-safe-speed/ Accessed on June 10, 2022.

3.4 Risk of Collision

Rule 7 of the COLREGs speaks of the risk of collision and states that:

(a). Every vessel shall use all available means appropriate to the prevailing circumstances and conditions to determine if risk of collision exists. If there is any doubt such risk shall be deemed to exist.
(b). Proper use shall be made of radar equipment if fitted and operational, including long-range scanning to obtain early warning of risk of collision and radar plotting or equivalent systematic observation of detected objects.
(c). Assumptions shall not be made on the basis of scanty information, especially scanty radar information.
(d). In determining if risk of collision exists the following considerations shall be among those taken into account:

(i). such risk shall be deemed to exist if the compass bearing of an approaching vessel does not appreciably change;
(ii). such risk may sometimes exist even when an appreciable bearing change is evident, particularly when approaching a very large vessel or a tow or when approaching a vessel at close range.

The author is of the view that an autonomous vessel should be able to comply with Rule 7 of the COLREGs, as the manufacturers will simply have to ensure that the vessel has radar equipment as required by Rule 7 (b). The manufacturers will also have to ensure that they have high tech equipment installed and programmed using Artificial Intelligence that can make assumptions where radar information is scanty.

3.5 Action to Avoid Collision

Rule 8 of the COLREGs addresses actions to avoid a collision and states that:

(a). Any action to avoid collision shall be taken in accordance with the Rules of this Part and shall, if the circumstances of the case admit, be positive, made in ample time and with due regard to the observance of good seamanship.
(b). Any alteration of course and/or speed to avoid collision shall, if the circumstances of the case admit, be large enough to be readily apparent to another vessel observing visually or by radar; a succession of small alterations of course and/or speed should be avoided.

In the case of the Evergreen Marine (UK) Limited v Nautical Challenge Ltd,[120] the lower court judge had to consider one of the allegations of fault made against ALEXANDRA 1, which was that she had approached too close to the end of the channel. The lower court judge asked the Elder Brethren the following question (paras 93–94):

[120] [2021] UKSC 6.

> Did good seamanship require ALEXANDRA I to keep a certain minimum distance from buoys No 1 so long as EVER SMART was still in the dredged channel and if so what was that distance?

Their advice was:

> Subject to a good aural and visual lookout, it would be reasonable and good seamanship for the Master of Alexander I to have approached the first pair of buoys keeping close to her own side of the entrance channel.

It is thus evident that applying good seamanship is important to avoid a collision at sea. The author is of the view that if the tenets of good seamanship that were discussed above can be pre-programmed into codes for an autonomous vessel to manipulate, then an autonomous vessel could comply with Rule 8 of the COLREGs.

3.6 Narrow Channels

Rule 9 of the COLREGs which addresses how vessels should operate in narrow channels have been a source of litigation when collisions have occurred at sea. Rule 9 states that:

> (a). A vessel proceeding along the course of a narrow channel or fairway shall keep as near to the outer limit of the channel or fairway which lies on her starboard side as is safe and practicable.
> (b). A vessel of less than 20 metres in length or a sailing vessel shall not impede the passage of a vessel which can safely navigate only within a narrow channel or fairway.
> (c). A vessel engaged in fishing shall not impede the passage of any other vessel navigating within a narrow channel or fairway.
> (d). A vessel shall not cross a narrow channel or fairway if such crossing impedes the passage of a vessel which can safely navigate only within such channel or fairway. The latter vessel may use the sound signal prescribed in Rule 34(d) if in doubt as to the intention of the crossing vessel.
> (e). (i). In a narrow channel or fairway when overtaking can take place only if the vessel to be overtaken has to take action to permit safe passing, the vessel intending to overtake shall indicate her intention by sounding the appropriate signal prescribed in Rule 34(c) (i). The vessel to be overtaken shall, if in agreement, sound the appropriate signal prescribed in Rule 34(c)(ii) and take steps to permit safe passing. If in doubt she may sound the signals prescribed in Rule 34(d)
> (ii). This Rule does not relieve the overtaking vessel of her obligation under Rule 13
> (f). A vessel nearing a bend or an area of a narrow channel or fairway where other vessels may be obscured by an intervening obstruction shall navigate with particular alertness and caution and shall sound the appropriate signal prescribed in Rule 34(e).
> (g). Any vessel shall, if the circumstances of the case admit, avoid anchoring in a narrow channel.

Defining what amounts to a narrow channel has been difficult, as the definition section of the COLREGs, that being Rule 3, does not provide any guidance. The

Bosporus,[121] the Rive Maas,[122] the River Parana,[123] the Eastham Channel in the Port of Liverpool,[124] the Mae Nam Chao Phraya,[125] The Danube,[126] and most recently, the pilot boarding area outside the port of Jebel Ali[127] have all been considered as narrow channels by case law. It appears that there is no specific length or breadth for a narrow channel. As such, in determining whether a particular area of water is considered a narrow channel, each case is decided on its own facts. The case of The Sedgepool provides some guidance, where Wilmer J stated:

> As I understand the law, one of the determining factors in deciding whether a given area is or is not within the Narrow Channel Rule is the way in which a seaman in fact regards it and behaves in it.

It has also been said that an area should only be considered as narrow if the state, port, or harbor authority declares it as narrow.[128] This declaration is usually published in navigational rules and documents so that seafarers are aware of it and as such, they would know how to act within such narrow channel.[129]

It is the authors' view that an amendment could be made to the COLREGs to clearly define what are the dimensions that would constitute a narrow channel. If this amendment was made, then it is the authors' view that these dimensions could be engineered into the software of an autonomous vessel, so that it would be able to detect that this body of water constitutes a narrow channel. Once this programming can be done, it is the authors' view that an autonomous vessel will be able to competently navigate a narrow channel.

3.7 Overtaking

It is important to also examine Rule 13 of the COLREGs which states that:

> (a). Notwithstanding anything contained in the Rules of part B, sections I and II, any vessel overtaking any other shall keep out of the way of the vessel being overtaken.

---

[121] The Elazig (1972) 1 Lloyd's Rep. 355.

[122] The Adolf Leonhardt (1973) 2 Lloyd's Rep. 318; and The Oldekerk (1974) 1 Lloyd's Rep. 95.

[123] The Martin Fierro (1974) 2 Lloyd's Rep. 203.

[124] The City of Leeds (1978) 2 Lloyd's Rep. 346.

[125] The Toluca (1981) 2 Lloyd's Rep. 548.

[126] The Satyam Padam (1985) 1 Lloyd's Rep. 338.

[127] Evergreen Marine (UK) Limited v Nautical Challenge Ltd, [2021] UKSC 6.

[128] Online document (Suppiah) "Departure from COLREGS infringement or good seamanship Departure from COLREGS infringement or good seamanship" https://commons.wmu.se/cgi/viewcontent.cgi?article=1116&context=all_dissertations Accessed on June 10, 2022.

[129] Online document (Suppiah) "Departure from COLREGS infringement or good seamanship Departure from COLREGS infringement or good seamanship" https://commons.wmu.se/cgi/viewcontent.cgi?article=1116&context=all_dissertations Accessed on June 10, 2022.

(b). A vessel shall be deemed to be overtaking when coming up with another vessel from a direction more than 22.5 degrees abaft her beam, that is, in such a position with reference to the vessel she is overtaking, that at night she would be able to see only the sternlight of that vessel but neither of her sidelights.
(c). When a vessel is in any doubt as to whether she is overtaking another, she shall assume that this is the case and act accordingly.
(d). Any subsequent alteration of the bearing between the two vessels shall not make the overtaking vessel a crossing vessel within the meaning of these Rules or relieve her of the duty of keeping clear of the overtaken vessel until she is finally past and clear.

Breach of Rule 13 of the COLREGs has resulted in collisions between vessels in the past. The rule continues to apply until all the danger of a collision is over, that is, when the overtaking vessel is past and clear.[130] The duty of the overtaking vessel is to keep clearly out of the way of the vessel being overtaken. If overtaking is occurring in a narrow channel, the overtaken vessel must take action to permit safe passing by giving the overtaking vessel special signals. A proposal to pass to starboard is signaled by two prolonged and short blasts, and two prolonged and two short blasts for passage to port. The overtaking vessel should not cross ahead of the other vessel but alter course or reduce speed to pass astern of the vessel that is being overtaken. The overtaking vessel is required to act at an early stage to avoid any confusion with the vessel that is being overtaken.[131]

The case of The Nowy Sacz[132] is illustrative of how a collision can occur because of improper overtaking. In this case, a collision occurred between two vessels called the Nowy Sacz and the Olympian. Both vessels were proceeding on about parallel courses in a northerly direction. The night was clear and the visibility good. At about 0245 h, when Olympian was bearing 25 degrees to 30 degrees abaft the starboard beam of Nowy Sacz, the second officer of Nowy Sacz was seeing the masthead light of Olympian but not her red light some three miles away. At the same time the second officer of Olympian had not yet observed any lights of Nowy Sacz. By 0300 h, when the bearing of Olympian from Nowy Sacz ceased to be more than two points abaft the beam, the second officer was seeing the red light of Olympian as well as her masthead lights. The second officer of Olympian also saw the masthead lights and the green lights of Nowy Sacz. Owing to the relative courses and speed of the two ships-the closing speed being between two and three knots—the time at which risk of collision arose was about 0330 h, when Nowy Sacz was on Olympian's port beam and appeared to be closing on a crossing course from port to starboard at an angle of 25 degrees to 30 degrees. On hearing a signal of one short blast from the Olympian which was the about one.

[130] Online document (Clawson Jr) "Overtaking or crossing? Don't assume what other ship will do" https://www.professionalmariner.com/overtaking-or-crossing-dont-assume-what-other-ship-will-do/ Accessed on June 12, 2022.

[131] Online document (Suppiah) "Departure from COLREGS infringement or good seamanship Departure from COLREGS infringement or good seamanship" https://commons.wmu.se/cgi/viewcontent.cgi?article=1116&context=all_dissertations Accessed on June 10, 2022.

[132] (1976) 2 Lloyd's Rep. 682.

Two cables away, the second officer of Nowy Sacz put her engines first half, then full astern and sounded three short blasts. Olympian's master put her engines to stand-by, ordered one short blast and put her wheel hard to starboard. Shortly afterwards at 0357 h a collision occurred between the stem of Nowy Sacz and the port quarter of Olympian at an angle of about 10 degrees. Brandon, J. rejected the contention that the situation was an overtaking one and held that it was a crossing situation. Nowy Sacz should have kept out of the way of Olympian. The latter as the stand–on vessel should have kept her course and speed. Accordingly, Nowy Sacz was three quarters to blame and Olympian one-quarter. On appeal, the court said:

> The overtaking rule applies before there is a risk of collision. However, this does not mean that it necessarily comes into effect as soon as the vessels are in sight of one another. The overtaking rules begin to operate as soon as it could properly be said that the overtaking ship was coming up with the overtaken ship. When exactly that will be may not always be easy to determine but we see no reason to suppose that it will be any more difficult that the decision as to when the situation involves a risk of collision.
>
> Although the court of appeal approved Brandon, J. that the risk of collision arose in this case at about 0337 hrs, it considered the situation as overtaking. Nowy Sacz was the stand-on ship and Olympian was the give way ship. It apportioned the blame; Nowy Sacz - one quarter and Olympian - three quarters.

It is the authors' view that an autonomous vessel, once it is fitted with high-definition sensors, should be able to detect the presence of another vessel and accurately determine the amount of distance that is required for an autonomous vessel to successfully overtake another vessel. From the case of the Nowy Sacz, it is evident that poor judgment/human error in the overtaking process was cause of the collision. The author is also of the view however that an argument could be made that only remote-controlled autonomous vessels should be able or should be allowed to overtake, since there would be a human remotely monitoring the situation. However, the author believes that both remote-controlled unmanned vessels and fully autonomous vessels should be allowed to overtake once the fully autonomous vessel is properly programmed to do so.

3.8 Head-On Situations

Rule 14 of the COLREGs illustrate how vessels should operate in head-on situations. Rule 14 states that:

> (a). When two power-driven vessels are meeting on reciprocal or nearly reciprocal courses so as to involve risk of collision each shall alter her course to starboard so that each shall pass on the port side of the other.
> (b). Such a situation shall be deemed to exist when a vessel sees the other ahead or nearly ahead and by night she could see the masthead lights of the other in a line or nearly in a line and/or both sidelights and by day she observes the corresponding aspect of the other vessel.
> (c). When a vessel is in any doubt as to whether such a situation exists she shall assume that it does exist and act accordingly.

A power-driven vessel is any vessel propelled by a machinery, which is different from a sailing vessel which is any vessel under sail, provided that propelling machinery, if fitted, is not being used.[133]

The author is of the view that an autonomous vessel would be considered a power-driven vessel and as such, Rule 14 of the COLREGs could arguably apply to autonomous vessels. Similarly, to the other rules of the COLREGs, breach of this rule has also caused collisions. This can be seen in the American case of The Amoco Cremona[134] where two vessels started on reciprocal courses off the fairway buoy at the entrance to the Galveston Channel and one of the vessels altered its course to port instead of starboard, in breach of the COLREGs which caused a collision between the two vessels.[135]

3.9 Crossing Situations

Rule 15 of the COLREGs addresses crossing situations. These rules apply where the overtaking and head-on rules (Rule 13 and 14) of the COLREGs do not apply.

Rule 15 states that:

> When two power-driven vessels are crossing so as to involve risk of collision, the vessel which has the other on her own starboard side shall keep out of the way and shall, if the circumstances of the case admit, avoid crossing ahead of the other vessel.

A crossing situation occurs where two vessels are likely, if each keeps their expected course, to arrive at the same point at or nearly the same time. This involves a risk of collision and thus Rule 15 illustrates how vessels should operate to eliminate this risk.[136]

Similarly to Rule 14, the author is of the opinion that an autonomous vessel would be considered a power-driven vessel and as such, it is arguable that Rule 15 of the COLREGs can also apply to an autonomous vessel.

3.10 Conduct of Vessels in Restricted Visibility

Rule 19 of the COLREGs delineates how a vessel should navigate in an area that has restricted visibility. Rule 19 states that:

---

[133] See Rule 3 of the COLREGs.

[134] (1983) A.M.C 1087.

[135] See also the case of The Ballylesson (1968) 1 Lloyd's Rep. 69.

[136] Online document (Suppiah) "Departure from COLREGS infringement or good seamanship Departure from COLREGS infringement or good seamanship" https://commons.wmu.se/cgi/viewcontent.cgi?article=1116&context=all_dissertations Accessed on June 10, 2022.

(a). This Rule applies to vessels not in sight of one another when navigating in or near an area of restricted visibility.
(b). Every vessel shall proceed at a safe speed adapted to the prevailing circumstances and conditions of restricted visibility. A power-driven vessel shall have her engines ready for immediate manoeuvre.
(c). Every vessel shall have due regard to the prevailing circumstances and conditions of restricted visibility when complying with the Rules of section I of this part.
(d). A vessel which detects by radar alone the presence of another vessel shall determine if a close-quarters situation is developing and/or risk of collision exists. If so, she shall take avoiding action in ample time, provided that when such action consists of an alteration of course, so far as possible the following shall be avoided:

  (i). an alteration of course to port for a vessel forward of the beam, other than for a vessel being overtaken;
  (ii). an alteration of course towards a vessel abeam or abaft the beam.

(e). Except where it has been determined that a risk of collision does not exist, every vessel which hears apparently forward of her beam the fog signal of another vessel, or which cannot avoid a close-quarters situation with another vessel forward of her beam, shall reduce her speed to the minimum at which she can be kept on her course. She shall if necessary take all her way off and in any event navigate with extreme caution until danger of collision is over.

The author is of the view that an autonomous vessel can be fitted with the appropriate algorithms to detect when visibility is poor and to take the necessary steps in compliance with Rule 19 of the COLREGs to avoid a collision.

4 Discussion

Autonomous vessels are currently becoming the new phenomena in the marine industry. However, it is the authors’ own view that it will not be an instantaneous expectation for that all traditional vessels will entirely automate especially if these traditional vessels are able to meet the demands of their intended voyage as it relates to seaworthiness. It is the authors’ view however that if the maritime industry does evolve to the point where majority of the vessels in operation are remotely controlled and/or fully autonomous, then a traditional manned vessel could arguably be deemed unseaworthy in the future. The author is further of the view that “The Portland Trader” case would be decided very differently in this modern era. This is so because since the year 2002, due to SOLAS regulations, vessels constructed after July 1, 2002, have a mandatory requirement that they should be fitted with a radar and other navigational aids. Additionally, it is very likely that an autonomous vessel will be able to satisfy the requirements of being cargo worthy. With the advancement of the technology of the vessel, it is expected that the holds of an autonomous vessel will be more robust than a traditional vessel and some systems such as the pipes will have sensors and/or will be fully automated. It is thus submitted that an autonomous vessel will satisfy the requirements of seaworthiness as it relates to cargo worthiness.

Thus, the author submits that eventually, a fully automated vessel may be the only type of vessel that can satisfy the requirement of seaworthiness in the future.

It is arguably also very possible for an autonomous vessel to satisfy the requirements of good seamanship. It is the authors' view that an autonomous vessel could be pre-programmed with specific commands that reflect the different clauses of the COLREGs. For example, it is very possible to preprogramme an autonomous vessel to consistently maintain a safe speed throughout its voyage to comply with Rule 6 of the COLREGs. Several modern manned vessels are already equipped with Electronic Nautical Chart Systems[137] linked to speed and depth sensors, along with Global Positioning System (GPS) and Automatic Identification System (AIS). If similar systems were implemented on an autonomous vessel, it is arguable that "safe speed" under Rule 6 of the COLREGs could easily be complied with by an autonomous vessel. The Evergreen Marine (UK) Limited case mentioned above illustrates once more how human error contributes to these collisions and makes the case stronger that autonomous vessels may be the solution to human error. Similarly, as it relates to Rule 5 of the COLREGs which requires vessel to keep a proper lookout, the author is of the view that a high-definition camera could function as a fictional "eye" and a highly tuned microphone/speaker could function as the fictional "ear" on an autonomous vessel. If this interpretation is taken, the author is of the view that it would be very possible for an autonomous vessel to comply with Rule 5 of the COLREGs. Further, the author is of the view that this wide interpretation should be supported, as it would replace the need for a human lookout which has been a cause of many collisions.

5 Conclusion

The author is of the overall view that the traditional concepts of seaworthiness and good seamanship are still applicable to autonomous vessels. As outlined throughout this chapter, a liberal approach can be adopted to the meaning of seaworthiness under the common law, international conventions and under the common standard charterparties. The author is of the view that seaworthiness should no longer be viewed from the perspective of a competent master and crew and is of the view that an autonomous vessel can be considered as seaworthy. The author however is of the view that the best course of action would be to develop and adopt international regulations which specifically address autonomous vessels given its novelty. As it relates specifically to standard charter parties and the applicability of the seaworthiness clause therein to autonomous vessels, the author posits that the best course of action would be to develop a standard charter party such as the "*Autoshipman*" which is currently being developed by BIMCO, for exclusive use by autonomous vessel.

[137] Tikanmäki et al. (2021), p. 130.

Further, the author is of the opinion that upon careful consideration of the concept of good seamanship and the provisions of the COLREGs that underpin the concept of good seamanship, that an autonomous vessel would be able to comply with the provisions. This is particularly true if the courts adopt a liberal approach to the interpretation of the COLREGs. The author finds support in this position in Rule 2 (b) of the COLREGs, which signals a warning to navigators to not be too rigid when interpreting the COLREGs and illustrates that navigators overarching duty is to uphold the tenets of good seamanship. Judges, although they do not legislate, have always had the discretion to interpret statute and conventions and manipulate these said statutes to the specific facts before them. In the case of in Black-Clawson International v Papierwerke Waldhof Aschaffenburg,[138] Lord Diplock said:

> It is for the Court and no one else to decide what words in a statute mean.

If a liberal approach is taken to the interpretation of the COLREGs, then the author is of the view that any court could view that it is applicable to an autonomous vessel and that the autonomous vessel can comply with the tenets of good seamanship and with the provisions of the COLREGs.

The author however believes the best approach would be to create an amendment to the COLREGs that would consider specific situations that would affect an autonomous vessel. Alternatively, regulations designed specifically for good seamanship and prevention of collisions for autonomous vessels could be drafted if a liberal approach to the interpretation or an amendment of the COLREGs are not viewed as favorable by regulators.

References

Books

Baatz Y (2014) Maritime law, 3rd edn. Informa Law by Routledge
Branch A, Robarts M (2014) Branch's elements of shipping, 9th edn. Informa Law by Routledge
Djadjev, I (2017) The Obligations of the Carrier Regarding the Cargo: the Hague-Visby Rules (1st edn), Springer
Force, R (2004) Admiralty and Maritime Law, (1st edn) Federal Judicial Center
Güner-Özbek M (2008) The carriage of dangerous goods by sea, 1st edn. Springer
Liu C (2009) Maritime transport services in the law of the sea and the World Trade Organization, vol 14. International Academic Publishers
Lüddeke C, Johnson A (1995) The Hamburg rules: from Hague to Hamburg via Visby, 2nd edn. Lloyd's of London Press
McDowell C, Gibbs H (1954) Ocean transportation, 1st edn. Mcgraw-Hill Book Company
Mitchelhill A (1990) Bills of lading: law and practice, 2nd edn. Springer
Rose F (2013) Marine insurance: law and practice, 2nd edn. Routledge

[138] [1975] AC 591.

Shackelford S (2020) Cyber war and peace: toward cyber peace, 1st edn. Cambridge University Press
Soyer B (2006) Warranties in marine insurance, 2nd edn. Cavendish
Tetley W (1978) Marine cargo claims, vol 1. Butterworths
Todd P (2016) Principles of the carriage of goods by sea, 1st edn. Routledge

Book Chapters

Tikanmäki I, Räsänen J, Ruoslahti H, Rajamäki J (2021) Maritime surveillance and information sharing systems for better situational awareness on the European maritime domain: a literature review. In: Digital transformation, cyber security and resilience of modern societies, 1st edn. Springer, pp 117–136

Journal Articles

Felski A, Zwolak K (2020) The ocean-going autonomous ship—challenges and threats Andrzej Felski and Karolina Zwolak. J Mar Sci Eng 8(1):41

Reforming the Law of the Sea for the Future of Automated Shipping

Callum Laffey

Abstract Autonomous ships, controlled by a shore-based operator (SBO) or by an artificial intelligence, are in development and it is anticipated will soon be deployed into the world's seas and oceans. But on deployment, first, Maritime Autonomous Vehicles (MAVs) must be able to access the sea and ocean areas, according to the law of the sea. Access is predicated upon compatibility with maritime safety conventions. The current rules are geared towards ships with personnel onboard, not for autonomous ships. MAVs' integration into merchant shipping will also require international private law and commercial law to be compatible with autonomous ships. This chapter will indicate the most significant barriers to autonomous ship integration, in the aforementioned areas of law, and highlight priorities in the current regulatory framework that ought to be continued in the regime governing autonomous ships, from the lens of commercial expediency, safety and the prevention of pollution. This chapter will recommend solutions for the International Maritime Organization which are designed to facilitate the adoption of autonomous ships in a merchant setting.

Keywords Autonomous ships · Law of the Sea · International Maritime Organization · English commercial law · Implementation agreement

1 Introduction

Autonomous ships are being developed in several countries and are being considered for broad deployment for merchant operations in the coming decades. For this chapter, the craft in question will be called Maritime Autonomous Vehicles (MAVs)[1] which covers vehicles controlled by a shore-based operator (SBO) or an

[1] The phrase is coined by Klein et al. (2020).

C. Laffey (✉)
University of Exeter, Law School, Exeter, UK
e-mail: cl719@exeter.ac.uk

K. Noussia, M. Channon (eds.), *The Regulation of Automated and Autonomous Transport*, https://doi.org/10.1007/978-3-031-32356-0_8

artificial intelligence (AI).[2] Part of the motivation for developing MAVs is the prospect of vehicles which are safer than conventional ships, by avoiding human fatigue which causes the majority of maritime accidents.[3] But there are difficulties with unmanned vehicles; the lack a crew means fires and routine repairs cannot be tackled while underway, potentially leading to an increased incidence of accidents.[4] If MAVs are to be used in the international carriage of goods by sea, such vehicles would require unrestricted access to the seas and oceans. This chapter will determine if MAVs fit into the law of the sea, which is necessary for MAVs' deployment in international shipping; it will highlight priorities in the law of the sea that must be maintained in relation to MAVs (Sect. 2). Ultimately, access will hinge upon MAVs' compatibility with international safety standards (Sect. 3); as access to the law of the sea is dependent on MAVs' compatibility with international safety standards, the solution to safety standards and the law of the sea barriers will be brought together (Sect. 4). This chapter will also determine if the 'master' role will change with the introduction MAVs and the SBO (Sect. 6). It will identify any barriers to MAV in international private law (Sect. 7) and English commercial law (Sect. 5); it will identify priorities for policymakers and developers in relation to MAVs' long-term deployment. The operability of the wreck and salvage regimes assists MAV development by ensuring journeys may be prosecuted in the same was as conventional ships; first, by ensuring MAVs are used safely, second, there is the opportunity to be salvaged before a marine casualty occurs, which threatens the safety of other ships and the marine environment. MAVs are an exciting technological innovation in the shipping industry, an industry which supports supply chains that benefits all our lives intensely, but MAVs should be employed only if safe for other sea users, the coastal state and the marine environment. It is critically important that the safety credentials of MAVs are prioritised in any research proposing to ease the integration of MAVs, including this chapter. To understand the safety credentials, it is hoped the topic will be taken up by the scientific and engineering literature and a definitive answer as to the viability of MAVs.

2 Law of the Sea

The primary question to be answered in this chapter is whether MAVs will fit into the current framework governing the law of the sea. If MAVs do not, MAVs will not be able to deploy in the various bodies of water necessary for international shipping. The other issues of MAVs' compatibility with maritime safety conventions and commercial law would be entirely academic.

---

[2]The SBO version is the more likely candidate for merchant vessels, at least for the foreseeable future.

[3]Katsivela (2020). Veal and Ringbom (2017).

[4]Wróbel et al. (2017).

This section will consider MAVs' access to the specific zones in the law of the sea, as found in the United Nations Convention on the Law of the Sea (UNCLOS), which serves as 'a constitution of the oceans'.[5] The law of the sea governs the high seas, the exclusive economic zone and the territorial sea of coastal states, and coastal states' internal waters. If MAVs are to be viable for merchant work—and they must be proven to be viable to be worthy of the billions of dollars of investment for development—MAVs would require guaranteed and unfettered access to the aforementioned areas. Therefore, for MAVs' integration, it is crucial MAVs have access to these zones in the same way conventional ships do. This section also considers the internal waters and ports of the coastal state; access to ports is necessary for the commercial viability of merchant vessels, especially ones in bulk carrier work.[6] This section will develop the novel view that MAVs' access to key areas could not be guaranteed under the law of the sea.

2.1 Internal Waters

The internal waters, including the ports of the coastal state, are defined by being landward of the baseline and naturally are part of the coastal state's territory. The coastal state has almost complete control. Consequently, it would appear to follow that if a coastal state has a desire to exclude MAVs from its internal waters, the state would have the lawful right to exclude MAVs from its territory, thus frustrating the integration of MAVs into the merchant shipping industry. Although this position seems counter-intuitive given the broad access to internal waters and ports merchant vessels experience on a daily basis for commercial trade. So, does the coastal state have a qualified or absolute right to exclude ships and vehicles from its internal waters?

The qualified argument follows from the *Aramco Arbitration* in 1958,[7] which ruled that there is right of access to ports 'as a great principle of public international law'.[8] The arbitrator speaking, in obiter, said that according to international law ports can only be closed when the 'then the vital interests of the state so require' and was premised, inter alia, on Article 16 of the Statute on the International Regime of Maritime Ports,[9] which *only* permits exclusion in deviation from other articles in the Statute in times of emergency or where it is in the vital interests of the state to exclude.[10] The validity of this ruling ought to be questioned. Particularly, the Statute

[5]United Nations Convention on the Law of the Sea (adopted 10 December 1982, entered into force 16 November 1994) 1833 UNTS 3 (known as "UNCLOS"). See Boyle (2005).

[6]Karlis (2018).

[7]Aramco Arbitration (1963), p. 27 ILR 117. Cited in Lowe (1977), p. 598.

[8]Summarised by De La Fayette (1996) and Lowe (1977), p. 598.

[9]Convention and Statute on the International Regime of Maritime Ports (adopted 2 December 1926, entered into force 26 July 1926) 58 LNTS 285 (known as "Maritime Ports Convention").

[10]The power to deviate from Articles 2 to 7 of the Statute.

only applies to the signatories and is premised on reciprocity.[11] It cannot be read as either reflecting customary international law at the time of drafting nor did state practice emerge from the Statute as the majority of states have not accepted it.[12] To become customary international law, there must be extensive and virtually uniform acceptance amongst states; the Statute has only been accepted by a few dozen states.[13]

There is ample authority to conclude the coastal state has *near*-absolute jurisdiction to exclude vessels in the domestic law of various countries. For example, England—and now the United Kingdom—has since at least 1276 held that it is part of a state's sovereignty to deny access to its waters[14] and has utilised this idea in recent years, for example, when it denied access to a Soviet whaling vessel to Hong Kong in 1972, then a United Kingdom colony.[15] Several other states have adopted the same position in their practices.[16] As de la Fayette correctly observed, it would be pointless to join the Statute if guaranteed access were customary international law already.[17] The *Aramco Arbitration* dictum must be rejected.

There is authority in international law against qualified jurisdiction. The International Court of Justice (ICJ) in *Military and Paramilitary Activities in Nicaragua*[18] recognised the state's ability to regulate its ports is part of its sovereignty, saying 'it is also by virtue of its sovereignty that the coastal state may regulate access to its ports'.[19] This broad discretion is reflected in the language of UNCLOS, which provides in Article 25 that the coastal state is permitted set conditions for entry and to 'prevent any breach of conditions to which the admission of those ships . . . is subject', which necessarily implies a right to refuse access based on non-compliance with those standards.[20] This competence is significant for MAVs' integration or lack thereof; any coastal state would be permitted to set a condition against unmanned ships, thus ensuring MAVs could not be used in such a state's internal waters or ports. The only qualifications are the state must receive ships in distress or *force majeure*[21] and that denials are based on the pervasive principle in international law of reasonableness, meaning the denial must non-discriminatory, proportional and not

[11] Statute, Art. 2.

[12] Lowe (1977); De La Fayette (1996).

[13] De La Fayette (1996) counts 40. *Columbian–Peruvian asylum case (Columbia v Peru) (*Judgment) [1950] ICJ 266, 276.

[14] De La Fayette (1996), p. 8.

[15] Ibid.

[16] Ibid, p. 9. Japan, Belgium. Norway, Chile. Canada.

[17] Ibid, p. 17.

[18] *Case Concerning Military Aid and Paramilitary Activities in and against Nicaragua (Nicaragua v United States)* (Merits) [1986] ICJ 14.

[19] Ibid, paras 212–213. De La Fayette (1996), p. 2.

[20] UNCLOS, Art. 25. Ibid, p. 3.

[21] *Case Concerning Military Aid and Paramilitary Activities in and against Nicaragua (Nicaragua v United States)* (Merits) [1986] ICJ 14. *Kate A Hoff (United States) v Mexico* (1929) 4 RIAA 44. Erik Molenaar (2015), p. 228.

an abuse of rights.[22] If states oppose MAV integration, there is no indication that their denial to permit such vessels into their ports would be unreasonable, especially if inspired by preventing pollution or marine casualties. Lastly, the weight of academic opinion is against the claim of guaranteed access.[23] It is submitted that there is no right of access to ports *generally* in international law. There may be specific treaties or agreement a state is a party to that requires access.

For example, it has been argued that there is a right of access to ports under the WTO by virtue of GATT Article 5[24] which a freedom of transit through the territory of each state via the most convenient route for transit.[25] Similarly bilateral agreements, such as Friendship, Commerce and Navigation agreements[26] which require access to each states' ports. The latter only applies reciprocally to the parties to the agreement, rather than generally to the world at large, so such agreements are of limited scope. On GATT Article 5, it is beyond the scope of this chapter to tackle this issue in depth, beyond to say there are yet no cases which have concluded that the article requires access to ports.[27]

After concluding that coastal states may deny access to its internal waters and ports, it is necessary to consider practical constraints on this power. Veal and Ringbom, for example, argue that the absolute jurisdiction exists for conventional ships as it does for MAVs.[28] This might suggest the power will be sparingly used in relation to MAVs. However, it is submitted that MAVs will face more opposition, as (a) such vessels either being unsafe or being perceived as such; and (b) there is political opposition to MAVs by states which lose out to de-crewing. Further, this competence has been used in the past to prevent ships with an increased risk from entering ports, namely ships carrying nuclear cargo.[29] Therefore, there is potentially motive and historical precedent for excluding MAVs from the coastal state's ports.

However, there are some practical controls on exclusionary coastal states. Namely, the possibility that if a state closes its ports to MAVs, the flags of MAVs may exclude the coastal state's flagged ships from its port as a form of mutually assured detriment.[30] It follows that if states act to exclude MAVs from their ports, the flag of the MAV could act to exclude the coastal state's flagged ships from their ports. This would deter states who intend to exclude MAVs from excluding such vessels from their ports, but such a process is deeply unattractive for international trade, potentially threatening supply chains, and must therefore be avoided.

[22] Veal and Ringbom (2017), p. 103. Jordan (2020), p. 297. Lowe (1977).

[23] De La Fayette (1996); Molenaar (2015); Lowe (1977), p. 621.

[24] De la Fayette (1996).

[25] General Agreement on Tariffs and Trade (adopted 30 October 1947, entered into force 1 January 1948) 55 UNTS 194 (known as "GATT"), Art. 5(2).

[26] De La Fayette (1996), p. 4.

[27] Molenaar (2015), pp. 237–239; De La Fayette (1996), pp. 19–21.

[28] Veal and Ringbom (2017), p. 103.

[29] Lowe (1977), p. 614.

[30] Adapting the argument of: De La Fayette (1996), p. 22.

Furthermore, it is arguable the retribution practices are inherently unreasonable and are thereby an affront to the reasonableness principle. Nevertheless, this practice is likely to be used to deter denials of access to ports, until there is consensus that such practices are unreasonable and unlawful in international law.

Significantly, there are limits on what the port state may permit to leave her ports. Under UNCLOS, a port state finding, upon its own initiative or reported to it by others, that a ship in its ports 'is in violation of applicable *international standards* relating to seaworthiness of vessels and thereby threatens damage to the marine environment' then it 'shall, as far as practicable' prevent it from leaving port until the ship is repaired.[31] This acknowledges unseaworthiness per se is a threat to the marine environment and, more importantly, that the port state is obliged to prevent unseaworthy vessels within its jurisdiction from leaving its ports. Under this provision, a coastal state would be required to block MAVs leaving its port if the MAV does not comply with international standards, even if the coastal state supports MAV integration. As a practical matter, coastal states which favour MAV integration may not be disposed to use their initiative to identify non-compliant MAVs, although the coastal state must still investigate a ship if reported to it by another state. If a diligent third state reports the MAVs for not complying with international standards, MAVs will not be able to leave the port; such a position would render merchant activities impossible. Section 3 on maritime safety standards will determine if there is a violation of applicable international standards.

2.2 *Territorial Sea*

Access to the territorial sea of coastal states, i.e. the sea 12 nautical miles from the coast state's baseline, is necessary if MAV are to be involved in the international carriage of goods by sea. Ships tend to incidentally enter states' territorial seas other than their flag's at some point in their voyage. This section will consider whether MAVs have the right of innocent passage in the territorial sea.

2.2.1 MAV as Ships

First is it necessary to identify if MAVs are 'ships.' UNCLOS guarantees the right to innocent passage only to 'ships of all States'.[32] Therefore, the right to innocent passage depends upon whether MAVs will be considered 'ships'. It is for this reason this chapter has avoided the more common 'Maritime Autonomous Surface Ships' which obviously presupposes that such vehicles are ships.

[31] UNCLOS, Art. 219 (emphasis added).

[32] UNCLOS, Art. 17.

UNCLOS does not provide a definition of what a 'ship' is. It is not in the *travaux préparatoires* nor the conventions UNCLOS replaced.[33] But it is clear the term is equivalent to vessel as the two are used interchangeably throughout UNCLOS.[34] While no definition is given in UNCLOS, at the time of drafting of UNCLOS, it was clear that AI-controlled ships and remote-controlled ships were well beyond the contemplation of the drafters.[35] The drafters were intending to cover conventional ships, i.e. ships manned with a crew and master. With this in mind, it must be carefully considered whether UNCLOS's rights, duties and privileges that apply to 'ships' extend to MAVs. There are reasons to suppose not, for example, MAV craft may be especially vulnerable to the perils of the sea, thus being unseaworthy in the strict nautical sense, and therefore MAVs may be unlikely candidates for ship status and the various privileges UNCLOS bestows upon ships.

Prima facie, the task of defining a ship is left to the flag state as UNCLOS does not define the term. The flag is given the right under UNCLOS to set the conditions for their granting nationality to ships.[36] Other conventions may give some guidance as to what is generally understood in international law to be a ship and will influence the conditions the flag will set, but there is no pervasive definition in international agreements because conventions differ greatly and the definition given to 'ship' or 'vessel' reflexively matches the purpose of the convention. Some conventions are intended to be rather broad and encompass every craft in the seas and oceans, such as MARPOL[37] which is intended to cover pollution from all sea-borne vehicles, and some rather narrow. Still, no convention explicitly requires manning, to the surprise of some academics.[38] However, when these conventions were drafted only conventional ship were contemplated, so drafters did not consider it necessary to limit the conventions to manned vessels.

To understand what this power to flag means, it is necessary to consider the rules of treaty of interpretation. Under the Vienna Convention of the Law of Treaties (VCLT)[39] Article 31, a treaty is to 'interpreted in good faith in accordance with the ordinary meaning given to the terms of the treaty in their context and in light of the object and purpose'. In addition, the treaty shall include the text and any agreement made by the parties in connection with the treaty, any subsequent agreement or practice, and any relevant rules of international law. Roughly, interpretive practice is

---

[33] Daum and Stellpflug (2017).

[34] Veal and Ringbom (2017), p. 102. Veal and Tsimplis (2017), p. 307.

[35] Veal et al. (2019), p. 27.

[36] UNCLOS, Art. 91.

[37] Protocol of 1978 relating to the International Convention for the prevention of pollution from ships, 1973 (adopted 17 February 1978, entered into force 2 October 1983) 1340 UNTS 1983 (known as "MARPOL"), Art. 2(4).

[38] Daum and Stellpflug (2017).

[39] Vienna Convention on the Law of Treaties (adopted 23 May 1969, entered into force 27 January 1980) 1155 UNTS 331 (known as "VCLT").

to look first at the text, then the context, and then subsequent materials.[40] This places a limit on the flag's ability to grant nationality as there is a duty to apply UNCLOS, and therefore 'ship' status, in good faith. This constrains a state's granting its flag to ships. For example, a flag could not list a plane as a 'ship' or 'vessel'; also, the flag could not label a pure missile a 'ship' or 'vessel' as such a designation would obviously be applying the convention not in good faith. To understand a term according to the interpretive rules of VCLT, the 'ordinary meaning' is used. The ordinary meaning of 'ship' does not necessarily require manning.[41] But there is a duty under Article 31 to consider a term 'in [its] context', meaning by reading the convention as a whole.[42] Van Logchem eloquently argued that MAV may not be considered 'ships' under UNCLOS as there are various parts of UNCLOS, i.e. in the context the term is used, where manning is explicitly contemplated.[43]

For example, under UNCLOS, the flag state must ensure 'the manning of ships'[44] and that 'each ship is in the charge of a master and officers who possess appropriate qualifications'.[45] It is also contemplated in Article 29, which says a warship under command of a commissioned officer and 'manned by a crew' of the regular armed forces. These references to manning suggest that 'ships' in UNCLOS, when read in their context, cannot be understood to include unmanned vehicles.

The rebuttal usually comes in the form of a purposive argument. Namely, the 'object and purpose' of UNCLOS is to be a comprehensive agreement which provides stability to the law and therefore UNCLOS must cover all maritime operations.[46] However, it has been advanced by van Logchem that the 'object and purpose' method is a secondary consideration and the main technique for treaty interpretation is the wording of the text. This is based on ICJ in *Admission of a State to the United Nations*, where it was said 'If the relevant words in their natural and ordinary meaning makes sense in their context, that is the end of the matter'.[47] He argues that the treaty must be read in whole to understand the meaning of the terms *in context* and in the whole ships within UNCLOS are understood to be 'manned'.

In response, an argument could be made that UNCLOS was intended to be restrictive to a craft of peculiar features, this would have resulted in a definition of 'ship' in the Convention rather than leaving it open.[48] But this comes under the same argument previously made: MAV were not in contemplation of drafters as this technology was not yet developed, so there was no cause to exclude them explicitly.

[40] Van Logchem (2021), p. 30.

[41] Oxford English Dictionary. Simpson (n.d.).

[42] VCLT.

[43] Van Logchem (2021).

[44] UNCLOS, Art. 94(3)(b).

[45] Ibid.

[46] See Ringbom (2019), pp. 161–162.

[47] Van Logchem (2021), p. 39. *Admission of a State to the United Nations* (Advisory Opinion) [1948] ICJ Rep 57. *Territorial Dispute (Libya v Chad)* (Merits) [1994] ICJ Rep 6.

[48] Veal et al. (2019), p. 28.

However, early signs are that the international community and the IMO do consider MAVs to be ships, for example in naming MAVs 'Maritime Autonomous Surface Ships' in the Regulatory Scoping Exercise and the IMO Interim Guidelines.[49] Furthermore, the purpose of the IMO is to serve the state members;[50] if states desire MAVs to be integrated, it could be argued the IMO ought to accommodate this desire.[51]

The denying of MAV ship status under UNCLOS may be destabilising and has been criticised as excessive lawyering.[52] Another rejection is based on the supposed evolutionary approach in international law. This is given some weight by the *Nicaragua v Costa Rica*[53] ICJ case, which has been suggested to imbue a new evolutionary approach into international law which could be used to expand UNCLOS.[54] In this case they had to determine if 'commerce' covered passenger travel, which it did not when the agreement was originally drafted. The court applied an evolutionary approach as the agreement was to be eternal and therefore an expansive definition of trade was the intention of the drafters. However, *Nicaragua* has been confined to the facts of the case, which is reflected in its limited application since, and is not of general application.[55]

The purpose of this section is not to say that UNCLOS could not cover MAVs, but to note that there are questions to be asked about whether 'ship' status may extend to unmanned vehicles and to highlight that there are limits of a purposive argument or the evolutionary approach. This will hopefully serve as a foundation for future researchers to explore this issue in more depth and it is likely that signatories to UNCLOS and member states of the IMO will prefer that the IMO and UNCLOS supervise MAV adventures, which are clearly 'maritime' in nature. Such a preference amongst states will emerge over time.

2.2.2 Unless Prejudicial

For a ship to have the right to unimpeded innocent passage, it must satisfy the requirements of Article 19 of UNCLOS. Under paragraph (1), it is provided that passage will be innocent 'so long as it is not prejudicial to the peace, good order or

[49] International Maritime Organization, "Interim Guidelines for MASS Trials" (14 June 2019) MSC.1/Circ.1604. Veal et al. (2019), p. 29.

[50] Convention on the International Maritime Organization, as amended (adopted 6 March 1948, entered into force 17 March 1958) 289 UNTS 3, Art. 1(a).

[51] Ringbom (2019), pp. 161–162.

[52] Kraska (2010), p. 64.

[53] *Military and Paramilitary Activities in and against Nicaragua (Nicaragua v United States of America)* (Merits) [1986] ICJ rep 14.

[54] Veal and Tsimplis (2017).

[55] Dixon et al. (2016), p. 88.

security of the coastal State'.[56] UNCLOS follows up with a list of activities in paragraph (2) which 'shall be considered to be prejudicial' to the coastal state:[57]

(a) any threat or use of force against the sovereignty, territorial integrity or political independence of the coastal State, or in any other manner in violation of the principles of international law embodied in the Charter of the United Nations;
(b) any exercise or practice with weapons of any kind;
(c) any act aimed at collecting information to the prejudice of the defence or security of the coastal State;
(d) any act of propaganda aimed at affecting the defence or security of the coastal State;
(e) the launching, landing or taking on board of any aircraft;
(f) the launching, landing or taking on board of any military device;
(g) the loading or unloading of any commodity, currency or person contrary to the customs, fiscal, immigration or sanitary laws and regulations of the coastal State;
(h) any act of wilful and serious pollution contrary to this Convention;
(i) any fishing activities;
(j) the carrying out of research or survey activities;
(k) any act aimed at interfering with any systems of communication or any other facilities or installations of the coastal State;
(l) any other activity not having a direct bearing on passage.

For the purposes of this section, it is not clear if the listed activities under paragraph (2) are the means to be considered prejudicial and thereby non-innocent or if paragraph (1) provides a ground per se and paragraph 2 provides an illustrative but not exhaustive list of examples. The general view is paragraph 2 provides the real test, and paragraph 1 is perfunctory.[58] After this is presumed, it is generally followed by a determination that MAV will not necessarily take part in any of the listed activities, so MAVs' will not be prejudicial, and thus MAVs could not be denied access to the coastal state's territorial sea.[59]

There are strong reasons to suppose that the activities are exhaustive. The two global superpowers of the time, the USA and the USSR, said so in joint-statement. However, it is submitted that the list cannot be exhaustive or paragraph (1) would be superfluous.[60] Indeed, there are hints that the specific activities existed are not the only grounds by the broad catch-all of sub-paragraph (l) 'any other activity not having a direct bearing on passage'; this suggests a non-exhaustive intention in the drafters.[61] Notably, the non-exhaustive position has been adopted by significant maritime states; for example, Japan adopted this position justify its exclusion of warships carrying nuclear weapons from its territorial sea, saying such passage was

[56] UNCLOS, Art. 19(1).
[57] Ibid, Art. 19(2).
[58] Evans (2018), pp. 645–646.
[59] See: Veal and Tsimplis (2017). Ringbom (2020b).
[60] Tanaka (2015), p. 542.
[61] UNCLOS, Art. 19(2)(h). Hakapää and Molenaar (1999), p. 132.

prejudicial to its territory, although the carrying of nuclear material is not a listed activity.[62]

If it is presumed that paragraph 1 per se provides a ground, MAVs might be prejudicial to the *security* of the coastal state as MAVs will be difficult to inspect and board,[63] both of which are vital tools for coastal state security. There may be solutions, such as the virtual visit, whereby the visiting vessel might observe the cyber infrastructure of the vessel remotely—which is equivalent to the right of visit.[64] However, this solution is so far only theoretical and cannot be relied upon to conclude that MAVs are necessarily capable of inspection. MAVs have additional security issues. For example, MAVs which are capable of being remotely controlled over the Internet might be more prone to hijacking by hackers and terrorists and thereby a threat to the coastal state's security; such vessels might be directed to collide with other ships, the coastal state authority ships, or a ports' infrastructure. Therefore, MAVs cannot be considered automatically *not prejudicial* to the coastal state's security as there are strong reasons to suppose such vehicles might be denied access to the territorial sea, in accordance with UNCLOS.

However, even if the listed activities are exhaustive, it is not clear MAV will still be involved in innocent passage inherently. It is correct that MAV will not necessarily be engaged in most of the activities under paragraph 2, such as exercising with weapons or spreading propaganda, but there are serious questions to be asked about whether such vessels will be involved in 'wilful and serious' pollution.[65] There are reasons to suppose that MAV will be more prone to polluting the seas, as such ships cannot carry out repairs or respond to fires,[66] which increases the risk of casualty and cargo being discharged into the ocean. Such discharge could amount to 'serious pollution', but what of 'wilful'?

It has been widely argued that 'wilful' must relate to deliberate acts. Mere negligent pollution, including the use of vehicles more prone to polluting,[67] will not do.[68] However, this might not be correct. Roscini argues that wilful should be understood not only as deliberately causing a certain result, but also as the deliberate and conscious taking of 'high risks', especially in carrying dangerous cargo.[69] The ordinary meaning of 'wilful' would seem to exclude negligence.[70] But Proelss et al. (2020) commentary on UNCLOS gives weight to Roscini's argument as the authors include it without criticism. Crucially, Roscini's definition is not mere academic conjecture, but has some merit. For example, various insurance contracts provide

[62] Tanaka (2015), p. 542, citing Kanehara (1999).
[63] Ringbom (2020b), p. 456.
[64] Schmitt (2017), R 46(10). Klein et al. (2020), p. 732.
[65] UNCLOS, Art. 19(2)(h).
[66] Wróbel et al. (2017).
[67] Which has not been made out yet in relation to MAV.
[68] Serdy (2018), p. 325.
[69] Roscini (2002), pp. 258–259.
[70] Oxford English Dictionary. Simpson (n.d.).

that recklessness is considered 'wilful'.[71] Furthermore, non-deliberate pollution has been considered 'wilful' by some states.[72] Therefore, the argument cannot be discounted without further consideration by the literature.

Going forward, it is possible that states will use sub-paragraph (h) to justify the exclusion of MAV from the territorial sea, especially if such vehicles are regularly involved in serious pollution and develop a reputation. It would seem this only covers retrospective pollution, i.e. after pollution has occurred, which could not be engaged where a vehicle merely poses the possibility of pollution, although this defeats the point of granting states the power to turn back ships involved in 'wilful and "serious pollution"', as the damage would already be done; for the provision to be effective, it must apply anticipatorily.

It has been argued that significantly much riskier vessels are 'innocent', so that MAV must be considered 'innocent'. For example, Jordan said nuclear ships have the right to innocent passage,[73] therefore MAVs must also as they are significantly reduced risk profile in comparison.[74] The provision of UNCLOS Jordan refers to says that in exercising innocent passage, the nuclear ship must carry documents and observe any special precautionary measures.[75] However, as is clear from the provision, UNCLOS does not say that such ships are automatically 'innocent'. Rather, it expresses conditions *when* such ships are 'exercising their right of innocent passage' and therefore after they have demonstrated they are 'innocent'.[76] The provision does not indicate such vessels are necessarily non-innocent and therefore MAV cannot be assumed to be non-innocent by comparison.

As might be anticipated, if the passage is non-innocent, the coastal state 'may take the necessary steps [. . .] to prevent' it.[77] Practically, it may be beneficial for MAV operators in the initial deployment of MAV period to gain the coastal state's consent before entering its territorial sea. This approach is also proposed by the IMO in the *Interim Guidelines* for testing MAVs.[78]

2.2.3 Local Laws

Outside of innocent passage, the coastal state's sovereignty over its territorial sea provides legislative jurisdiction (called 'local laws') under UNCLOS Article 21. The coastal state's potential use of local laws in relation to MAV will be explored. To exercise this right, the laws must relate to the listed subject areas and must be in

---

[71] Howse (2020). Gard Rules 2019 (Rule 72).

[72] Van Dyke (1996), p. 384.

[73] UNCLOS, Art. 23.

[74] Jordan (2020), p. 297. See Ringbom (2020b), p. 454; Van Dyke (2002), p. 77.

[75] UNCLOS, Art. 23.

[76] Ibid, Art. 23.

[77] Ibid, Art. 25(1).

[78] Interim Guidelines for MASS Trials (2019), para 1.2.3.

conformity with UNCLOS and other rules of international law. The local laws may relate to:[79]

- 'safety of navigation and regulation of maritime traffic'
- 'protection of navigational aids and facilities,' etc., cables and pipelines
- 'the conservation of the living resources of the sea'
- enforcement of fishery laws
- 'the preservation of the environment of the coastal States and prevention, reduction and control of pollution'
- marine scientific research and hydrographic surveys
- protection of customs, fiscal, immigration or sanitary laws

This legislative prima facie provides grounds for supposing coastal states may produce laws that prevent or deter MAVs in their territorial sea by creating laws as to the safety of navigation or preservation of the environment. But the local laws may not relate to 'the design, construction, manning or equipment' (CDEM), unless the laws 'are giving effect to generally accepted international rules' (GAIRS).[80] It is highly likely that any law aimed at prohibiting MAV deployment in the territorial sea would ipso facto relate to manning or design, so there would be no means for a coastal state to prevent MAVs deployment of the territorial sea through the local laws competence, unless the state is giving effect to GAIRS. MAVs' compatibility with GAIRS will be considered in the maritime safety portion of this chapter (Sect. 3).

It should be noted there are no CDEM limitations when determining if MAVs are 'prejudicial', so that provides a much more likely barrier for MAVs to be excluded from the territorial sea than local laws.[81]

2.3 Exclusive Economic Zone

The further the ship gets from the coastal state, the less control the coastal state has over the ship. One might suspect therefore that 200 nautical miles from the coastal state's baseline the coastal state would have no jurisdiction over the ship and autonomous ship, however, this area forms the costal state's exclusive economic zone (EEZ) and the coastal state has 'sovereign rights' over living and non-living resources in the EEZ.[82] However, the coastal state has general jurisdiction under Article 56 to the creation of artificial islands, marine scientific research, and 'protection and preservation of the marine environment'.[83] It is likely that MAV will

[79] UNCLOS, Art. 21.
[80] Ibid, Art. 21(2).
[81] UNCLOS, Art. 19.
[82] Ibid, Arts. 56 and 57.
[83] Ibid, Art. 56.

have access to these areas, if MAVs are ships as ships have navigational rights in the area,[84] although the environmental competence of the coastal state poses some interesting questions. It is vital that MAVs deployed for international trade in goods have access to the EEZs of states, as international shipping journeys will necessarily cross into the EEZs of other states.

There are potentially barriers for flag states to deploy MAVs. Particularly, Article 28 says states must have 'due regard' for the rights of the coastal state; if MAVs' deployment interferes with the rights of the coastal state, for example in creating accidents that limit the coastal state's ability to exploit its resources, then these states would be in breach of Article 28. This is not to conclude that MAVs will not interfere with other ship users, but that flag states must be conscious of this duty before allowing MAVs' deployment in the EEZ of other states.

The marine scientific research competence of the coastal state might appear to be a potential barrier for MAVs, as marine scientific research in the EEZ must gain prior permission from the coastal state; it may be incorrectly concluded that MAVs are operate under the marine scientific research heading. Although MAVs are experimental, and their initial use is in the form of research conducted in the marine area, MAVs' experimental initial use would not qualify as marine scientific research; this term applies where the target object of the research is the marine environment *itself* rather than vessels.[85] For example, a glider which is collecting hydrographic data would have the marine environment itself as its object, but research into how a glider operates nautically in the ocean would not. Therefore, coastal state permission would not be required under this coastal state competence.

2.3.1 Environmental Justification

As mentioned in relation to the territorial sea, there may be a serious risk of pollution where MAVs are used, as such vessels may be involved in an increased number of marine casualties, which will pollute the oceans. As such, the coastal state's jurisdiction over the environment of the EEZ provides a potential barrier for MAVs' deployment in those zones.

Although the literature has rejected these grounds, it is possible it may be used to exclude MAVs from the EEZ. Ringbom argues the coastal state's jurisdiction in relation to pollution is unlikely to be relevant to MAVs.[86] However, if MAVs are at an increased risk of causing casualties in a coastal state's EEZ, MAVs must pose a pollution risk to the EEZ. Should MAVs develop a track record of polluting coastlines, states will be anxious to exclude such vessel from their waters. This approach has been rejected by a purposive analysis, which contends that the EEZ

[84] Ibid, Arts. 57 and 58. See Sect. 2.2.1.

[85] Stephens and Rothwell (2015), p. 578.

[86] Ringbom (2020b), p. 457.

was created under UNCLOS to recognise the economic rights of the coastal state, nothing additional.[87] But pollution harms the coastal state's economic interests; pollution harms the living resources which the coastal state has an interest in exploiting. The suggested power to exclude MAVs arguably meets the purpose of the EEZ when added to the law of the sea. Additionally, it is contended in the literature that the coastal state's jurisdiction over the EEZ environment is 'narrow',[88] presumably as it does not set out the rights of the coastal state, so the possibility to exclude ships which pollute is limited.

However, the lack of qualification to 'jurisdiction', i.e. it is not called a prescriptive or legislative jurisdiction in UNCLOS, suggests a broad meaning and arguably covers acts which are prejudicial to the coastal state's sovereign rights. Further, it ought to be accepted that coastal states have the right to protect their living and non-living resources, and its populace's health and well-being, by restricting high-risk vessels from its EEZ. This position, rather than being heretical, has been advanced by several states.[89] For example, Canada began this trend in 1970 by asserting unilateral control over navigation in the 100-mile zone surrounding its arctic territory.[90] Similarly, Spain in 2003 excluded single-hulled tankers from its EEZ over fears of a repeat of the 2002 *Prestige* oil spill.[91]

This position is arguably justified by appeal to the precautionary principle, as articulated in the Rio Declaration.[92] It allows states to '[a]pply preventive, precautionary and anticipatory approaches so as to avoid degradation of the marine environment'.[93] Environmental protection has a long history in international law. Prior to the Rio Declaration, it was found in the no harm principle from *Trail Smelter*[94] as well as the idea of custodianship from 1893 *Pacific Fur Seals Arbitration*.[95] Under the principle, the key goal is prevention, followed by repair and latterly compensation. A similar concept, the duty to prevent, has been found to be customary law.[96] Indeed, the precautionary principle is arguably present in various parts of UNCLOS.[97] For example, UNCLOS Article 220 permits detention and proceedings where there is 'clear objective evidence' of a violation which either causes 'major damage' or *threatens* it.[98] UNCLOS's allowing enforcement actions for violations that pose a '*threat* of major damage' entails a precautionary approach.

[87] Klein (2006), p. 317.

[88] Hakapää and Molenaar (1999), p. 145.

[89] Roscini (2002), p. 255. For example, Oman, Iran, Egypt, Guinea, Malaysia.

[90] Oxman (2006), p. 849.

[91] Serdy (2018), p. 326.

[92] "Report of the United Nations Conference on Environment and Development" (Rio De Janeiro 3–14 June 1992) (12 August 1992) UN Doc A/CONF.151/26, Agenda 21, Ch 17.

[93] Van Dyke (2002). Ibid, para 17.22.

[94] *Trail Smelter Arbitration (United States v Canada)* (1941) 3 RIAA 1905.

[95] *Pacific Fur Seals Arbitration* (1893) 1 Moore's Int'l Arb Awards 755.

[96] *Pulp Mills on the River Uruguay (Argentine v Uruguay) (Judgment)* [2010] ICJ 14.

[97] Van Dyke (2005). Van Dyke (2002), p. 88.

[98] UNCLOS, Art. 220(6).

The precautionary principle should influence our reading of what 'jurisdiction' means in relation to the EEZ. Although the precautionary principle has not become customary international law,[99] it has been found to be an interpretive aid when making sense and interpreting treaties.[100] The principle may be used to interpret UNCLOS, such as the right to exclude MAVs that pose a risk to the marine environment of the coastal state. To adopt the purposive approach, it would be obscene if coastal states would be obliged to watch on as ships pollute its marine environment without having the means to expel those vessels. This priority to prevent pollution is reflected in the potential liability for the flag; flag states might be liable when they fail to supervise flagged ships properly, especially where there is a pattern of behaviour.[101] In relation to ultrahazardous cargo, this can involve strict liability.[102] This liability suggests flags must be cautious in flagging MAVs, but also highlights how seriously the task of preventing pollution is taken in international law.

However, exclusion of MAVs from the EEZ would be contrary to the doctrine of freedom of navigation that all ships have in the EEZ. If access is denied, MAV shipowners might try gain access by applying for prior authorisation before entering, but this model is seriously burdensome.[103] It is expected this would seriously infringe freedom of navigation. It is recommended coastal states which oppose MAVs in their EEZ express their disagreement via peaceful means, although the lack of international rules compels actions to defend one's marine environment,[104] rather than unilaterally requiring consent and expelling ships which do not gain consent by the threat of force. The right to exclude ships which endanger the coastal state's territory, as justified by the precautionary principle and arguably its interpretive effect on the generic jurisdiction of the coastal state over pollution in the EEZ, should be clarified by the IMO immediately. It is preferable this clarification provide for the right to exclude, reflecting the perversity of denying the coastal state a means to expel polluting vessels.

Additional methods for protection of the environment of the EEZ exist under UNCLOS. Under Article 220, where there is 'clear objective evidence' of major damage or the threat of major damage to the coastline or interests of the coastal state,

---

[99] *Pulp Mills on the River Uruguay (Argentina v Uruguay) (Judgment)* [2010] ICJ 14, para 164. WTO, *EC Measures Concerning Meat and Meat Products (Hormones)* (16 January 1998) WT/DS26/AB/R and WT/DS48/AB/R (Appellate Body Report), paras 120-25; WTO, *EC Measures Affecting the Approval and Marketing of Biotech Products* (29 September 2006) WT/DS291/R (Panel Report), para 7.89. Redgwell (2018), p. 685.

[100] Redgwell (2018), pp. 685–686.

[101] Kirk (2015), p. 520. *Responsibilities and obligations of States with respect to activities in the Area* (Advisory Opinion, 1 February 2011) ITLOS Reports 2011, 10 [131].

[102] Van Dyke (2002), p. 57. *Trail Smelter Arbitration (United States v Canada)* (1941) 3 RIAA 1905; *Corfu Channel case (United Kingdom v Albania)* (Judgment) [1949] ICJ 4.

[103] Van Dyke (2002).

[104] UNCLOS, Art. 279. Van Dyke (2002), pp. 88–89.

following a breach of international standards,[105] the territorial sea or EEZ, the coastal state may detain and bring proceedings against ships and therefore MAVs,[106] but must allow the vessel to proceed when that information is secured.[107] In respect of the flag state (even in relation to the EEZ), the coastal state must notify the flag of any proceedings.[108] Similarly, proceedings of the coastal state must terminate when the flag state initiates its own, unless the proceedings concern major damage caused to the coastal state or the flag state has repeatedly disregarded its enforcement obligations.[109] In practice, enforcement will be left to the flag state, unless there is proof of substantial harm or there is a threat of significant harm.[110] But the coastal state's refraining from bringing proceedings de facto may not be continued by states which oppose MAVs' integration.

The coastal state has additional powers, beyond the generic jurisdiction outlined above. Under Article 22, the coastal state is granted prescriptive jurisdiction to create laws and regulations, in conformity with international standards, concerning pollution.[111] The power to create laws could effectively prevent MAVs' operation in the EEZ if MAVs are incompatible with environmental international standards. Should MAVs not be compatible with such standards, the coastal state may enforce its laws by boarding, inspecting, and arresting the and by bringing judicial proceedings for violations,[112] although the flag must be notified of the vessel's arrest and detention, and the detention must promptly end when security is provided.[113] Whether MAVs are compatible with environmental standards is beyond the scope of this chapter, but it will be considered by the Marine Environment Protection Committee following the completion of the IMO's Regulatory Scoping Exercise.[114]

2.3.2 Self-Defence

The doctrine of self-defence has been used in relation to environmental protection, in a time when the precautionary principle was not fully articulated. Self-defence is an aspect of customary international law that permits a state's use of force in defence of itself. Self-defence must be exercised reasonably, meeting the requirements of

[105] Ibid, Art. 220(6) and (3).

[106] Ibid, Arts. 220 (6), 228(1), 226(1)(c).

[107] Ibid, Art. 220 (7).

[108] UNCLOS, Art. 231.

[109] Ibid, Art. 228.

[110] Andreone (2015), p. 176. Ibid, Art. 220 (5)-(6).

[111] UNCLOS, Art. 211(5).

[112] Ibid, Art. 73.

[113] Ibid.

[114] Klein et al. (2020), p. 727.

necessity and proportionality.[115] Historically, the ability to exclude ships owing to the threat to the environment has been justified by appeal to self-defence; the rationale being that the effects of the pollution are just as serious as a military act.[116] This approach was used by the UK in 1967 over the *Torrey Canyon* grounding in the English channel, leading to bombing of the vessel.[117] If MAVs are perceived as posing a threat to a coastal state, the coastal state may appeal to this outdated doctrine to prevent MAVs from entering its EEZ.

However, it is unlikely this ground is legally sound today. Self-defence is understood to require an 'armed attack', as demonstrated by the wording of the self-defence doctrine in the UN Charter, which also reflects customary international law.[118] Therefore, it is unlikely the doctrine would apply to MAV integration which does not pose a threat of armed attack.

2.4 High Seas

In the areas beyond the EEZ, the high seas, there is freedom of navigation as it is understood the area belongs to all. UNCLOS provides the flag states 'the right to sail ships flying its flag on the high seas'[119] and it is submitted that MAVs will have access to the high seas,[120] as there is no coastal state to restrict MAVs. Further, if the flag state considers MAVs 'ships' and UNCLOS is understood to cover MAVs as 'ships', MAVs' freedom to sail the high seas will be acknowledged by other states.[121] Given there is no other to exercise control, the flag state would have exclusive jurisdiction over the ship in the high seas,[122] which may only be infringed if a ship is engaged in unauthorised broadcasts, shipment of narcotic or psychotropic drugs, or the vessel is without a flag or has a false flag.[123]

In the high seas, no state is permitted to claim sovereignty over the area[124] and it is a common resource shared by all.[125] As there is no other state to exercise jurisdiction in the high seas, it is likely the freedom of navigation practically subsists if MAVs are not considered ships. Rather, the loss of ship status would mean the flag

---

[115] Charter of the United Nations and Statute of the International Court of Justice (adopted 26 June 1945, entered into force 24 October 1945) 1 UNTS 16 (known as "UN Charter"), Art. 51.

[116] Roscini (2002), p. 258.

[117] Van Dyke (2002), p. 89.

[118] UN Charter, Art. 51. Brownlie (1963), Ch 13. Greenwood (2011).

[119] UNCLOS, Art. 90.

[120] Ibid, Art. 86.

[121] Serdy (2018), p. 329.

[122] Veal and Ringbom (2017). UNCLOS, Art. 92.

[123] UNCLOS, Art. 110.

[124] Ibid, Art. 89.

[125] Guilfoyle (2015), p. 203.

would lose exclusive jurisdiction over the ship and any state could impose its jurisdiction on the vehicle.[126] This is occurred in the Subic Sea incident in 2016, where China detained an American unmanned underwater vehicle (UUV) in Philippines' EEZ and said that the vehicle was not a ship and, therefore, did not have the protection of exclusive jurisdiction of the flag, the United States.[127] The loss of exclusive jurisdiction would not only be a great inconvenience for MAV deployment, involving delays while the vessel is visited, but it would allow foreign states to tamper with the devices or copy the device's intellectual property, including the designs, technical equipment and operational code. The loss of exclusive jurisdiction would also risk exposing MAVs to piracy, as operators grow accustomed to allowing visits in the high seas without the means to verify the *bona fides* of the visitors.

MAVs' flags may be constrained in their ability to allow MAVs to sail in the high seas. As with the EEZ, the freedom of navigation must be exercised with 'due regard for the interests of other States in their exercise of the freedom of the high seas'.[128] If MAVs are proven to be a hazard to other ships,[129] MAVs' use would arguably infringe other states' freedom of navigation interests in the high seas; other states' ships may avoid an area entirely for risk of collision with a MAV. If so, flags who permit MAVs' deployment in the EEZ will breach this duty. It is not possible to know the safety characteristics of MAVs at this stage, as they have not been widely deployed, but future developers should bear in mind their responsibility to other sea users before allowing MAVs into the high seas and coastal waters.

As to the exclusive jurisdiction, the flag state must exercise this jurisdiction 'in administrative, technical and social matters'.[130] Pursuant to this, the flag state is required to take measures, taking into account international instruments over the 'construction, equipment and seaworthiness of ships', 'the manning of ships' and training, use of signals and communications and prevention of collisions.[131] The flag must ensure proper surveying and the vessel is equipped with the proper navigational equipment.[132] These rules created by the flag must be as stringent as GAIRS and may exceed GAIRS.[133] To flag a vessel, there must also be a genuine link between flag and ship.[134]

Some maritime law associations have anticipated that the flag state may struggle to show a 'genuine link' to a MAV[135] presumably as there is no crew over which to exercise jurisdiction; or because the operator may not be in the flag state's

[126] Ringbom (2020b).
[127] Veal et al. (2019), p. 28.
[128] UNCLOS, Art. 87(2).
[129] As argued by China in relation to the Subic Sea incident.
[130] UNCLOS, Art. 94(1).
[131] Ibid, Art. 94(3).
[132] Ibid, Art. 94(4)(a).
[133] Ringbom (2020b), p. 443.
[134] UNCLOS, Art. 91(2).
[135] CMI (2018), p. 5. Accessed on 05/01/2021. Brazil and Denmark.

territory.[136] The 'genuine link' criterion might suggest that other states can interfere with the flag state's exclusive jurisdiction by disputing the flag's claim of a 'genuine link', but this is incorrect. In the *M/V Saiga* case, the International Tribunal for the Law of the Sea (ITLOS) indicated the requirement it is 'to secure more effective implementation of the duties of the flag', 'not to establish criteria by reference to which the validity of the registration of ships in a flag State may be challenged by other States'.[137]

If the flag fails to supervise, other states would not be able to challenge the flag's jurisdiction, nor impose its own jurisdiction in the high seas on ships under the defective flag.[138] Therefore, if a country concludes that deploying MAVs is inherently reckless and a dereliction of flag state supervision, that country would not be able to reject the flag state's supervision. Such states may only report the facts to the flag to investigate the vessel.[139] However, mandatory auditing under the Instruments Implementation Code provides a level of supervision of flag states and encourages flag states comply to with international standards, which would extend to supervising MAVs or rejecting MAVs for flagging if MAVs are not compliant with GAIRS.[140]

2.4.1 Provisions Requiring Manning

In the parts of UNCLOS setting out the flag state's duties, there are several provisions which either explicitly or implicitly contemplate manning. This section will identify these issues and offer solutions.

In exchange for granting nationality, the flag is required to ensure 'that each ship is in the charge of a master and officers who possess appropriate qualifications' for the vessel type[141] and that the personnel are familiar with rules relating to safety of life at sea, prevention of collisions, pollution and communications.[142] Furthermore, the flag must ensure, inter alia, 'the manning of ships'. In common parlance, the reference to manning suggests at least *some* crew is on board. However, UNCLOS provides that the duty is to ensure manning through measures '*as are necessary* to ensure safety at sea'.[143] Therefore, if MAVs are safe with no manning, it would not

[136] Van Logchem (2021). Daum and Stellpflug (2017).

[137] *M/V 'Saiga' (No 2) (Saint Vincent and the Grenadines v Guinea)* (Judgment, 1 July 1999) ITLOS Reports 1999, 10 [83]. Guilfoyle (2015), p. 215.

[138] Veal et al. (2019), p. 25.

[139] UNCLOS, Art. 94(6).

[140] International Maritime Organization, "IMO Instruments Implementation Code (III Code) (4 December 2013) IMO Res A.1070.

[141] UNCLOS, Art. 94(5).

[142] Ibid, Art. 94(4)(c).

[143] UNCLOS, Art. 94(3).

be necessary for the flag *to take measures* that require manning.[144] To avoid the potential for dispute, the MAVs' ability to comply with this requirement with zero crew should be clarified at the IMO level.

As to the requirement of being 'in the charge of a master',[145] there is no definition provided in UNCLOS for 'master'. For context, English law understands a master as the person in 'command or charge' over a ship.[146] It is submitted that the SBO is in command of the ship and takes on this role, as the SBO is performing the same navigational role.[147] However, a ship controlled by an algorithm would have no master and would thereby be incompatible with UNCLOS. If member states wish for fully autonomous ships to be utilised in the high seas, UNCLOS would have to be amended.

2.4.2 Criminal Competence Relevant for MAV?

The flag state is responsible for supervising ships flying its flag in the high seas to avoid complete lawlessness in international waters.[148] The lack of connection between the autonomous ship and the flag state creates new problems. Presently, when the ship is in the high seas the flag's criminal law applies; e.g. the flag has penal jurisdiction over collisions in the high seas.[149] However, the flag's exclusive criminal jurisdiction is not suitable for MAVs controlled by a shore-based operator (SBO). The arrangement is intended to apply to a crew in the middle of international waters, where no other state has jurisdiction, leaving only the flag state as the state with the necessary nexus to regulate the vessel. SBOs would not be in the high seas, but on a country's land territory, and there is no reason to suppose the flag is better placed to control the crew and investigate potential criminality in the lead-up to a collision than the country in whose territory the shore-based control centre (SBCC) is based. Therefore, UNCLOS must be amended to reflect the different circumstances for MAVs and switch criminal jurisdiction to the country where the SBCC is located.

2.4.3 Duty to Render Assistance

The duty to render assistance to those lost at sea is a noble requirement. As Cockburn CJ said in *Scaramanga v Stamp*, 'The impulsive desire to save human life in peril is one of the most beneficial instincts of humanity, and is nowhere more salutary in its

144 Veal and Tsimplis (2017), p. 319.

145 UNCLOS, Art. 94(b).

146 Merchant Shipping Act 1995, s 313.

147 Veal et al. (2019), p. 36. from Carey (2017), p. 210.

148 Guilfoyle (2015), p. 209.

149 Ibid, Arts. 94, 92, and 97.

results than in bringing help to those who, exposed to the destruction from the fury of the winds and waves, would perish if left without assistance'.[150] The law does not only protect this desire but compels it nationally and internationally. It is compelled in the Safety of Life at Sea Convention and, significantly, under UNCLOS.[151]

UNCLOS compels a master of a ship to render assistance 'in so far as he can do so without serious danger to the ship, the crew or passengers' to 'any person found at sea in danger', and 'to proceed with all possible speed to the rescue of persons in distress'.[152] Similarly, the master has a duty after a collision, such as exchanging details of the name of the ship, port of registry, and nearest port at which she will call.[153] MAVs ability to complete both tasks should be considered.

On the first, MAVs' obligation would upon who is the 'master' of an unmanned vehicle. It is probable the SBO would be the master, so the duty to render assistance would be placed on that individual.[154] However, the master's ability to assist is more limited than a master on a conventional ship. The quintessential life-saving tasks of throwing a life-saving buoy or pulling an unconscious body from the water, and performing CPR would be impossible,[155] until humanoids robots populate the ship.

However, UNCLOS does not indicate that the assistance must be physical, only that it must be given. It requires the assistance to be provided by the master *in so far as he can do so*. Therefore, the limitation of the master in question is considered in relation to this duty; therefore, if the SBO could only assist non-physically, this would satisfy the duty. Non-physical could be nevertheless significant, for example, in the relaying the radio signal and acting as a communication hub or deploying a life raft.[156] In some respect, MAVs may provide superior assistance in some respect by engaging in risky rescue efforts a manned ship would avoid, and could potentially meet the priority to prevent loss of life at sea in an enhanced manner.[157]

Although the literature assumes the second requirement would be easy for a MAV to satisfy, it causes more issues. The SBO may employ a relay on the ship to communicate via VHF radiocommunications to communicate with another ship,[158] but if this fails, the SBO nor could they not board the ship and relay the information in person. Instead, the SBO could send it to the other ship electronically. But if the MAV loses connectivity, there is a risk the SBO may not be aware of the collision to

---

150 (1880) 5 CPD 265, 304 (Cockburn CJ).

151 The International Convention for the Safety of Life at Sea, as amended (adopted 1 November 1974, entered into force 25 May 1980) 1184 UNTS 276 (known as "SOLAS"), Ch V, Reg 33. UNCLOS, Art. 98.

152 UNCLOS, Art. 98(1).

153 Ibid, Art. 98.

154 Carey (2017), p. 211.

155 International Convention on Maritime Search and Rescue, as amended (adopted 27 April 1979, entered into force 22 June 1985) 1405 UNTS 118 (known as "SAR Convention").

156 Veal and Ringbom (2017), p. 104. Jordan (2020), pp. 298–299. Veal and Tsimplis (2017), p. 330. Carey (2017), p. 213.

157 Carey (2017), p. 213.

158 AAWA (2016), p. 40.

know to give over information. To overcome this, the information could be transmitted between ships automatically through the Automatic Identification System or AIS which ships currently deploy in accordance with international law.

What about fully autonomous ships? Such ships would not have a 'master' and the duty, as it stands, would have no legal force over them.[159] However, it is submitted that the fully autonomous ship ought to be under a duty to assist nonetheless and the language of UNCLOS ought to be amended to reflect this. The reasoning in favour of this can be found in the 1870 case of *Ferguson v Hutchinson*,[160] where an English steamer collided with a coble and the steamer failed to attempt to render assistance to the crew of the other ship. In legal proceedings, the steamers' crew argued they were under no duty to render assistance under national law as cobles were not ships and the duty did not arise. As Blackburn J said, it would be a 'monstrous consequence' to permit the steamers to watch their fellow seafarers drown without moving a muscle based on a legal technicality.[161] One hundred fifty years later, this lesson must be learnt and there must be a same priority to save life; it would be monstrous if fully autonomous MAVs observed from afar the loss of human life without offering help, especially as assisting would present little to no risk to the MAV in question. To ensure this duty is maintained, UNCLOS must be amended to ensure fully autonomous ships have a duty to assist.

In contrast, the argument would not apply for crewed ships to render assistance to uncrewed MAV ships in distress as there are no human lives to be saved.[162] This is matched by the text of UNCLOS which requires assistance for 'any person found at sea in danger of being lost' and 'to proceed with all possible speed to the rescue of persons in distress'.[163] Therefore, there is no legal or moral duty to assist a ship without a crew.

3 Maritime Safety Conventions

As highlighted in the previous sections, MAVs' ability to operate in the territorial sea, to be flagged, and to leave ports, will depend on MAVs' ability to comply with generally accepted international rules and standards (GAIRS). This section will determine if MAVs may comply with such standard in relation to Safety of Life at Sea (SOLAS),[164] Convention on the International Regulations for Preventing

159 UNCLOS, Art. 94(4)(b). Veal and Ringbom (2017), p. 104.

160 Ferguson v Hutchinson (1870-71) LR 6 QB 280.

161 Ibid, 292.

162 Carey (2017), p. 213.

163 UNCLOS, Art. 98.

164 The International Convention for the Safety of Life at Sea, as amended (adopted 1 November 1974, entered into force 25 May 1980) 1184 UNTS 276 (known as "SOLAS").

Collisions at Sea (COLREGs),[165] and the International Convention on Standards of Training, Certification and Watchkeeping for Seafarers (STCW).[166] Some international standards, namely as to the environment, are beyond the scope of this chapter and must considered in further research. As will be demonstrated, there are many parts of the international standards which present difficulties for MAVs; this section will highlight the difficulties and determine if amendments to the standards are required. It will also establish the priorities set out in the international standards which ought to continue to apply to MAV and ought to be included in the regulatory regime governing MAVs.

3.1 SOLAS

The first version of the Safety of Life at Sea Convention was devised following *The Titanic* disaster in 1912 and its focus is increasing safety of life sea, which might appear to be excluded with MAVs, but other sea users are carrying lives who may be imperilled by unsafe practices aboard MAVs. Furthermore, SOLAS can be useful for ensuring safe operations, such as avoiding accidents which are harmful to the commercial enterprise and key for avoiding loss of cargo and ships. Consequently, SOLAS is a vitally important convention and MAV comply with it fully.

The first step in determining MAV compatibility is answering whether SOLAS applies to unmanned vehicles. SOLAS sets out that it applies 'only to ships engaged on international voyages'[167] and it is highly likely that MAV which are used for international trade would satisfy this requirement.[168] If this were not so, it would be necessary to amend SOLAS to ensure it applies to such vehicles given the crucial role SOLAS plays in ensuring safety. However, there are compatibility issues in several the provisions, which require manning.

For example, under Chapter II-2, Regulation 13, there is a requirement of a means of escape to allow persons onboard to disembark, requirement of access to a lifeboat, and clearly marked escape routes. This exists only to protect onboard persons. MAVs may technically comply by having clear safe routes so that a hypothetical person might escape, but it may prima facie be considered redundant for MAVs which will not have a crew physically on board. However, this position would disregard the potential for manning in certain conditions, such as when a maintenance crew is on board while the vessel is underway to carry out repairs in

[165] The International Regulations for Preventing Collisions at Sea (adopted 20 October 1972, entered into force 15 July 1977) 1060 UNTS 16 (known as "COLREGS").

[166] The International Convention on the Standards of Training, Certification and Watchkeeping for Seafarers, as amended (adopted 7 July 1978, entered into force 28 April 1984) 1361 UNTS 2 (known as "STCW").

[167] SOLAS, Ch 1, Reg 1.

[168] Van Hooydonk (2014).

emergency situations. It would be perverse if such people were trapped on the vessel without a means of escape; therefore, a means of escape must be required onboard for MAVs, so long as occasional manning is possible. Furthermore, it is expected that MAV will allow the boarding of the ship by distressed persons and provide such persons with refuge.[169]

In the same chapter, there are duties as to ensuring training of seafarers on board and that seafarers take part in drills. There are fire drills,[170] which would be prudent for an SBO crew to simulate and practise their reaction times and abilities, particularly in relation to how to navigate the vessel and communicate with rescue workers and shore authorities. However, such SBCC-based drills would not meet this requirement as presently the drills must be carried out *onboard*.[171] The flag state may exempt a ship if the alternative arrangements meet the fire safety objectives and the functional requirements of the prescribed arrangement.[172] It remains to be seen if the industry will develop the technology to make MAV safe from fires or provide the means to mitigate fires in a safe and reliable way without a crew on board.

Under SOLAS, the master must receive real-time insight as to stability information to make decisions. It says the master will be 'supplied with such information . . . as is necessary to enable him by rapid and simple processes to obtain accurate guidance as to the stability of the ship'.[173] It is anticipated this will be met by a shore-based operator as master, but the connectivity with the ship must be sufficient to give sufficiently real-time to permit 'rapid and simple processes' to acquire the information.[174] Initial simulation studies suggest just 6-second delays in feedback to the controller can result in collisions and lead to a casualty; developers of MAV must ensure that the SBO is given real-time information to meet this regulation and the objective behind it.[175] As for the connectivity needed, a few megabits per second (Mbps) is needed to upload each Light Detection and Ranging (LiDAR) and high definition (HD) video feeds to the SBCC, meaning at least 6 Mbps is needed in upload capacity for remote-controlled ships to give the SBO this information.[176]

Simple 4G connection might meet this as it can provide 100 Mpbs in download speeds,[177] but it would not work for international voyages as both 4G and 5G which have limits of around 100 km, even using balloon-based antennas to extend range.[178] A 100 km or 54 nautical mile limit would significantly limit MAVs' range; such

[169] NB: part of economic benefit from MAV is reduction in accommodation size to house more cargo: see Katsivela (2020), p. 240.

[170] SOLAS, Ch II-2, Regulation 15-16.

[171] Ibid, Ch II-2, Regulation 15-6, 2.1.1.

[172] Ibid, Ch II-2, Regulation 17.

[173] Ibid, Ch II-1, Reg 5-1.

[174] If the processes are rapid, it stands to reason the acquiring of the information must also be rapid.

[175] Perera and Batalden (2019).

[176] Chae et al. (2020), p. 9.

[177] DNV (2015), p. 13.

[178] Höyhtyä (2019), p. 88.

vessels could not leave the EEZ of a state, which is 200 nm from shore. To maintain connectivity beyond this range, satellites are routinely used and will likely be increasingly used in the coming years as both MAV and conventional ships acquire greater connectivity to the shore—for seafarer comfort and entertainment, and to ensure robust communications with shore management. SpaceX's Starlink will use 12,000 satellites at roughly 340 miles from the earth's surface to provide Internet access to isolated areas, including ships, and has average speeds of 87.25 Mbps and uploads of 13.54 Mbps.[179] This suffices the upload requirements of HD video and LiDAR. However, the viability of these satellites on a large scale might be doubted by the overcrowding of space.[180]

Furthermore, good cyber security will be necessary to ensure ships are not left stranded in busy international straits or used to collide intentionally with other vessels or the shore. This security protocol is best achieved by international prescription, likely as an annex SOLAS rather than individual companies or SBOs.

3.1.1 Vessel Design

SOLAS contains design and constructions standards, some of which will not be relevant for MAVs which lack a crew and will need amendment to support MAVs' integration, and some which will be relevant to MAVs. This section will go through some of the compatibility barriers and highlight some of prescriptions useful for MAV development.

Under SOLAS, the bridge design must be ergonomic and meet an array of requirements, for example, it must give the master a full appraisal of the situation.[181] Significantly, the requirement of a 'bridge' should be removed for MAV, as MAVs will be no crew on board to operate the ship from that location.[182] It is likely the SBCC will not be considered the 'bridge', as the *Oxford English Dictionary* specifies a 'bridge' is onboard the vessel.[183] However, a similar standard should be met for the SBCC to allow the SBOs to have a full appraisal of the placement and condition of the ship. This is likely more important given the lack of feedback the SBO, for example, it is unlikely the SBO will feel vibrations onboard the ship, which may indicate collision with the seafloor, in the same way a master on a conventional ship would. This deficit should be accounted for by providing augmented visual and audio cues on the SBCC compared to that of a conventional master's bridge to ensure a full appraisal as set out by SOLAS.

The specific requirements for alarm systems mandated by SOLAS are not directly applicable to MAVs. For example, under SOLAS, the engineers' alarm which must

[179] Roulette (2021); Dellinger (2021).

[180] Kwan and Henley (2021).

[181] SOLAS, Ch 5, Reg 15.

[182] Veal and Tsimplis (2017).

[183] Simpson (n.d.).

be operated from the engine control room, or other locations, and be audible in the engineering accommodation.[184] The lack of engine control room would mean a MAV would breach this requirement; this provision would require amendment to ensure MAVs are compatible with SOLAS, but it is important that alarm systems which were obligatory for conventional ships are fed into the SBO's system to alert the SBO to issues. This would ensure at the international level the SBO has situational awareness and can navigate the ship safely, rather than leaving the MAVs alarm systems completely to the discretion of manufacturers and developers.

The requirement of a safety centre under SOLAS is likely not appropriate for MAVs. For passenger ships, there must be a safety centre for passengers in the bridge or adjacent to it, which is a place for passengers go to in emergency situations.[185] However, it is likely that autonomous ships will not be used in the movement of passengers, as passengers expect assistance and hospitality from the crew, so this requirement as a practical matter need not be applicable to MAV, until they carry passengers.[186]

There are other parts which expressly mandate manning. For example, it is required that '[e]very ship carry personnel' qualified in distress and safety radiocommunications,[187] which could not be satisfied by a qualified SBO as SOLAS establishes the personnel must be carried on the ship. This provision must be amended if MAVs are to be deployed on merchant voyages, but it should also be noted that the SBO's qualifications should cover emergency radiocommunications to permit effective communication with the shore and other ships, as such communications are likely to continue to be over a radio and fulfilled by radio relay back to the SBCC built on the ship. The rules relating to physical pilot transfer will probably not be relevant for MAV.

Under SOLAS, the pilot transfer arrangements require a procedure for the pilot to 'embark' and 'disembark', amongst other things, and requires a pilot ladder.[188] This clearly contemplates the pilot physically coming on board and it may continue to be necessary, even if there is no conventional bridge on board, for example, if a pilot will use a small navigation booth onboard the MAV to carry out the piloting activities. But it is more likely that the pilot will perform their role remotely.[189] Technical compliance would be possible, but would be illogical: it would be a redundancy and the pilot ladder would become a potential security risk as it would provide a means for in-person hijackers to board the vessel. This provision must be removed once it becomes clear what the pilot practices are in relation to MAVs. One certainty is that pilotage in some form will continue, partly for their local knowledge

[184] SOLAS, Ch II, Reg 38.

[185] Ibid, Ch II-2 Reg 23.

[186] Ibid.

[187] Ibid, Ch IV, Reg 16.

[188] Ibid, Ch V, Reg 23.

[189] Davies (2020).

which avoids catastrophes, but especially as these local monopolies are incredibly well-financed and have political support.[190]

The requirement to switch to manual steering poses fewer issues than might be immediately anticipated from the wording. Under Chapter V, Regulation 24, the ship needs to be able to switch from automated to manual steering immediately in certain weather conditions. However, this switch refers to moving from autopilot back to 'manned' control. The presence of a control unit on MAVs allowing persons on board to manually steer should suffice.[191] Further, this priority should be upheld and a similar requirement ought to be adopted for MAVs; SBOs ought to switch from automatic steering to manual in difficult weather to ensure heightened situational awareness in risky circumstances.

There are data recording requirements which will be much easier for MAVs to accomplish and exceed. Under Chapter V, Regulation 20, voyage data records must be stored, which may be accomplished by MAVs more easily than conventional ships as the SBO's entire interaction, included what information the SBO receives from sensors and cameras, and the SBO's digital inputs to the ship, is digitally stored and will be available for analysis by accident investigators and the technology developers.[192]

The duties as to watchkeeping in SOLAS often specify onboard activities with which MAVs cannot comply. Under SOLAS, it is required that where helicopter facilities are used, two crew must be firefighting trained and equipped at all times when a helicopter is being used.[193] An SBO and their crew have little need for a fireproof overall and a hose; this duty is clearly intended for persons serving on board. SBOs could use automatic firefighting processes, but this equivalent should be set out in an amendment for helicopters to be safely used on ships. Helicopters will invariably be used from time to time even on an unmanned MAV, for example, to bring on emergency crew for repairs or to evacuate distressed persons who have been saved and offered refuge.

3.1.2 Sufficiently and Efficiently Manned—Master's Authority

The requirement for a sufficiently and efficiently manned ship, which is echoed in UNCLOS, poses obvious issues for MAVs. Under Chapter V, Regulation 14, SOLAS provides 'all ships shall be sufficiently and efficiently manned'. However, as Veal et al. contend, zero manning should be efficient and sufficient if the safety goal of SOLAS is met, i.e. MAV with zero are just as safe as conventional ships carrying the characteristic complement of crew.[194] An extensive study of

[190] Davies (2020).

[191] Van Hooydonk (2014). Daum and Stellpflug (2017).

[192] See Van Hooydonk (2014), p. 415.

[193] SOLAS, Ch II-2, Reg 18.

[194] Veal et al. (2019).

accident data, once MAVs are deployed and data is created, would inform policymakers and states if MAVs are equally safe relative to conventional ships, but this will emerge in time and should not be presumed because MAVs face increased risk in several areas. This interpretation accords with the *Oxford English Dictionary* which provides that manning means providing with a crew,[195] without specifying that the crew is on board the ship. Similarly, for the adjective 'manned', it means supplied with people, which is an SBO would meet; the SBO and their assistants would be 'supplied' to and serve a MAV. Therefore, there is no reason to suppose personnel must serve on board per SOLAS, but this position ought to be clarified in the form of an addition to SOLAS. There is a similar duty to that found in UNCLOS of rendering assistance,[196] which is substantially the same as UNCLOS and has been dealt with above.[197]

The master's position must be secure under SOLAS. Under SOLAS, decisions of the master are not constrained by the company,[198] which requires the owner or charterer to respect the master's discretion to make decisions necessary as regards safety at sea and protection of the marine environment. The master's decisions are privileged because they have more situational awareness than the owners and company. But this privilege is at odds with the master as servant of the shipowner and a servant who may be dismissed at any time.[199] While SBOs will be exposed to more supervision by the owners, given the potential physical proximity of the SBCC to the company, it is submitted that this duty must extend to the SBO who has a direct insight into the voyage, more expertise into the operation of the craft in question than their employers, situational awareness and has a duty to ensure the safety of the vessel.

When speaking of issues which must be amended, such measures are often not an absolute bar to MAV or unmanned ships as SOLAS extensively permits the flag state to permit exemptions where (a) it is reasonable to do so and (b) a similar level of safety is demonstrated. It may also be permitted through the use of equivalents.[200] For example, under Chapter II-1, Regulation 55 certain designs may be deviated from where 'the alternative design and arrangements meet the intent of the requirements concerned and provide an equivalent level of safety'; the alternative measure will be subject to an engineering analysis, evaluation and approval of the design.[201] Consequently, if MAVs do not comply with these rules, they might be exempted under the equivalents method so long as MAVs meet the intended level of safety of the regulations. This is a wide-ranging power that could permit MAV development, even contrary to the wording of the regulations within SOLAS, but it is a short-term

[195] Simpson (n.d.).

[196] SOLAS, Ch V, Reg 33.

[197] At Sect. 2.4.1.

[198] SOLAS, Ch IV, Reg 34-1.

[199] Cartner et al. (2009), para 8.8.

[200] Baughen and Tettenborn (2021), p. 21.

[201] Parts C, D, E, and G.

solution. Stability will be found using amendments directed at MAV operations specifically. Nevertheless, there are several parts of SOLAS which will continue to be useful, for example, the hull construction, the coating of ballasts, waterhead bulkheads, fire protection regulations are relevant to MAV in the same way they are nautically proper and safe conventional ships.[202]

3.1.3 ISM Under SOLAS

The International Safety Management (ISM) Code,[203] as adopted under SOLAS Chapter XI, plays an important role in reducing accident by eliminating management-side issues that lead to accidents and pollution to the marine environment and elsewhere.

Under the Code, the company and the ship must meet the requirements of the ISM Code and possess a Document of Compliance,[204] which is issued to every company that meets the Code's requirements.[205] A copy of Document of Compliance must be kept on board so they can 'produce it' for verification.[206] With MAVs, a physical copy might be kept on board in a prominent place so that it may be easily discovered by the inspecting authority. However, this does not meet the verb 'produce' which suggests the master has a role in finding it and displaying it. The master may 'produce' a digital copy, although this would not be kept on board. Therefore, if a digital means is to be permitted, an amendment to the ISM Code will be necessary.

As to the specific requirements of the Code, there is a duty to 'continuously improve safety management skills of personnel ashore and aboard ships'.[207] The onboard personnel will be under no duty, but this provision will continue to be useful to the MAV's crew in an SBCC, who are personnel ashore. Further, the ISM Code mandates a 'designated person' who provides a link between the Company and onboard and direct access to highest level of management and ensures safety and pollution standards are met, and guarantees the ship is given adequate resources.[208] In relation to the connection between MAV's master and the company's highest level of management, it may be easier to achieve as the SBO may be in the same office building as management and the company, thereby aiding communication.[209] However, this will not necessarily be the case as SBOs might be located in a different

[202] Baughen and Tettenborn (2021).

[203] IMO, "International Management Code for the Safe Operation of Ships and for Pollution Prevention (International Safety Management (ISM) Code" (17 November 1993) resolution A.741(18), as amended.

[204] SOLAS, Ch XI, Reg 3.

[205] Ibid, Ch XI, Reg 4.

[206] Ibid, Reg 4. ISM Code, Reg 14.1.

[207] ISM Code, Reg 1.2.3.

[208] Ibid, Reg 4.

[209] Karlis (2018), p. 126.

country to the company; therefore, the role of designated person should continue to be mandated to ensure an efficient means of communication and ensure management-side problems do not risk the safety of the MAV.

Under the Code, the master is responsible motivating crew to comply with company policy and the company should clearly define the master's role in their policies.[210] With MAVs, it will continue to be important for the crew, i.e. the SBO's assistants, to understand the role of the SBO as master and appreciate the division of labour, thus ensuring critical decisions are made quickly without lengthy discussion about who has authority to make the decision. Furthermore, the SBO, as supervisor of the assistants, ought to take on the role of motivating their assistants to comply with company policy—preferably this supervision role should be established at an international level as opposed to individual ad hoc measures of companies and manufacturers.

3.2 COLREGs

The International Regulations for Preventing Collisions at Sea (COLREGs), commonly called the 'rules of the road',[211] are used to prevent ships colliding into each other. The effectiveness of the regimes relies on each ship knowing or having a reasonable expectation of how other ships in their proximity will behave, and it is crucial this continues to apply for MAV. This section will determine if MAVs comply with the rules.

The scope of COLREGs' is broad; it applies to 'all vessels upon the high seas and in all waters connected therewith by seagoing vessels'.[212] The definition of vessel is incredibly broad, as it 'includes every description of a water craft, including non-displacement craft, WIG craft and seaplanes, used or capable of being used as a means of transportation'.[213] It would apply to MAV as a water craft used and capable of being used as a means of transportation. As mentioned above, there is a strong reason for applying COLREGs to MAVs, as other vessels must have a reasonable expectation of how vessels within their vicinity, including autonomous vessels, will behave to avoid collisions.

It has been suggested MAVs may struggle with COLREGs contain subjective elements. Under Rule 2, it is set out that compliance with the rules will not exonerate the vessel from the requirements of the ordinary practice of seamen or the special circumstances of the relevant case and it allows departure from the Rules where it is necessary to avoid danger.[214] It is anticipated the SBO, initially with seagoing

[210] ISM Code, Reg 5.2.

[211] Katsivela (2020), p. 244.

[212] COLREGs, Rule 1(a).

[213] Ibid, Rule 3(a).

[214] Ibid, Rule 2(2).

experience, would be able to rely on their expertise to form a subjective evaluation to depart from the Rules where it is necessary for safety. But it is less clear how a fully autonomous MAV would comply with this element. Portions of the literature suggest this discretion could not be properly exercised by an AI,[215] but this may be wrong. In simulation studies, AI have been demonstrated to act in non-compliance with COLREGs to avoid collisions,[216] suggesting that a level of subjective evaluation may be possible, especially following more intense research and development. Whether this competence emerges will depend on technological advancements in the coming decades, which are impossible to predict. Within COLREGs, there are additional parts which require human sentience, for example Rule 14 which speaks of a vessel being 'in any doubt';[217] such a feeling would be impossible for a software to have. MAVs without a master onboard may prima facie be considered 'not under command'.

Under Rule 18, it is provided that a power-driven vessel shall keep out of the way of vessels 'not under command' (1) or 'a vessel restricted in her ability to manoeuvre' (2), which has led to concerns that MAV may be 'not under command' and, consequently, causing hardship for conventional ships. However, COLREGs provides some clarity as to what the phrase 'not under command' means. It speaks of 'a vessel which through some exceptional circumstances is unable to manoeuvre' as required and is therefore unable to keep out of the way of other vessels.[218] In practice, it usually refers to equipment loss, as in *Mendip Range v Radcliffe*.[219] By the definition, it covers exceptional circumstances, which a MAV's ordinary mode of travel cannot qualify as,[220] although it may cover MAVs with lost connection to the shore or with a breakdown in the operational code. Alternatively, a MAV may be considered a 'vessel restricted in her ability to manoeuvre' with the same obligation that other power-driven ships keep out of the way. But again, COLREGs provides a definition, which specifies this limitation is from the 'nature of her work' and includes a non-exhaustive list of activities which qualify, such as laying underwater cables and dredging.[221] It is clear it is the type of work that limits the manoeuvrability under this Rule, rather than the ship's design. Therefore, power-drive ships will not be under the burdensome duty to keep out of the way of MAVs per se.

The most widely discussed issue in COLREGs, in relation to MAVs, is the requirement to keep a lookout by 'sight and hearing' under Rule 5 which tends to be understood as contemplating a human element—sight and hearing connote

[215] Veal et al. (2019).

[216] Perera and Batalden (2019).

[217] Daum and Stellpflug (2017).

[218] COLREGs, Rule 3(f).

[219] [1921] 1 AC 556. Case cited in Carey (2017), p. 209.

[220] Veal and Tsimplis (2017).

[221] COLREGs, Rule 3.

human perception.[222] A compute program, extending to an AI, could not be said to 'see' or 'hear', although a ship controlled by an SBO likely would (assuming that there are listening devices transmitting sound and cameras relaying image to the SBO).[223] To be effective as a 'watch', this must involve a real-time feed to the SBO;[224] therefore, to satisfy Rule 5 MAVs with an SBO must possess reliable connectivity with minimal latency in the connection.

The SBO's watch by sight and hearing is not radically different from modern watchkeeping practices, as equivalent technology has been used to incorporated to meet watchout requirements, most notably radar and VHF radiocommunications.[225] However, nothing short of MAVs' equivalent safety will be accepted by regulators and it will not be accepted by other sea users. There have been some doubts that the SBO could perform sight and hearing watch, however.[226] A clarification of the Rules, and the SBO's ability to perform, would offer some stability for states enforcing the regulations and for operators and manufacturers seeking to comply with them.

3.3 STCW

The International Convention on Standards, Training, Certification and Watchkeeping for Seafarers (STCW), as might be expected, contains rules prescribing watchkeeping, and this inevitably troubles MAVs which lack the means to perform some of the types of watch listed therein. This part will highlight some of the issues found in the STCW in relation to MAVs and explain why MAVs' compliance with STCW will not problematic.

Under STCW, the officer of the watch (OOW) must not leave the bridge, the bridge being on board the ship.[227] The SBO as OOW cannot be *on the bridge* to satisfy this requirement. Nonetheless, it is submitted that SBO ought to be under a comparable duty not to leave the SBCC until relieved to ensure that MAVs are properly supervised at all times. There are additional watches dependent on cargo. For hazardous cargoes on a ship, there must be continuous watch maintained in port,[228] which being in port could be achieved by bringing on a crew once or immediately before a ship is berthed. A more probable solution would be not to use MAVs for hazardous cargoes owing to the inability of the crew to supervise

[222] Veal et al. (2019).
[223] See Daum and Stellpflug (2017) who argue MAVs meet Rule 5.
[224] Veal and Tsimplis (2017).
[225] Veal et al. (2019).
[226] Baughen and Tettenborn (2021), p. 22.
[227] STCW, Ch VII/2, para 3-1, para 17.
[228] Ibid, Ch A-VIII/2, Part 5.5 and General 105. Baughen and Tettenborn (2021), p. 18.

properly such cargo while underway. There are additional watches MAVs cannot comply with.

An engineering watch is required under STCW and part of the duty of the watch is to be immediately available for engineering work in machinery spaces,[229] which the distant SBO may perform by being available for engineering-related tasks capable of being accomplished from ashore, but the SBO and their crew would not be immediately available to work in machinery spaces. MAVs would inherently be non-compliant with this part of STCW. However, the utility of an engineering watch is reduced for MAVs, as the crew cannot perform many of the engineering-related repairs, but there must be a duty to be aware of repair and engineering issues on the part of the SBO, as there is a need to know the nautical limitations on the vessels in terms of avoiding exacerbating a breakdown, and an awareness of the need to call for a repair crew personnel to be parachuted in.

Some watches may be met by the SBO. The STCW requires a radio watch must be maintained,[230] which pertains to listening for communications from other ships and the shore. An SBO could meet this obligation by having a radio relay from the ship transmitted to the SBCC; it is vital that the SBO uses this means to communicate to ensure that they are aware of other sea users, to avoid collisions, to relay their own information to the shore and other ships, and be alerted when persons are lost at sea.

The quality of the watch is important. Authorities require the master to ensure watchkeeping is adequate, taking into account prevailing circumstances.[231] The main duty of the OOW under the STCW is safety of navigation,[232] which involves passage planning, checking a course and positioning, asking for support, contacting the master and not leaving the bridge;[233] the master must also complete a passage plan that covers hazards.[234] Passage plans, including those concerning hazards, could be completed by the SBO without any great difficulty.

As MAVs could not comply with certain watchkeeping requirements, it is necessary to determine if the STCW binds MAVs. Straightforwardly, STCW attaches specifically to seafarers: 'The Convention shall apply to seafarers serving on board seagoing ships . . .'.[235] As there are no seafarers *on board* seagoing ships, there are no individuals for STCW to attach to and STCW could have no obligatory application to remote-controlled or fully autonomous MAVs.

However, there are domestic enactments bringing STCW into domestic law which may use broader language; for example, the United Kingdom's domestic

[229] Ibid, Ch A-VIII/2.

[230] Ibid, Ch A-VIII/2.

[231] Schelin (2020). Ibid, Ch VIII/2, Section 2.

[232] STCW, Ch A-VIII/2, Part 4, para 10.

[233] Schelin (2020), p. 268.

[234] STCW, Ch A-VIII/2, Part 2, paras 3 and 5.

[235] Ibid, Art. III. Veal and Tsimplis (2017), p. 322.

enactment says it applies to 'sea-going ships', without the need for seafarers.[236] MAVs, as sea-going ships, could not comply with the various duties under STCW and therefore the wording of the UK incorporation of STCW into domestic law would provide a significant barrier to MAVs' integration in this key maritime state. The enactment may be simply revised by a domestic amendment. Despite the inapplicability of STCW to MAVs, STCW could continue to be useful internationally for MAVs' regulation, such as the requirement of a constant watch and its provisions mandating training and qualifications of seafarers. Both priorities should be maintained for MAVs. Particularly, STCW's training provisions ought to be replicated in the regime for SBOs, as to ensure proficiency comparable to that of a licensed sea-going seafarer on a conventional ship.[237] It cannot be accepted that those without the proper qualifications or seafaring experience can be entrusted to oversee this perilous activity.[238]

4 What Is the Solution?

The solution to these problems at the International Maritime Organization (IMO) level could be in the form of guidelines and soft law, amendments, an implementation agreement or a MAV Code. This part will analyse the strengths and weaknesses of each method and prescribe a given method that will work best for MAVs' integration in the medium- and long-term. Work has already been commissioned in this area by the IMO. The Regulatory Scoping Exercise is the IMO's first effort and considers whether the IMO conventions are compatible with MAVs:[239] first determining whether (a) maritime autonomous surface ships (MASS) could operate by equivalences set out in the text of the instrument, (b) interpretations need to be developed, (c) amendment of the instrument is needed, (d) or a new instrument is needed.

In the meantime, alternative ad hoc measures might be used to permit MAVs' deployment, such as bilateral agreements by pro-integration states or the of the existing framework unamended.[240] These approaches are useful for the short-term, particularly as the IMO process is both busy and slow. But it has been suggested that the IMO's work looking to change the law of the sea is inappropriate and is characterised by excessive lawyering that threatens the stability the current legal order provides.[241] This is partially correct as numerous overlapping legal

[236] Merchant Shipping (Standards of Training, Certification and Watchkeeping) Regulations 1997 (Reg 3).

[237] See Veal and Tsimplis (2017), p. 323.

[238] Fan et al. (2020), p. 9.

[239] Maritime Autonomous Surface Ship (MASS) in the language of the IMO.

[240] Veal and Tsimplis (2017); Carey (2017).

[241] Kraska (2010).

instruments could threaten the definitive balance UNCLOS has struck in the law of the sea, but there needs to be clarity in this area as legal uncertainty leads to disputes that may turn hot. Disputes have already begun, for example, the aforementioned 2016 Subic Sea incident which involved military actors. Therefore, the stability of the legal order is threatened rather more by inactivity. The prospect of IMO regulation is not far-fetched. While traditionally the IMO has been slow to develop new rules,[242] it has taken a more proactive in recent years, such as developing the oily water separator rules before the technology had been developed.[243] It is hoped this proactivity continues and that there is an appetite for an IMO-level solution for a medium- or long-term rather than ad hoc measures. The next section will go through the options.

4.1 *Soft Law*

Soft law in the form of guidelines, potentially passed by the IMO general assembly, could clarify the law as regards MAVs in a non-legally binding manner. Soft law has the advantage of being able to be constructed and adopted at pace, allowing quick adjustments for emerging technological trends and industry practices; it often benefits from the specificity that binding agreements ordinarily avoid.[244] Soft law guidance could initially be simple, such as clarifying several terms, e.g. recognising the SBO as the master in all the IMO instruments, and followed up with more complex guidelines as industry best practices emerge.

Soft law has utility in the short-term. It can mitigate the risk of confrontation between states by forging a mutual understanding between parties, but it has limits. It is unlikely industry will be willing to invest billions and transform their fleets to autonomous vessels based on mere guidelines without binding legal force.[245] This is logical; without legally binding rules, states which oppose MAVs may quietly accept guidelines, but object to MAVs use in their territorial sea at a later point, especially if there are safety or pollution concerns that emerge at a later date.

Soft law has limited use as an interpretive tool in relation to changing the language of the IMO instruments. As mentioned in Sect. 2.3.1, general principles have emerged from soft law and developed into well-recognised interpretive tools in international law, such as the precautionary principle, which led to the International Court of Justice recognising sustainable development as an interstitial norm.[246] Consequently, UNCLOS fisheries rules have been modified by the precautionary

---

[242] George (2021).

[243] Kenney (2018), in MARPOL Annex I in 1983.

[244] Stemre (2018).

[245] Ringbom (2020a).

[246] Boyle (2005), p. 573. Gabčíkovo-Nagymaros Project (Hungary/Slovakia) [1997] ICJ 7, 140.

principle, as found by ITLOS in *Southern Bluefin Tuna*.[247] Guidelines can therefore slowly emerge and change how treaties are understood.[248] But an interpretation could not emerge in this way if it is directly contrary to the terms of UNCLOS. As demonstrated in Sect. 2, there are parts of UNCLOS which directly do not contemplate MAVs, especially for fully autonomous ships. As Soyer correctly notes, guidelines could not solve this, as VCLT Article 31(1) provides a term must be interpreted in good faith; it would necessarily be contrary good faith to apply a term inconsistent with the intention of the drafters.[249] MAVs need a more solid legal foundation to ensure compatibility with the law of the sea, which will only emerge from a legally binding effort.

This is not to suppose that soft law will not be useful to the development of MAVs. Guidelines' specificity may still be useful for the medium- and long-term, especially in the form of codes of practice[250] which quicky respond to new technology and disseminate best practices across the spectrum of the shipping industry, and, crucially, will ensure MAVs are used in a safe manner.

4.2 Amendment Process

In some form, a change of the language of the law of the sea will be needed. This part will consider individual amendments to UNCLOS to ensure compatibility with the terms of the convention. Despite UNCLOS's being in force for nearly 30 years,[251] there has been no amendment to it.[252] This is little surprise as it took ten years to negotiate UNCLOS, which involved balancing the interests of all the IMO member states and international governmental organisations.

Despite the lack of amendment to UNCLOS, there is an amendment procedure under Article 312, which sets out that any state party may propose to Secretary-General of the United Nations, and the Secretary-General shall convene a conference if not less than half of states respond favourably within 12 months. As part of the process, the proponents must take every effort to achieve consensus and there will be no vote until all other efforts have been exhausted. This bias towards consensus ensures amendments are difficult to pass, as states which oppose the amendment for national reasons can simply block the process; this bias towards consensus is useful as it reduces the risk of fragmented legal orders as different states adopt different

[247] Boyle (2005), 573. Citing: *Southern Bluefin Tuna Cases* (Provisional Measures) (1999) ITLOS Nos 3 and 4. Paras 77-79. Judges Laing at 16-19, Treves at para 9.

[248] Allen (2018).

[249] Soyer (2020), p. 171.

[250] Bartlett (2018).

[251] Entered into force 1994.

[252] Van Logchem (2021).

rules, thus creating disunity in the rules governing a global enterprise. UNCLOS also contains a simplified process, but this has a more extreme version of this bias, relying on no opposition for an amendment to pass.[253] The lack of amendment to date signifies the difficulty of the process. Lastly, reflecting the Vienna Convention on the Law of Treaties which provides a state is only bound to a treaty with their consent,[254] under UNCLOS an amendment would not bind parties who do not sign up for it.[255] Therefore, if a state opposes MAVs' integration, that state would not be bound to accept the text of the amendment. MAV integrationists must provide compelling evidence, particularly as regards safety and pollution, if they hope to form consensus for any amendment to UNCLOS.

The likelihood for gaining consensus to bring in amendments is unlikely, owing to the ten years of diligent and careful negotiation required to conclude the successful UNCLOS agreement,[256] but there is a demonstrated appetite for some regulation in this area, as evidenced by the IMO's, and therefore IMO's member states, commissioning the Regulatory Scoping Exercise and passing the Interim Guidelines for MASS. Consequently, there are early indications that amendments may come to have significant support.

However, the process of amending individual parts would be incredibly laborious, as there are multiple articles of UNCLOS which need to be amended and a vote organised for each proposed amendment.[257] Furthermore, there are jurisdictional questions to be answered in relation to an amendment, particularly concerning whether the IMO is competent to regulate MAVs if MAVs are considered 'non-ships'.[258] The Canadian Maritime Law Association offered some insight in their response to the CMI, saying this issue is decided upon one's subjective belief in the role of IMO in relation to UNCLOS: is the IMO to set standards or merely reflect standards established in UNCLOS.[259] It is submitted that this issue will be determined by what member states accept in practice,[260] but the IMO is the institution best suited to govern craft in the maritime area. IMO member states' supporting the IMO's commissioning of the Regulatory Scoping Exercise suggests states consider MAVs within the IMO's competence.

---

[253] UNCLOS, Art. 313.

[254] VCLT, Art 34.

[255] UNCLOS, Art. 316.

[256] Boyle (2005).

[257] Ringbom (2020a).

[258] Van Logchem (2021).

[259] Ibid, p. 59. Canadian MLA response to the CMI.

[260] CMI (2018). Eleven MLAs said inconsistencies in UNCLOS could be resolved by the IMO. Argentina, Britain and Germany said Article 94 could be clarified by the IMO. China, Finland and Spain said it would need a framework convention which could be done by the IMO.

4.3 New Convention

A new convention may be proposed for MAV, in parallel to UNCLOS, which could be advantageous in avoiding the laborious process of individual amendments and providing a bespoke regime for MAV,[261] ensuring harmonisation of the law.[262]

However, UNCLOS was time-consuming to develop. It would be incredibly unlikely that a MAV version of UNCLOS could be negotiated and gain the same level of success in securing accessions, at least not before the technology is developed and already deployed.[263] For this reason, the new convention method is only academically attractive and it would likely be opposed by states which supply a disproportionate number of seafarers to the global shipping economy.[264] Further, the bespoke regime is a misnomer as conventions aiming to garner broad international support lose specificity to ensure success, and therefore the convention could offer little more insight than an amendment to UNCLOS. Furthermore, it would risk a limited number of signatories and therefore risk a fragmented regulatory regime internationally.[265] Instead an implementation agreement or MAV Code is more achievable and therefore preferable.

4.4 Implementation Agreement or MAV Code

An implementation agreement, which is an agreement concluded subsequent to completion of the treaty and provides rules about the implementation of the treaty,[266] is the most likely solution for MAVs. Within this, the parties to the agreement can set a certain interpretation out, such as all references in the instruments to ‘master’ being expanded to cover the SBO, in a legally binding way without going through the troublesome process of individual amendments. Implementation agreements have normative force because they aid the effectiveness of the original treaty, which is a key objective of international law.[267]

There is ample reason to suppose such agreements would work for the MAVs and law of the sea, namely they have already been successfully used in relation to UNCLOS. The Agreement Relating to the Implementation of Part XI and the 1995 UN Fisheries Agreement both succeeded, and there is another in the works, in the form of the 2019 Biodiversity in Sea Beyond the National Jurisdiction.[268] Such

[261] Bork et al. (2008).

[262] Ibid, p. 319. Katsivela (2020).

[263] See Soyer (2020).

[264] Ibid, p. 168.

[265] Stemre (2018).

[266] Kojima and Vereshchetin (2013).

[267] Ibid.

[268] Although it was frustrated by COVID.

agreements are successful because states find implementation agreements easier to stomach than other legally binding agreements; this makes such agreements easier to pass than an amendment or a new convention,[269] and also improves the chances of gaining consensus.

Agreements can be significant and fundamentally change how treaty or convention works. For example, the Deep Seabed Agreement[270] affects UNCLOS as whole as it is taken to revise the text of UNCLOS; Article 2(1) of the agreement expressly provides that the language of the agreement prevails over UNCLOS. Additionally, it is presumed to signatories of the Convention have acquiesced.[271] In contrast, the Straddling Fish Stocks Agreement[272] is expressly subordinate to UNCLOS under Article 4 of the agreement. Therefore, an implementation agreement would meet the needs of MAVs' integration in providing a clarification of UNCLOS, even if contrary to the wording of the treaty text. However, as this chapter has hopefully set out, it is not only UNCLOS but several instruments at the IMO which pose difficulties for MAVs.

There are issues with SOLAS and potentially in MARPOL and the environmental instruments. It is preferable that all these barriers to MAVs' integration are tackled in one broad agreement, which amends several provisions in several conventions in one go,[273] as occurred when IMO regulated voyages to the polar regions through the Polar Code.[274] This method for MAV integration has wide support[275] and is economical: a binding clarification that the SBO is the 'master' is needed in several conventions and a Polar Code would settle this in one.[276] Furthermore, while having the character of an implementation agreement and therefore more easily attracting consensus, the amendment to the maritime safety convention could benefit from the tacit acceptance procedure found in several conventions, binds a state to a technical amendment unless they explicitly opt out.[277] This process could ensure a less fragmented regime to govern MAVs. The success of the Polar Code ought to be emulated with a Maritime Autonomous Vehicle Code.

---

[269] Eder (2019).

[270] Agreement relating the Implementation of Part XI of the Convention on the Law of the Sea (1994). Agreement relating to the Implementation of Part XI of the United Nations Convention on the Law of the Sea of 10 December 1982 (adopted 28 July 1994, entered into force 16 November 1994) 1836 UNTS 3.

[271] Boyle (2005), p. 565.

[272] Agreement for the Implementation of the Provisions of the United Nations Convention on the Law of the Sea of 10 December 1982 relating to the Conservation and Management of Straddling Fish Stocks and Highly Migratory Fish Stocks (adopted 4 August 1995, entered into force 11 December 2001) 2167 UNTS 3.

[273] Eder (2019).

[274] International Code for Ships Operating in Polar Waters (Polar Code) (21 November 2014) resolution MSC.385(94).

[275] Ibid, 55.

[276] Ibid.

[277] Stemre (2018).

5 Carrier and Insurance

For MAV to be viable for international carriage of goods, their operation must be compatible with international commercial law; MAVs will either be used to carry goods in a bill of lading contract or be chartered under a charter contract, and will need to be insured. Such contracts must comport with the needs of MAVs. This section deals with the barriers found in bill of lading regulations, under the Hague-Visby Rules and the Hamburg Rules, and considers how the contractual terms in charterparties and insurance law match with the needs of MAVs. The commercial law analysis will be primarily from the perspective of English law.

5.1 Bill of Lading Contracts

5.1.1 Hague Rules

The Hague Rules[278]—or Hague-Visby Rules as adopted under English law—[279] present a key challenge to MAV integration owing to the duty of seaworthiness under Article III and the duty to ensure proper manning, equipping and supplying the ship.[280] This attaches before and at the commencement of the voyage and operates as an overriding obligation.[281] If MAVs fail to be seaworthy, the shipowner may be exposed to great liabilities without the use of the limitation found in the Rules[282] and the exemptions, which arise from: the default or neglect of the master and crew, fire, an Act of God, war, act of public enemies, and latent defects.[283] To understand if a ship is seaworthy in broad terms, a ship will be considered so if a reasonably prudent owner, had they have known the relevant facts to and circumstances of the ship, would have allowed the vessel to go to sea.[284]

MAVs may fail this duty in many respects, in the same way as conventional ships: if the vehicles commence their voyages with a hole in the hull, or if their engines are not fit for purpose, or if there is a serious fire at the commencement of the voyage. But there are new considerations for MAVs, which impacts a MAV's ability

[278] International Convention for the Unification of Certain Rules of Law relating to Bills of Lading (adopted 25 August 1924, entered into force 02 June 1931) UKTS 021/2004: CM 6259 (known as "Hague Rules").

[279] Carriage of Goods by Sea Act 1971, s 1. Protocol to amend the International Convention for the unification of certain rules of law relating to bills of lading, signed at Brussels on 25 August 1924 (adopted 23 February 1968, entered into force 26 June 1977) 1412 UNTS 127 (known as the "Visby Protocol", combining with the Hague Rules to create the Hague-Visby Rules).

[280] Hague-Visby Rules, Art. III(3)(b).

[281] Maxine Footwear Co v Canadian Government Merchant Marine [1959] AC 589.

[282] Hague-Visby Rules, Art. IV.

[283] Ibid, Art. IV(2).

[284] The Eurasian Dream [2002] EWHC 118 (Comm); [2002] 1 Lloyd's Rep 719.

to go to go sea compared to a conventional ship, namely the MAV's IT infrastructure will be key to whether it is fit for the voyage. For example, a latent defect in the code controlling the MAV would potentially be a breach of the seaworthiness duty, if the defect existed before or at commencement of the voyage, as the vehicle would not be fit to go to sea. As this would breach the overriding obligation, a carrier would not be able to excuse the defect by the latent defect exemption under the Rules.[285] Therefore, it is crucially important that MAVs comply with Article III and it is important that MAVs' developers are aware of the additional aspects that will become more relevant to autonomous vehicles' seaworthiness. It is expected that MAVs will from time to time fail to be seaworthy just as conventional ships do, but developers should strive to ensure that MAVs are not inherently unseaworthy or especially prone to unseaworthiness, for example, through a poorly developed, non-updated or inadequately tested operational code and IT infrastructure.

Under Article III, there is a duty on the carrier to ensure a vessel is properly manned, which presents obvious difficulties for a vessel without a crew onboard. But this provision does not prescribe a level of manning, nor does it say manning must be onboard the vessel.[286] This is supported by the dictionary definition of 'manning' which speaks of a crew being supplied, but not necessarily supplied to serve onboard a vessel.[287] Instead, proper manning rather speaks to a level of manning necessary for a safe and secure voyage; the manning requirement was not inserted by maritime labour unions to ensure employment for seafarers, but rather the purpose is to ensure a ship can safely sail and carry the cargo as agreed under the carriage contract.[288] There is judicial authority to support this position; it was acknowledged by Sellers LJ in *The Hongkong Fir* that a numerically deficient crew may nevertheless be safe.[289] Whether MAV may be safe will depend on the technology that emerges and safety practices devised by manufacturers and stakeholders; unfortunately, it cannot be determined at this time.[290]

In addition to the quantity to manning, the duty concerns competence of the crew.[291] For example, in *The Makedonia*, inefficient engineers at commencement rendered the ship unseaworthy.[292] In *Hongkong Fir*, the seafarer in question was addicted to drink and this was found to make the ship unseaworthy.[293] One incident

---

[285] A fire caused by the vessel's unseaworthiness prevented the reliance of the fire exemption: Maxine Footwear Co v Canadian Government Merchant Marine [1959] AC 589. See Karlis (2018).

[286] Baughen (2021), p. 83.

[287] Oxford English Dictionary, Simpson (n.d.).

[288] Stevens (2020), p. 250.

[289] The Hongkong Fir [1962] 2 QB 26, 55 (Sellers LJ) as cited in Carey (2017), p. 204.

[290] It is hoped science and engineering will explore this technology in more depth, so it may be known how safe this technology is in practice.

[291] The Star Sea [1997] 1 Lloyd's Rep 360. The Eurasian Dream [2002] EWHC 118 (Comm); [2002] 1 Lloyd's Rep 719. The Hongkong Fir [1962] 2 QB 26.

[292] The Makedonia [1962] 1 Lloyd's Rep 316.

[293] The Hongkong Fir [1962] 2 QB 26.

of negligence will not be enough to render a crew incompetent;[294] instead, incompetence refers disabling want of knowledge or skill[295] and is generally found where there is an ineffective management system.[296] It is rarely found in recent years,[297] in part owing to it the distaste in the exemption being pleaded in disputes.[298] Crew incompetence from lack of knowledge was found in *Standard Oil v Clan Line*[299] which involved lack of manual instructions and a lack of knowledge of how to operate safety systems was found in *The Eurasian Dream.*[300] As it relates to MAVs, there is a risk that SBO's lack of training could render the SBO incompetent and therefore the MAV unseaworthy.[301]

This is in part because the current understanding of a competent crew is seafaring experience. While the first SBOs will likely be masters with decades of seafaring experience, and indeed extensive training in how to operate the SBO safety systems, as MAV increased and conventional ships diminish, SBOs of-the-future will have less opportunity to gain sea-going experience and therefore would be considered to be lacking competence.[302] However, future SBOs will have experience *of an SBO*, potentially as assistants of SBOs with seagoing experience, and will have the necessary competence to supervise a remotely controlled vessel. Though, SBOs could further demonstrate competence by completing courses in STCW and COLREGs required courses.

As mentioned in relation to a faulty code, the design and construction of the vessel affect a ship's seaworthiness; ships usually demonstrate seaworthy design and construction by meeting the international technical standards set out.[303] Likewise, it ought to be ensured that MAVs meet the same nautical standard of conventional ships, such as hull thickness and design. The seaworthiness duty will likely be more intense for MAVs and to a higher standard. As it stands, 'easy fixes' that may be ameliorated quickly while the vessel underway do not render a vessel unseaworthy.[304] However, MAVs will not have a crew on board to conduct easy fixes, repair cracks in the metalwork, or deal with minor flooding, which means minor issues that are acceptable for conventional ships could render MAVs unseaworthy. MAVs' developers must ensure increased construction and design quality to reduce the incidence of breakdowns and ensure effective inspection of the

[294] Steel v The State Line Steamship Company (1877) 3 AC 72.
[295] See Stevens (2020), p. 248 citing The Clan Gordon [1924] AC 100, 120-121 (Lord Atkinson).
[296] The Eurasian Dream [2002] EWHC 118 (Comm); [2002] 1 Lloyd's Rep 719.
[297] Stevens (2020), p. 248.
[298] Carey (2017), p. 205.
[299] Steel v The State Line Steamship Company (1877) 3 AC 72.
[300] The Eurasian Dream [2002] EWHC 118 (Comm); [2002] 1 Lloyd's Rep 719.
[301] Wróbel et al. (2017).
[302] Karlis (2018).
[303] Stevens (2020), p. 244.
[304] Quoted in Stevens (2020), p. 247.

MAV before every voyage is commenced to guarantee a MAV does not commence her voyage with minor issues that render her unseaworthy.

There are aspects with which MAVs will comply without difficulty. The master's passage plan is part of the seaworthiness duty.[305] An SBO from shore could devise such a plan prior to the voyage and amend it as needed while the MAV is underway, although it is not clear if fully autonomous ships could complete passage plan, as the technology has not been demonstrated. The role of the programmer in MAVs' operations will be much more intense than in conventional ships, as computational updates and bugs will be fixed by programmers continuously.

Under the Rules, seaworthiness is non-delegable, meaning the carrier's assigning the task to independent contractor (in this case, a programmer) would not extinguish the carrier's liability in relation to failures in that contracted work.[306] A failure in uploading the software, even if assigned to a programmer, would squarely fall as the responsibility of the shipowner. However, this only applies to programmers contracted after the ship entered the shipowner's sphere influence of the carrier. For example, in *The Happy Ranger*, the shipowner was held not to be liable for the shipbuilder's errors over cranes as they occurred before the cranes had entered the shipowner's orbit; but the shipowner was found liable after entered orbit for failing to proof test the cranes.[307] This case demonstrates that defects in the MAVs' software from the ship's manufacturer will not be the shipowner's responsibility, but still could attract liability by failing to proper tests. However, with machine-learning AI, developed over time and might make the ship unseaworthy unless undetectable by due diligence, under Article IV(2)(p). With navigation software added after purchase, this would likely come down to a service contract with a programmer or software company; the independent contractors' defective work would still make the ship unseaworthy.[308] Therefore, shipowners ought to be aware that the carrier is liable for defective software upgrades contracted out to a third party.

The next issue is whether the exemptions and the grounds for excluding liability apply to autonomous vehicles. MAVs deployers will rely on the exemptions to ensure they are not exposed to additional liabilities relative to conventional ships; it is crucial to identify if they apply. First, the exemption for the neglect or default of the master.[309] This provision is sufficiently broad to cover the SBO as it expressly covers pilots 'or the servants of the carrier in the navigation or in the management of the ship'. An SBO is certainly a servant and involved in the navigation of the ship.

The Rules provide an additional duty under Article III(2) of properly and carefully loading and caring for the goods when received. This is partly equivalent

[305] The CMA CGM Libra [2020] EWCA Civ 293; [2020] All ER (Comm) 1072.

[306] The Muncaster Castle [1961] AC 807.

[307] The Happy Ranger [2006] EWHC 122 (Comm); [2006] 1 Lloyd's Rep 649. Baughen (2021), pp. 84–85.

[308] Baughen (2021), p. 85. The Muncaster Castle [1961] AC 807.

[309] Hague-Visby Rules, Art. 2(4)(a).

to the cargoworthiness duty, i.e. the vessel is fit to carry the goods in question, as found in *Maori King v Hughes*.[310] In the *Maori King* case, it was held refrigerators were held to be warranted in a contract to carry meats. The standard for proper caring and carrying is determined by whether there is a sound system in place for caring for the cargo,[311] which will be found if the practice reflects industry standards.[312] This is a significant issue for MAVs. While the loading portion could be handled by the carrier's port agent,[313] the ongoing duty to care could not be met by an agent and there is no crew onboard to meet it. The MAV would lack crew to inspect cargo, observe the goods to ensure they are not being contaminated or damaged, or ensure key machinery supporting the cargo in transit is working. It would be impossible for a remote operator, for example, to repair a mechanically broken refrigerator. However, it remains to be seen if technology or industry practices develop to ensure fail-safe machinery to carry cargo such as refrigerators or provide mechanisms for ventilation of holds to ensure cargo is not contaminated.[314] Developers should be conscious of the need to ensure cargoworthy merchant vehicles and devise a solution to the aforementioned issues prior to MAV deployment. Amendment to the Rules may be considered to ensure MAVs can be deployed, but removing the duty to care for cargo approach would defeat the purpose of the carriage cargo, i.e. to carry cargo without causing its loss.

The Rules provide a documentary duty to sign the bill of lading, attesting to the quantity, weight, quality, etc., which primarily lies with the master.[315] As the SBO is master, this duty would fall on them. However, there are practical issues in the SBO signing the bill of lading. First, the SBO would not be able to sign a physical bill of lading, but instead would be limited to an electronic bill of lading.[316] There is no indication in the Rules that the bill of lading cannot be electronic.[317] Second, and more fundamentally, the SBO could not inspect the cargo at loading to sign the document. Therefore, it would likely fall to the port agent, who has the means of inspecting the goods upon loading, to be responsible for singing the bill of lading. This is permitted under the Rules, as it may be performed by the 'agent of the carrier'.[318]

Lastly, the master has powers over cargo which relate to dangerous goods if not disclosed to the carrier, which permits rejecting dangerous goods and discharging

[310] [1895] 2 QB 550.

[311] The Albacora [1966] 2 Llyod's Rep 53. Carey (2017), p. 205.

[312] Volcafe Ltd v Compania Sud Americana de Vapores SA [2016] EWCA Civ 1103, [72] (Flaux J). Ibid.

[313] Carey (2017), p. 213.

[314] Baughen (2021), p. 86.

[315] Hague-Visby Rules, Art. 3(3).

[316] Baughen (2021), p. 89.

[317] Ibid.

[318] Hague-Visby Rules, Art. 3(3).

goods or rendering said goods innocuous without compensating the shipper.[319] This is important for MAVs which are less capable of carrying dangerous cargo safely and must be able to reject dangerous cargo. Under the Rules, where dangerous goods have been brought on a ship without the carrier's consent, the master may discharge or render that cargo innocuous at the shipper's expense.[320] The port agent can act to reject dangerous cargoes at loading, as the individual supervising loading, but it will not be possible for an SBO of a vessel underway to discharge goods or render them innocuous. This is an increased risk for carriers, therefore, but the liability for damage from the dangerous cargoes lies with the shipper. It is incumbent upon the shipper to disclose fully all goods shipped on an autonomous vehicle to avoid causing a casualty and, more compellingly, bearing the financial consequences of a ship and her cargo being lost owing to the shipper's fault.

5.1.2 Hamburg Rules

A parallel international regime for carriage of goods by sea is provided by the Hamburg Rules,[321] which has significantly fewer signatories. It has similar duties to the Hague Rules and therefore similar barriers to MAVs' use of the Rules. This part will identify the potential barriers to MAVs in the Hamburg Rules.

Under the Rules, the carriage contract is carriage from one port to another.[322] Therefore, any MAVs involved in such operations will attract the operation of the Rules. The duties of the carrier are set out in Article 4. These duties extend to 'the servants or agents' of the carrier[323] and would cover the SBO. The carrier will be liable occurred while the goods are in the carrier's care, unless the carrier proves their servants or agents took all measures reasonably required to avoid the occurrence.[324] The period of responsibility is ongoing, from the time the goods are put in the carrier's charge at the port of loading, during carriage and at the port of discharge.[325]

The criterion of 'in the charge' poses issues for the SBO. According to the Rules, the carrier will be liable when put in the charge of themselves, or the carrier's servants and agents. It is unclear if goods put onto a ship controlled by an SBO would be sufficient to put the goods in the SBO's 'charge'. The SBO is responsible for the vessel and the goods, but 'charge' in common parlance connotes physical proximity which the SBO will not have with the goods. A clarification of the

319 Ibid, Art. IV(6).

320 Ibid.

321 United Nations Convention on the Carriage of Goods by Sea (adopted 31 March 1978, entered into force 1 November 1992) 1695 UNTS 3 (known as the "Hamburg Rules").

322 Hamburg Rules, Art. 1(6).

323 Ibid, Art. 1.

324 Hamburg Rules, Art. 5.

325 Ibid, Art. 4.

meaning of this provision ought to be adopted to accommodate MAVs' unrestricted commercial utilisation in states which have adopted the Hamburg Rules. However, as mentioned above, the goods will likely be loaded under the port agent's supervision when MAVs are used. Agents are expressly provided for under this provision and their receipt would involve the necessary physical proximity contemplated under this article.

The carrier will be liable if goods are lost or damaged while in their care, unless the carrier shows that they, their servants or agents took all measures reasonably required to avoid the loss.[326] This is not confined to onboard servants, so a carrier may avoid loss by proving the SBO as navigator took all reasonable measures, for example, by observing prudent navigational practices. For fully autonomous ships, there would be no defence through the AI as it is not a servant or agent; however, the carrier could still avoid liability by showing *they* took all measures reasonably required concerning the AI-controlled MAV.

The same defences and limitations apply to carrier's limitation and defences apply under the Rules, if the individual acted within the scope of their employment,[327] but will not apply if 'it is proved that the loss, damage or delay in the delivery resulted from an act or omission of the carrier done with the intent to cause such loss, damage or delay, or recklessly and with knowledge that such loss, damage or delay would probably result'.[328] It is submitted that this extends to an SBO acting within their scope of employment.

Similar to the Hague Rules, the Hamburg Rules provides that in cases of dangerous goods, where a shipper's fails to notify carrier of the nature of goods, the master has the power to unload the goods or destroy them without payment to the shipper.[329] The Rules do not specify who can perform the unloading or destroying,[330] although it ought to apply to the SBO and, more sensibly, the port agent.

Under the Rules, the carrier must issue the bill of lading on demand of the shipper,[331] but the Rules say the document may be signed by a person with the authority to sign from the carrier,[332] which could extend to the port agent, who will likely be granted such authority. There is the possibility that the 'carrier or other such person' may place a reservation on the bill of lading, inter alia, if 'he had no reasonable means of checking such particulars' such as the quantity and number.[333] The SBO may be inclined to use this reservation extensively to account for their lack of physical proximity and means of inspecting the cargo loaded. However, this

---

[326] Ibid, Art. 5.

[327] Ibid, Art. 7(2).

[328] Ibid, Art. 8(2).

[329] Ibid, Art. 13.

[330] Ibid.

[331] Ibid, Art. 14.

[332] Hamburg Rules, Art. 14(2).

[333] Ibid, Art. 14.

would render the singing of the bill of lading redundant and it is more sensible that port agent performs this task instead of the SBO.

5.2 Charter-Party

The other type of shipping contracts concern charterparties, which are commonly in the form of voyage, time and bareboat charters. Unlike bill of lading contracts, there is no international mandatory regime for charterparties and instead it is governed by domestic contract law. Therefore, this part will consider English law in area by looking at two quintessential charterparty duties: the charterer's duty to nominate a safe port and the master's duty not to deviate from the nominated course.

Under a time charter, the master is to follow the orders of the charterer as to the ports of call. To permit this, the charterer offers the shipowner and master an indemnity; the charterer warrants to nominate only safe ports.[334] It is likely this duty would be the same in MAV with little amendment. This contractual duty will continue to be significant to MAVs as MAVs ought to be employed to safe ports; there is the same risk of sinking from unsafe weather conditions[335] or colliding into the port owing a laxity of hierarchy in the port's organisation[336] as affect conventional ships. However, if ports require vessels that enter to be manned by personnel onboard, MAVs may be forced to load crew shortly before entering port to comply with the conditions of entry. It is arguable this would make the port unsafe for a ship, based on *The Alhambra*,[337] where the ship chartered was too large for the port in question and the ship had to lighten the vessel at another location to access the port. This was deemed to render the port unsafe. The boarding of crew would be a significant burden if mandated, and it would deny MAVs some of their economic benefits, in that the temporary crew would need accommodation that would take up cargo space.[338]

However, it is submitted that the offloading of cargo is a much more intensive activity than positioning a crew onboard for a brief period of time; therefore, the latter requirement is not comparable and such a requirement to 'crew up' should not render a port unsafe. But this has yet to be decided by a court, and it could be fairly argued that the loading of a temporary crew is comparable to lightening, in that the port in question is not fit for the vessel under her normal configuration. Judicial determination of this issue is crucial.

---

[334] The Houston City [1956] AC 266.

[335] The Stork [1955] 2 QB 68.

[336] The Marinicki [2003] EWHC 1894 (Admlty).

[337] (1881) 6 PD 68, 72.

[338] Carey (2017), p. 203.

Ports may be unsafe as there is a risk that MAVs are confiscated at the nominated at the port,[339] for example, by a country opposing MAV integration this could render the port unsafe. However, it must be balanced against the seaworthiness duty, where a ship will be unseaworthy if it cannot satisfy the legal requirements of ports anticipated on the journey.[340] The balancing of the duties—the charterer's safe port duty and the shipowner's seaworthiness duty—has not been determined and needs special consideration at to the construction of the contract, given both parties were presumably aware MAV might face such obstacles when they struck the bargain, and it cannot be concluded one party has more blame than the other.

The second quintessential duty under a charterparty is the duty on the master not to voluntarily deviate from nominated voyage.[341] Significantly, breach of this duty will engage the anachronistic doctrine of fundamental breach; this doctrine has been in extinguished for general English law of contract, but continued for deviation. Therefore, any deviation, however slight, will go to the root of the contract and defeat any exclusion clauses or limitations.[342] In short, it will have the effect of displacing the contract entirely.[343] There was some doubt following the quasi-deviation case law, but it is clear that the House of Lords authority in *Hain Steamship v Tate & Lyle* has not been reversed by the quasi-deviation case law, nor did the House of Lords cases, which ended fundamental breach for general contract law, apply to deviation, which is *sui generis*.[344]

Within deviation, there are exceptions, such as saving a life,[345] which does not extend to property.[346] This would equally apply to the SBO who would not be permitted to deviate from the course set, but would be emboldened and indeed required to save a life. In relation to fully autonomous ships, it is not clear if such could 'voluntarily' deviate, given such devices presently lack intentionality or the ability to will. A more detailed study is needed to determine if AI will be treated as having a level of intentionality sufficient to do anything 'voluntarily', which will depend on technological advancements and social attitudes as regards AI advancements. Additionally, further research is needed to look at MAVs' compatibility with the key standard charterparty forms.[347]

---

[339] Ogden v Graham (1861) 121 ER 901.

[340] The Ellie and The Frixos [2008] EWCA Civ 584. The Madeline [1967] 2 Lloyd's Rep 224. The Derby [1985] 2 Lloyd's Rep 325.

[341] Rio Tinto v Seed Shipping (1926) 24 LL Rep 316.

[342] Hain Steamship v Tate & Lyle [1936] 2 All ER 597.

[343] The MJ Sur [2018] EWHC 1673 (Comm) [87], (Carr J). Ibid.

[344] Hain Steamship v Tate & Lyle [1936] 2 All ER 597. The Antares [1987] 2 Lloyd's Rep 424. Suisse Atlantique v NC [1967] 1 AC 361. Photo Production Ltd v Securior Transport [1980] AC 827. The MJ Sur [2018] EWHC 1673 (Comm).

[345] See Sect. 2.4.3.

[346] Scaramanga v Stamp (1880) 5 CPD 295.

[347] See Baughen (2021).

5.3 Insurance

This section will consider insurance cover from an English law perspective. If MAVs are to be used in international trade, affordable and certain insurance cover is a necessity. This part will tackle whether MAVs are compatible with the duties and terms found in English insurance law as pertains to shipping, primarily found in the Insurance Act 2015, and also to what extent MAVs are compatible with the protection and indemnity (P&I) Clubs policies. There are other routes to indemnity, such as flag states providing the indemnity as occurred in the initial deployment of nuclear-powered ships,[348] but a more scalable model is needed for broad adoption of MAVs.

5.3.1 Carriage Insurance

Unlike the old legislation, the Marine Insurance Act 1906 which provided that a breach of the disclosure duties voided the policy and breach of a promissory discharged the insurer from the point of breach,[349] the current legislation under the Insurance Act 2015 is significantly less draconian. The marine implied warranties continue to apply. However, breach of these duties will be suspensive, i.e. coverage would be lost until it 'becomes essentially the same as that originally contemplated by the parties'.[350]

As the implied warranties subsist from the Marine Insurance Act 1906, there is an implied warranty that the commencement of each stage the ships is seaworthy in respect of such preparation or equipment for the purposes of that stage.[351] Each stage has different requirements. For example, in *Oliverson v Loughman*,[352] a contract for New Orleans to Liverpool, a ship injured by worms in mud of Mississippi, would be unseaworthy for the sea voyage, but was fit for the particular stage of the journey. The repairs were carried out before the sea voyage stage. It is unlikely that stages will impact MAVs any differently than conventional ships; the different risk profile for MAVs is in cyber risks, which will be nearly identical at each stage of a voyage.[353]

As for correcting breaches of the insurance warranty of seaworthiness, some breaches may be corrected, such as breakdowns in the connectivity by regaining connection or the operating system might be corrected by updates underway, but other breaches, such as mechanical issues at commencement, will likely be repairable. Therefore, shipowners and operators ought to be exceedingly cautious when

[348] Veal and Tsimplis (2017).

[349] Marine Insurance Act 1906, ss 17 and 33.

[350] Insurance Act 2015, s 10.

[351] Marine Insurance Act 1906, s 39(3).

[352] Oliverson v Loughman (1815) 105 ER 862.

[353] Wang (2020).

deploying MAVs to ensure MAVs are mechanically sound, as the owner will be unable to correct it.

To meet the warranty, a vessel is expected to meet the perils of the sea anticipated on the contracted voyage.[354] Different vessels have different expectations. For example, a floating dock faces a lower standard.[355] If MAVs have reduced capabilities, this relative seaworthiness duty may save them. However, this cannot be correct. Part of the reason a floating dock has different standards to a ship is because of the nature of journeys it is expected to endure are much less risky and complex than voyages ordinary ships participate in. Unlike floating docks, MAVs would be deployed on international voyages and should be treated to the same standards as conventional ships.

Replacing the duty to replace all material facts under the Marine Insurance Act 1906, there is a duty of fair presentation under the Insurance Act 2015.[356] To comply with this, the information must be so provided reasonably clear and accessible for the prudent insurer, it must include every material presentation, and it must be substantially correct and made with good faith.[357] Importantly, the information is material if it would influence the judgement of a prudent insurer in determining whether to take the risk and, if so, on what terms.[358] A breach will only be remedied if the insurer can prove that they would not have entered the contract or would have entered the contract on different terms.[359] Furthermore, there is no duty to disclose the information to the insurer if the insurer knows the information, ought to know it or is presumed to know it.[360]

For this radically new technology, it is unclear what information an insurer would find 'material', as insurers have yet to develop standard practices or models for calculating the risk. Furthermore, it is unclear what an insurer is ought to know or is presumed to know, again considering the lack of standard practice in relation to this new type of shipping platform. To ease the burden on the insured, it necessary that insurers develop guidance documents to reflect on the materiality of information relating to MAV marine adventure, which will in turn simplify the claims process for insurers and will ensure policies are more certain and provide confidence to early deployers of this technology. Confidence is a necessary precondition for investment into this technology.[361]

[354] Steel v The State Line Steamship Company (1877) 3 AC 72.

[355] Cantiere Meccanico Brindisino v Janson [1912] 3 KB 452.

[356] Marine Insurance Act 1906, s 18; Insurance Act 2015, s 3.

[357] Insurance Act 2015, s 3.

[358] Ibid, s 7.

[359] Ibid, s 8.

[360] Ibid, s 3.

[361] See for more detailed study: MacDonald Eggers (2021).

5.3.2 Protection & Indemnity (P&I)

This next section briefly considers P&I Clubs and ability for MAVs to be covered by Clubs policies as well as some practical issues as to cascading risk. Clubs are useful for covering incredibly large sums through the Pooling Agreement, reinsurance and overspill mechanisms.[362] For example, it was P&I Clubs which paid for *The Ever Given*'s release from Egyptian authorities, for a reported $600 million, following the 2021 grounding that blocked the Suez Canal.[363] It is crucially important for MAVs' developers to come under Club rules or they will face this liability individually, which might deter any serious investment in fleets of MAVs or significant use of MAVs.

Under the Gard Rules, Clubs cover 'ships' which does not mention manning.[364] This likely applies to MAVs and Clubs will likely not create barriers for MAVs as, fundamentally, Clubs want to maintain relevance by offering insurance policies for new maritime technologies.[365] Clubs cover an assortment of risk, from on-board, collisions, loss or damage to floating objects, pollution, wreck removal, incidental liabilities, and legal costs—all of which will be relevant for MAVs.[366] A condition of this cover is the vessel must meet classification society rules, which extends to equipment and manning amongst others.[367] However, as submitted throughout this work, manning is related to safety of manning levels, not an absolute prescription that some crew are on board; therefore, MAVs will comply with this requirement if MAVs demonstrate safety.

Going forward, it is anticipated that Clubs, which primarily focus on ship-board activities and matters, will begin to regulate and specific the shore-side of the operation to ensure standards, which is where key decisions will be taken. However, this switch to the shore needs to be considered as a matter of urgency to guarantee that MAVs are covered, but also that MAVs engage in activities in a safe which do not expose Clubs to cascading risks.[368] Currently, Clubs operate on the basis that they can cover high economic impact but low frequency, but with MAVs there is the possibility of multiple ships in an autonomous fleet and across different companies possess the same software and the same technical issue; this could lead to dozens of marine casualties, with catastrophic outcomes, in the same timeframe. If this were on the scale of *The Ever Given*, it could mean Clubs could not pay out the sums. Clubs must carefully consider whether their maximum coverage, reportedly of $8.3 billion USD,[369] is sufficient for MAV technology.

---

[362] Howse (2020).

[363] Jankowicz (2021).

[364] Gard Rules 2019 (Rule 1.1), cited in Howse (2020), p. 198.

[365] Howse (2020).

[366] Ibid, p. 194.

[367] Gard Rules 2019 (Rule 8).

[368] See Viljanen (2020).

[369] Howse (2020), p. 195.

6 The Changing Role of the Master

At current, the master or the captain of the ship, has a role as the chief navigator and supervisor of the crew. Inevitably, removing the crew and the master from the ship drastically changes the master's role. The master's present powers, which are extensive, are justified as both the company and flag state have significant geographic distances between the ship and themselves; therefore, the master is the only authority to control the crew's actions. This will change with MAVs, as there will always be a country close the SBCC. Presently, the master is the God, King and Constitution on the High Seas,[370] but this grand role is no longer justified. Although he master's role of navigator will subsist for remotely controlled MAVs, given the SBOs will make key navigational decisions akin to the master of a conventional ship. For MAVs controlled by an AI, it remains to be seen if a program can possess the intentionality to be considered equivalent to a 'master'.

The master's liabilities must be considered in relation to MAVs. Prior to the development of MAVs, there were concerns about the broad liabilities the master has compared to the limited power.[371] In the past, the grand role was justified as the master had absolute control over the ship while it was underway, leading to commensurate liability. The modern-day master has less control: the company sets policies, changes orders while underway, can make crewing, and agents interfere with the powers of the master while at port and near the coast. In contrast, the master's liabilities continue to be significant: first, as a fiduciary duty to the ship; second, the master's criminal liability, for example, the master will be prosecuted in English law for sailing a dangerously unsafe ship.[372] The fiduciary responsibility is not justified in relation to MAVs; this was developed when the master was far away from the shipowner and the shipowner had few means to supervise the master or has reason to question the honesty of the master with the ship or goods.[373] But this cannot be justified nowadays. With MAVs, the SBO's handling of the goods can be digitally stored; any dishonest dealing is easily detected. By comparison, the criminal liability, at least as to sailing a dangerously unsafe ship, is of vital importance and must continue when the master operates a MAV for the sake of safety.

Similar to the outdated fiduciary role, there are powers which will be redundant. For example, the master has incredible agency powers as an agent of necessity which emerges primarily where the master is not able to communicate with the shore manager or shipowner.[374] Presently, these agency powers are rarely used as communications with the shore are generally reliable;[375] therefore, there is no reason that a MAV controlled by an SBO, who has even more reliable connectivity with the

[370] Cartner et al. (2009), Ch 1.

[371] Gold (2004).

[372] Cartner et al. (2009), Ch 1. Merchant Shipping Act 1995, s 98.

[373] See reasoning in Morse v Slue (1672) 1 Ventris 238; 86 ER 159.

[374] Cartner et al. (2009), Ch 8.

[375] Vojković and Milenković (2020).

shipowner as a land-based person, ought to have these powers. The supervisory function is no longer justified.

Under most domestic laws, the master has a supervisory competence; they are responsible for supervising officers and ensuring the crew complies with company policies and the laws of the flag and any coastal state; they supervise winchmen, write up and inform the charterer of damage to cargo,[376] and must consent to hoses or hatches being used on the ship, even under charter. The general supervision of assistants should fall to the SBO, who should make key navigational and operational decisions, but some supervisory powers will be lost. The master will not be present onboard the ship to observe winchmen or detect damage to cargo.[377] Furthermore, the master supervises the pilot and at all times the master is responsible for the pilot's navigational decisions.[378] But this piloting regime makes little sense with an SBO, who has limited ability to supervise the work of the pilot or ability to intervene immediately to prevent an accident. There needs to be serious reflection on the SBO's responsibility for the pilot; but it is likely the liability would be more easily insured by the shipowner who has coverage with P&I clubs as opposed to port authorities.

MAVs may not be able to fulfil their duty to prosecute certain criminal offences under the law of the sea. For example, it has been reported that the flag state must ensure that the master prosecutes slavery, piracy, narcotics and unauthorised broadcasts.[379] This would be difficult for a MAV to do without a crew, as MAVs would have no means of detaining the vessel or the crew engaged in the qualifying illicit acts. However, it is clear that MAVs would be under no such duty. UNCLOS specifies that the flag state, rather than shups under its flag, have a duty to cooperate in relation to vessels engaging in the illicit activities and gives the flag the competence to prosecute such vessels.[380] Rather than this responsibility being placed on any vessel, UNCLOS says only a warship or any duly authorised ships 'marked and identifiable as being on government service' may perform this role. Indeed, the advice for the master in relation to piracy is to report and take pictures,[381] rather than physically intervening. A MAV's SBO could offer similar support, of reporting and taking videos, in the prosecution of piracy and the other illicit activities.

As to the supervision, there is a disciplinarian function. In English law, the master is empowered to restrain anybody on board 'as long as it appears necessary to him or expedient . . . for good order and discipline'.[382] There is little reason to continue this practice with an SBO; there will certainly be no cause for the SBO to send their fellow office workers to a cell to instil discipline, when ordinary police and

[376] Cartner et al. (2009), Ch 8.

[377] This could fall to the port agent.

[378] Cartner et al. (2009), Ch 13.

[379] Ibid, Ch 5. Based on UNCLOS, Arts. 98–100, 108–109.

[380] UNCLOS, Art. 100.

[381] See Best Management Practices – BMP5 (2018).

[382] Merchant Shipping Act, s 105.

regulatory authorities are available to prosecute and imprison the shore-based personnel involved in criminal acts.

Lastly, some obscure duties ought to remain, at least in principle. For example, in an emergency, it is convention the master is the last to leave ship.[383] This comports with the master's duty to oversee the emergency efforts, which is best achieved from onboard the ship. A similar duty should apply to the SBO; while the SBO is not onboard the vessel, they should continue their presence in the SBCC during a crisis. This is achievable as SBO has no cause to leave during an emergency situation; their life would not be imperilled. But they should offer continued supervision, with their enhanced situational awareness, for the duration of the crisis, until there is an appropriate opportunity to be relieved. Policymakers ought to include such a requirement in soft law guidelines and in binding additions to SOLAS, or the MAV Code.[384]

7 Private Law Issues

This section deals with quintessential private law issue: arrest, wreck, maritime lien, salvage and limitation. The literature as it stands is under-developed and it is hoped this section will provide some much-needed attention to this area of law and act as a launchpad for further research in this area. As with commercial law, international private law must fit MAVs if MAVs are to be effectively integrated into the shipping industry. This section will highlight possible issues and priorities, offer solutions.

7.1 Arrest

Arrest of ships is governed by the Arrest Convention.[385] The purpose of the Arrest Convention is to regulate how a vessel with liabilities in one state might be arrested by another state to secure the payment of those liabilities. Without this, the other state and the creditors, including parties which have experienced losses owing to the vessel in question, would have limited means of security payment of liabilities. Therefore, it is necessary for good order of society that MAVs should be subject to the Arrest Convention. The first issue is whether the Arrest Convention would apply to MAVs.

The Convention applies to a ship and says a ship is 'any vessel' flying the flag of a contracting state.[386] 'Any vessel' is intentionally broad and would likely apply to

[383] Cartner et al. (2009), para 9.1.4.

[384] See Sect 4.4.

[385] International Convention relating to the arrest of seagoing ships (adopted 10 May 1952, entered into force 24 February 1956) 439 UNTS 193 (known as the "Arrest Convention").

[386] Arrest Convention, Arts. 1 and 8.

MAVs. Indeed, given the need for the convention to apply to MAVs, there is no reason to suppose it would not.

The Arrest Convention covers damage by one ship to another, loss of life or personal injury 'in connexion with the operation of any ship'.[387] A MAV will be a 'ship' and any collision or other relevant act will be '*in connexion with* the operation of any ship', therefore attracting liability.[388] If MAVs did not attract this liability, it would mean MAVs could enter seas committing collisions with limited recourse for the victims of these collisions, which would be inequitable; any reading of the convention to this effect would go directly against its purpose.

Some authors suggest that MAVs will be less prone to accidents, as most accidents are attributable to humans and human fatigue which will not be an issue with unmanned ships.[389] However, SBOs are humans too and therefore flawed; there will continue to accidents caused by the traditional issue of human errors and fatigue, but MAVs will be at increased risk types of accidents, such as ones caused by the SBO's reduced situational awareness. Therefore, MAVs will continue to be avoid in collisions and other acts in connexion with the operation of a ship. However, it is unlikely that MAVs' operators will create *such* liabilities, it their office, from their personal injury or loss of life. Ordinary domestic rules for any other type of land-based employee should apply to such injuries, not the Arrest Convention.

As mentioned in the UNCLOS chapter relating to the difficulty of visits (Sect. 2.2.2), there will be difficulties in arresting a vessel without a crew. The solution may be in the shipowner handing over its operational code to the Admiralty Marshal to allow the Marshal to take control of the vessel and potentially restrict the shipowner from reacquiring access.[390] Furthermore, additional measure such as physical cables attached to the vessel once it is berthed could ensure that the arrested MAV could not get underway even if the shipowner reacquires access to the vehicle. Thus, the practical difficulty of MAV arrests has several solutions, but there needs to be further research into the practical viability of these methods from a technical perspective before their efficacy may be presumed.

7.2 Wrecks

As collisions will continue to occur with MAV, it is highly likely wrecks will continue also. The Nairobi Convention[391] requires that such wreck should be removed to prevent pollution to the environment and free up busy waterways for

[387] Ibid, Art. 1.

[388] Ibid, Art. 1(1)(b).

[389] Zanella (2020).

[390] Soyer and Tettenborn (2021), pp. 69–70.

[391] Nairobi International Convention on the removal of wrecks (adopted 18 May 2007, entered into force 2015) UKTS 030/2016: CM 9315 (known as the "Nairobi Convention").

other vessels, which is a priority that must continue in relation to MAV wrecks. Particularly, the convention is useful as it obliges the master and operator of a ship to notify the affected state[392] following a maritime casualty, that the ship was involved in, that resulted in a wreck.[393] Further, should the affected state find the wreck to be a hazard, the convention requires that the shipowner remove the wreck under the state's supervision, or the state may remove it at the ship's registered owner's expense, with limited exceptions.[394] The first question is whether the convention applies to MAVs.

The convention applies to a sea-going vessel of any type, which is sufficiently broad to incorporate a sea-going MAV.[395] Under the convention, a wreck is a maritime casualty, a sunken or stranded ship, any part thereof, and any object lost at sea from a ship which is stranded, sunken or adrift.[396] If MAVs were not considered 'vessels' under the convention, prima facie, it may appear to be a wreckable object, under the 'any object' provision may attract the operation of the convention and ensure, inter alia, that the registered owner of the wreck is obliged to remove it from the water. But it is clear the object must *come from a ship*, not be the putative ship; therefore, 'any object' would not be an alternative means for attracting the operation of the convention. However, MAVs meet the purpose of convention and ought to be within its scope. MAV shipowners, who profit from the use of the vehicle type, should be responsible financially for their vehicle and the removal of its wreck.

Whether the master and operator's duty to notify is applicable to MAVs will depend on the functionality of vehicle. There is a potential issue of, should SBOs are not considered masters, could the SBO[397] be responsible as an 'operator'? This is not likely. The operator is the organisation or person acting as the owner or manager, rather than the navigator per se.[398] MAVs may comply with this provision by virtue of the SBO informing the operator of the ship who would then be bound to notify the affected state. An SBO might be disinclined to do this, so the rules should be clarified to indicate that an SBO is bound to inform the operator or is bound to inform the affected state. A similar problem emerges for fully autonomous ships controlled by an AI; AIs are likely not 'masters' and therefore such MAVs would not be a duty to inform. The AI might inform the operators of the vessel who could in turn inform the affected state, but there must be a fail-safe way of the AI informing to ensure that warning is given to the affected state as soon as possible, thus reducing the environmental harm caused by the wreck. However, the fail-safe means of

[392] In whose waters the wreck is located.

[393] Nairobi Convention, Art. 5.

[394] Ibid, Arts. 9–10.

[395] Ibid, Art. 1.

[396] Ibid.

[397] That is, 'shore-based operator'.

[398] Baughen and Tettenborn (2021), p. 18.

informing should be solved by a convention dealing with design and functionality of the ship, such as SOLAS, rather than under the Nairobi Convention.

7.3 Lien

This part will briefly deal with the maritime lien. This is a claim held by the master or crew that runs *in rem*, which ensures certain payments are made even if the ship is sold. This is covered primarily by the domestic law of a state.[399] For example, only three countries signed up 1926 Convention on Maritime Liens and Mortgages, and a handful have signed up to the 1993 Convention on Maritime Liens and Mortgages, which entered into force in 2004. This section is a brief overview of this area of law and its compatibility with MAVs, but it is hoped it will provide some clarity. It considers maritime liens in English law. The two maritime liens which will be most relevant to MAVs are ones concerning claims for damage relating to a ship and claims for wages.

To qualify under the first, there must be 'damage done by a ship' which covers damage caused in the navigation or by the management of a ship; the ship must be the actual or noxious instrument in it, damage sustained is to a person or property external to a ship.[400] If a MAV is considered a ship, then it will be capable of attracting this liability both as a causer of damage and as having damage caused to it by another ship. The applicability of this lien for MAV operations would be useful to ensure liabilities are paid to the wronged party.

The less viable lien, in relation to MAVs, is the second, covering the master's and crew's wages and disbursements.[401] This should apply if the SBO is the 'master', which it is submitted will apply; it is less clear whether the SBO's assistants will be considered crew. Although the lien will apply to MAVs, there is little reason it should. The primary reason the wage lien is justified for conventional ships is owing to the difficulty in the seafarers' suing a shipowner or operator for wages where that individual is not in the same state as the seafarer. This arrangement is less likely with a shore-based SBO, who will more likely have a national association with the shipowner and crew and there will probably be a company registered the country the SBCC is based to deal with SBCC administration and crew pay. If this is the case, there is less need for a maritime lien for crew and wages of MAVs.

[399] See Baughen (2019).

[400] The Rama [1996] 2 Lloyd's Rep 281.

[401] Merchant Shipping Act 1995, s 41. For example, The Ever Success [1999] 1 Lloyd's Rep 823. Baughen and Tettenborn (2021), 18.

7.4 Salvage

As wrecks and collisions will continue to occur for MAVs, the need for salvage will persist also. This is governed internationally by the Salvage Convention.[402] These rules, in encouraging salvage by establishing an effective legal regime, assist in reducing loss of cargo and ships, as well as limiting damage to the environment; therefore, it is important that these rules continue to apply for MAVs. The initial question is whether the Salvage Convention applies to MAVs and what would the impact be.

Amongst its many provisions, the convention establishes the master's ability to conclude salvage contracts on behalf of the owner.[403] It covers MAVs, as it applies to vessels; where a vessel is defined as 'any ship or craft, or any structure capable of navigation'.[404] MAVs, even fully autonomous versions, will certainly be fit within the broad definition of 'any structure capable of navigation'.

As to the ability to conclude salvage contracts on behalf of the owner of the vessel, the contract may only be annulled or modified if the contract had been entered into under undue influence or if the terms of the contract are inequitable; or if the payment agreed is excessive or too small.[405] This power attaches to 'the master', which would include the SBO, but the utility of this provision is limited for MAVs. There is less need for the SBO of a MAV to have this power compared to a conventional ship. The power is designed for situations in which the navigator of the ship is out of communication with the management and shipowner and must secure the services of a salvor urgently; hesitancy could lead to the loss of the ship. However, if an SBO has sufficient connectivity to contact a salvor, they will be able to raise the shipowner or operator, especially if the SBCC is in the same location as the shipowner or management.

In contrast, for fully autonomous ships controlled by an AI, it is feasible that the ship will lose contact with the shore in extreme weather conditions, in which case the management and shipowner would not be able to conclude salvage contracts on behalf of the vessel. It is unclear what the appropriate solution would be in this situation; perhaps it is absurd futurism to conclude an AI could have this capacity to contract, but it is beyond the scope of this work to determine if AI could have legal personality. In ordinary circumstances, the shipowner and management could conclude salvage on behalf of the AI-controlled MAV, presuming the requisite connectivity is maintained with the ship and shore. Therefore, it is not necessary to adopt a detailed solution in the Salvage Convention for fully autonomous ships.

MAVs will continue to rely on salvage contracts. Unlike the duty to render assistance which does not apply to MAVs as there are no lives to save, salvage

[402] International Convention on Salvage (adopted 28 April 1989, entered into force 14 July 1996) 1953 UNTS 165 (known as the "Salvage Convention").

[403] Ibid, Art. 6.

[404] Ibid, Art. 1.

[405] Salvage Convention, Arts. 6 and 7.

contracts are explicitly not designed for saving lives; there is no incentive for life salvage under the convention.[406] Therefore, there is no reason the convention should not apply. However, there are questions about the scope and how far it extends to programmers who troubleshoot and save a broken-down MAV.[407] Such salvage is crucial for safety and commercial reasons, but it has yet to be judicially determined; it is important that this position is clarified under domestic law or by an amendment to the convention, so potential salvors are motivated to assist the salving of MAVs and have certainty that there is a valid salvage contract under the convention.[408]

7.5 *Limitation*

This section will briefly deal with the issue of limitation. This is a vastly significant topic with large consequences for maritime operators seeking to deploy MAVs, conventional sea users who share the seas with MAVs, and importantly software developers and programmers who could be exposed to extraordinarily large liabilities without a limitation. Consequently, limitation is important for ships; it encourages socially beneficial activities by ensuring the maritime adventure is insurable and the risk is certain.[409] It will play an equally significant role in the adoption of MAVs.

This section will determine if maritime limitation will apply to MAVs and if any parts of limitation pose present a barrier for MAVs, particularly (a) whether limitation covers the MAV vessel and (b) if it does, whether it extends to the SBO. This will focus on the Limitation Liability for Maritime Claims Convention (LLMC).[410] The LLMC says that the limitation applies to the shipowner and, for the purposes of the LLMC, the shipowner includes the manager or operator of a 'seagoing ship'.[411] It is submitted that MAVs engaged in seagoing activities will meet this definition. There is ample reason the limitation ought to apply, given both conventional ships and MAVs are engaged in broadly the same activity, for the benefit of society.

The limitation may be used be used 'whatever the basis of liability may be', including loss of life or personal injury or damage to property, losses from delays, contractual-based losses, losses from raising or removal of a ship, the cost of rendering harmless cargo of a ship.[412] If a claim is made in relation to a person's act, neglect or default and the shipowner is responsible for that person, that person

[406] Ibid, Art. 16.

[407] Soyer and Tettenborn (2021), p. 67.

[408] Ibid, p. 67.

[409] Song (2020).

[410] Convention on Limitation of Liability for Maritime Claims (adopted 17 November 1976, entered into force 1 December 1986) 1456 UNTS 221 (known as "LLMC").

[411] Ibid, Art. 1(2).

[412] Ibid, Art. 2.

may use the limitation, with certain exceptions.[413] This would extend to the SBO's operations, as the shipowner will be liable for the SBO's actions. It is crucial SBOs that may limit their liability; to attract qualified people to the job who are not deterred by prohibitive potential liabilities. However, if the SBO were not covered by the LLMC, the SBO may still have their liability limited. Similar protection may be provided for by the shipowner in the form of a *Himalaya* clause for the benefit of the SBO;[414] or the shipowner could undertake to pay the SBO's liabilities, as has occurred in the United States.[415] Although these solutions are effective, it is preferred that SBOs gain limitation by international rather than ad hoc national standards, to ensure claims are limited in each jurisdiction the vessel may visit.

8 Conclusion

This chapter has identified the barriers created in the law of the sea in relation to MAV and suggested solutions, focusing on public and private international law, and English commercial law. It found that if MAVs are considered ships and covered by UNCLOS, there may be barriers to entry to the port states (which has absolute jurisdiction), the territorial sea and the EEZ. In relation to the territorial sea, MAVs may be declined the privilege of innocent passage as such craft may be prejudicial to the coastal state. First, MAVs may pose a security threat to the coastal sate, primarily from a potential vulnerability to hijacking. Second, MAVs may be prejudicial to the coastal state through the risk of pollution. In relation to the EEZ, the coastal state has a generic jurisdiction over this zone in relation to pollution, which may—using the precautionary principle as an interpretive aid—permit the coastal state to exclude vessels which pose a threat to the coastal state's sovereign interests through pollution. In the high seas, MAVs' potential lack of ship status poses a risk of detention by states hostile to MAV integration. The status of MAVs must be clarified immediately by the IMO to avoid disputes amongst states and to clarify the rights of MAVs in the territorial sea. MAVs' ability to leave port states, enter the territorial sea of different states and their ability to flagged depends on their meeting international standards, which MAVs will not meet without amendment to those standards.[416] The environmental protection standards pose a likely barrier to MAVs, but that topic is beyond the scope of this chapter and must be considered by further research. To justify amendment to UNCLOS and the maritime safety conventions, MAVs must demonstrate a comparable level of safety to conventional ships; this also must be demonstrated for MAVs to integrate into English commercial law and for the shipowner to meet the classic duty of seaworthiness. It is not clear if MAVs can offer a comparable

[413] Ibid, Art. 1.

[414] Baughen (2021), p. 90.

[415] Cartner et al. (2009), para 14.2.1.

[416] Although the use of equivalents may be used by the flag state.

level of safety; this must be considered carefully in the engineering and scientific literature. MAVs will not easily meet the basic requirements of a bill of lading contract, that is, the signing the bill of lading. But it is likely that the port agent will meet this requirement in bill of lading contracts through an expanded role as supervisor of in the loading and discharging of cargo and issuer of the bill of lading. While the SBO will take on the role of master as regards navigation, many of the master's powers and duties will no longer be relevant, such as the power to imprison crew to instil discipline. As already stated, for MAVs to be deployed under the law of the sea, there must be an amendment to the IMO's legal instruments, which could be achieved by an implementation agreement or a MAV Code akin to the Polar Code, which is preferred as it has the benefit of amending the terms of several conventions in one effort.

References

Books and Journals

Allen C (2018) Determining the legal status of unmanned maritime vehicles: formalism vs functionalism. J Mar Law Com 49:477–514

Andreone G (2015) The exclusive economic zone. In: Rothwell D et al. (eds) The Oxford handbook of the law of the sea. OUP, pp. 159–180

Bartlett M (2018) Game of drones: unmanned maritime vehicles and the law of the sea. Auckland Univ Law Rev 24:66–91

Baughen S (2019) Maritime liens. In: Greenberg D (ed) Westlaw Insight

Baughen S (2021) Unmanned vessels and international conventions for the carriage of goods by sea. In: Soyer B, Tettenborn A (eds) Artificial intelligence and autonomous shipping: developing the international legal framework. Bloombury, London, pp 81–98

Baughen S, Tettenborn A (2021) International regulation of shipping and unmanned vessels. In: Soyer B, Tettenborn A (eds) Artificial intelligence and autonomous shipping: developing the international legal framework. Bloombury, London, pp 7–24

Bork K et al (2008) The legal regulation of floats and gliders—in quest of a new regime? Ocean Dev Int Law 39:298–328

Boyle A (2005) Further development of the law of the sea convention: mechanism for change. ICLQ 54:563–584

Brownlie I (1963) International law and the use of force by states. OUP, Oxford

Carey L (2017) All hands off deck? The legal barriers to autonomous ships. JIML 23:202–219

Cartner J et al (2009) International law of the shipmaster. Informa, London

Chae CJ et al (2020) A study on identification of development status of MASS technologies and directions of improvement. Appl Sci 10:4564–4582

Daum O, Stellpflug T (2017) The implications of international law on unmanned merchant vessels. JIML 23:363–374

Davies M (2020) Pilotage of autonomous and remotely-controlled ships. In: Ringbom H et al (eds) Autonomous ships and the law. Routledge, London, pp 279–294

De La Fayette L (1996) Access to ports in international law. Int J Mar Coast Law 11:1–22

Dixon M et al. (2016) Cases & materials on international law, 6th edn. OUP, Oxford

Eder B (2019) Unmanned vessels: challenges ahead. LMCLQ 45:47–56

Evans M (2018) The law of the sea. In: Evans M (ed) International law, 5th edn. OUP, Oxford, pp 635–670
Fan C et al (2020) A framework to identify factors influencing navigational risk for maritime autonomous surface ships. Ocean Eng 202:article 107188
Gold E (2004) From privilege to peril—the shipmaster's current legal rights and responsibilities. WMU J MA 3:51–66
Greenwood C (2011) Self-defence. In: Wolfrum R (ed) Max Planck institute of public international law. OUP, Oxford
Guilfoyle D (2015) The high seas. In: Rothwell D et al. (eds) The oxford handbook of the law of the sea. OUP, pp. 203–225
Hakapää K, Molenaar E (1999) Innocent passage—past and present. Marine Policy 23:131–145
Howse T (2020) P&I perspectives. In: Ringbom H et al (eds) Autonomous ships and the law. Routledge, London, pp 193–206
Höyhtyä M (2019) Connectivity manager: ensuring robust connections for autonomous ships. 2019 2nd ICoIAS, 86–90
Jordan S (2020) Captain, my captain: a look at autonomous ships and how they should operate under admiralty law. Indiana Int Comp Law Rev 30:283–317
Kanehara A (1999) The Japanese legal system concerning innocent passage of foreign vessels (1990–1998). Jap Ann Int Law 42:90–110
Karlis T (2018) Maritime law issues related to the operation of unmanned autonomous cargo ships. WMU J MA 17:119–128
Katsivela M (2020) Unmanned vessels and regulatory concerns. JIML 26:239–252
Kenney F (2018) Global regulation of ships: the future of development and implementation at the International Maritime Organization. Tul Mar Law J 42:259–268
Kirk E (2015) Science and the international regulation of marine pollution. In: Rothwell D et al. (eds) The Oxford handbook of the law of the sea. OUP, pp. 516–534
Klein N (2006) Legal limitations on ensuring Australia's maritime security. Melb J Int Law 7:306–338
Klein N et al (2020) Maritime autonomous vehicles: new frontiers in the law of the sea. ICLQ 69: 719–734
Kojima C, Vereshchetin V (2013) Implementation agreements. In: Wolfrum R (ed) Max Planck institute of public international law. OUP, Oxford
Kraska J (2010) The law of unmanned naval systems in war and peace. Subsea Vehicl 5:44–64
Lowe AV (1977) The right of entry into maritime ports in international law. San Diego Law Rev 14: 597–622
Macdonald Eggers P (2021) Maritime autonomous surface ships: marine insurance response to risks. In: Soyer B, Tettenborn A (eds) Artificial intelligence and autonomous shipping: developing the international legal framework. Bloombury, London, pp 139–160
Molenaar E (2015) Port and coastal states. In: Rothwell D et al (eds) The Oxford handbook of the law of the sea. OUP, Oxford, pp 228–303
Oxman B (2006) The territorial temptation: a siren song at sea. Am J Int Law 100:830–851
Perera L, Batalden BM (2019) Possible COLREGs Failures Under Digital Helmsman of Autonomous Ships. Oceans 2019 - Marseille 1–7
Proelss A et al (2020) United Nations Convention on the law of the sea: a commentary. Bloomsbury Collections, Munchen
Redgwell C (2018) International environmental law. In: Evans M (ed) International law, 5th edn. OUP, Oxford, pp 675–716
Ringbom H (2019) Regulating autonomous ships—concepts, challenges and precedents. Ocean Dev Int Law 40:141–169
Ringbom H (2020a) Developments, challenges, and prospects at the IMO. In: Ringbom H et al (eds) Autonomous ships and the law. Routledge, London, pp 56–68
Ringbom H (2020b) Legalizing autonomous ships. Ocean Yearb 34:429–460

Roscini M (2002) The navigational rights of nuclear ships. Leiden J Int Law 15:251–265
Schelin J (2020) Manning of unmanned ships. In: Ringbom H et al (eds) Autonomous ships and the law. Routledge, London, pp 261–278
Schmitt M (2017) Tallinn Manual 2.0 on the international law applicable to cyber operations. CUP, Cambridge
Serdy A (2018) Public international law aspects of shipping regulation. In: Baatz Y et al (eds) Maritime law, 4th edn. Routledge, Oxford
Simpson J (n.d.) Oxford English Dictionary. OUP, Oxford
Song M (2020) Moving forward by looking back. In: Ringbom H et al (eds) Autonomous ships and the law. Routledge, London, pp 222–239
Soyer B (2020) The future of autonomous shipping - the regulatory challenge. In: Soyer B, Tettenborn A (eds) Artificial intelligence and autonomous shipping: developing the international legal framework. Bloombury, London, pp 163–183
Soyer B, Tettenborn A (2021) Autonomous ships and private law issues. In: Soyer B, Tettenborn A (eds) Artificial intelligence and autonomous shipping: developing the international legal framework. Bloombury, London, pp 63–80
Stemre T (2018) The IMO, the polar regions, and global ocean governance: newly accessible maritime environments. In: Attard D et al (eds) The IMLI treatise on global governance: volume III: the IMO and global governance. OUP, Oxford
Stephens T, Rothwell D (2015) Marine scientific research. In: Rothwell D et al. (eds) The Oxford handbook of the law of the sea. OUP, pp. 559–581
Stevens F (2020) Seaworthiness and good seamanship in the age of autonomous vessels. In: Ringbom H et al (eds) Autonomous ships and the law. Routledge, London, pp 243–260
Tanaka Y (2015) Navigational rights and freedoms. In: Rothwell D et al. (eds) The Oxford handbook of the law of the sea. OUP, pp. 536–559
Van Dyke J (1996) Applying the precautionary principle to ocean shipments of radioactive materials. Ocean Dev Int Law 27:379–397
Van Dyke J (2002) The legal regime governing sea transport of ultrahazardous radioactive materials. Ocean Dev Int Law 33:77–108
Van Dyke J (2005) The disappearing right to navigational freedom in the exclusive economic zone. Marine Policy 29:107–121
Van Hooydonk E (2014) The law of unmanned merchant shipping—an exploration. JIML 6:403–423
Van Logchem Y (2021) International law of the sea and autonomous cargo 'Vessels'. In: Soyer B, Tettenborn A (eds) Artificial intelligence and autonomous shipping: developing the international legal framework. Bloombury, London, pp 25–62
Veal R, Ringbom H (2017) Unmanned ships and the international regulatory framework. JIML 23: 100–118
Veal R, Tsimplis M (2017) The integration of unmanned ships into the lex maritima. LMCLQ 43: 303–335
Veal R et al (2019) The legal status and operation of unmanned maritime vehicles. Ocean Dev Int Law 50:23–48
Viljanen M (2020) Insuring autonomous vessels. In: Ringbom H et al (eds) Autonomous ships and the law. Routledge, London, pp 207–221
Vojković G, Milenković M (2020) Autonomous ships and legal authorities of the ship master. Case Stud Transp Policy 8:333–340
Wang F (2020) The warranty of seaworthiness and cyber risk of unmanned ships. JBL 4:311–325
Wróbel K et al (2017) Towards the assessment of potential impact of unmanned vessels on maritime transportation safety. Reliabil Eng Syst Saf 165:155–169
Zanella TV (2020) The environmental impacts of the 'Maritime Autonomous Surface Ships' (MASS). Veredas do Direito 17:367–384

Websites

AAWA (2016) Remote and Autonomous Ships: The next steps. https://www.rolls-royce.com/~/media/Files/R/Rolls-Royce/documents/customers/marine/ship-intel/aawa-whitepaper-210616.pdf. Accessed on 7 August 2022

CMI (2018) Summary of responses. https://comitemaritime.org/wp-content/uploads/2018/05/Summary-of-Responses-to-the-CMI-Questionnaire.docx. Accessed on 7 August 2022

Dellinger AJ (2021) Elon Musk's Starlink Is Quietly Expanding, But Speeds Are Getting Worse. https://www.forbes.com/sites/ajdellinger/2021/12/23/elon-musks-starlink-is-quietly-expanding-but-speeds-are-getting-worse/?sh=ded4fc424197. Accessed on 7 August 2022

DNV (2015) Ship Connectivity. https://www.dnv.com/Publications/ship-connectivity-28107. Accessed on 7 August 2022

George R (2021) Wind . . . or worse: was pilot error to blame for the Suez blockage? https://www.theguardian.com/environment/2021/apr/03/wind-or-worse-was-pilot-error-to-blame-for-the-suez-blockage. Accessed on 7 August 2021

Jankowicz M (2021) Ever Given insurers say Egypt's $600 million claim is still too much, as authorities continue to impound the ship. https://www.businessinsider.com/even-given-insurers-egypts-600-million-claim-still-too-big-2021-5?r=US&IR=T. Accessed on 7 August 2022

Kwan R, Henley J (2021) China berates US after 'close encounters' with Elon Musk satellites. https://www.theguardian.com/science/2021/dec/28/china-complains-to-un-after-space-station-is-forced-to-move-to-avoid-starlink-satellites. Accessed on 7 August 2022

Roulette J (2021) SpaceX moves to beam Starlink internet into trucks, boats, and aircraft. https://www.theverge.com/2021/3/8/22319761/spacex-starlink-fcc-internet-cars-boats-aircraft. Accessed on 7 August 2022

Part III
Road Vehicles: Road Automated/Autonomous Transportation Systems

Automated Vehicles, Liability, and Insurance

Sara Landini and Francesco La Fata

Abstract This chapter aims to investigate the problems that algorithmic automation of vehicles poses in the area of civil liability and liability insurance.

The first part frames the types of automatic vehicles by identifying the different risk profiles. The second part deals with the ethical issues that arise with respect to automated choice. The third part deals with the problem of civil liability in the event of an automated action. The fourth part is dedicated to insurance coverage to find relief from the damage caused by the automated vehicle.

Keywords Automation · Ethical issues · Insurance

1 Automated Vehicles

1.1 What Is an Automated Vehicle

As well known, the concept of automation applied to vehicles is strictly related to the use of algorithms. It refers to systems that display intelligent behavior by analyzing their environment and taking actions—with some degree of autonomy—to achieve specific goals.[1] Only in 1979 Stanford Cart was able to move without human intervention using an image processing algorithm called The Cart's Vision

This chapter was written by Prof. Sara Landini, Full Professor Business Law, University of Florence, and Dr. Francesco La Fata, Researcher in Economic Law, at the University of Florence. Particularly, paragraphs 1, 2, 3, 4, and 6 refer to Prof. Sara Landini, while paragraph 5 refers to Dr. Francesco La Fata.

[1]Nof (2009), pp. 13–52.

S. Landini · F. La Fata (✉)
University of Florence, Firenze, FI, Italy
e-mail: sara.landini@unifi.it; francesco.lafata@unifi.ti

K. Noussia, M. Channon (eds.), *The Regulation of Automated and Autonomous Transport*, https://doi.org/10.1007/978-3-031-32356-0_9

Algorithm, that was inspired by the Blocks World planning method. It consisted in the reduction of the image to a set of edges.[2]

SAE (Society of Automobile Engineers) International's On-Road Automated Vehicle Standards Committee, along with experts from industry and government, provided an information report defining key concepts related to the increasing automation of on-road vehicles. Central to this report are six levels of driving automation: 0 (no automation), 1 (driver assistance), 2 (partial automation), 3 (conditional automation), 4 (high automation), and 5 (full automation).

According to forecasts, there will be as many as 10 million fully driverless cars on roads around the world in the coming years. Important numbers for a technology that grows faster than expectations. This is why understanding the future of driverless cars is not easy. Governments around the world are discussing the regulations to be implemented to allow driverless vehicles to be present in our cities. As we will see in the next paragraphs, it is necessary to understand whether it is necessary to write new rules or it is sufficient to adapt the current legislation. An increase in connected devices is expected globally with the IoT (Internet of Things) which will also have a huge impact on road transport networks. Vehicle-to-vehicle and vehicle-to-infrastructure communication will allow driverless cars to leverage data in real time, making mobility for city residents much more enjoyable and easier. Furthermore, it seems likely that the majority of future driverless cars will be fully electric vehicles (FEVs). Cities will therefore need to increase investment in charging infrastructure to support the growing share of automated and non-automated electric vehicles.

We are witnessing a revolution in road traffic which requires rethinking civil liability and its insurance.

1.2 Level of Automation

To define the cause of action in case of an automated choice, it is important to consider the above-mentioned levels of automation. Generally, and not only on vehicles, we can distinguish the levels of automation as follows:

- at level 1 the human operator acts and turns to the computer to implement his/her actions;
- at level 2 the computer helps the human operator by determining the options;
- at level 3 the computer suggests options and the human operator can choose to follow the recommendation;
- at level 4 the computer selects the action and the human operator decides if it should be done or not;
- at level 5 the computer selects the action and implements it if the human operator approves the selected action;

[2] Simon (1979), p. 10.

- at level 6 the computer selects the action and informs the human operator who can cancel the action;
- at level 7 the computer does the action and inform the human operator;
- at level 8 the computer does the action and inform the human only if the human operator asks;
- at level 9 the computer does the action and informs the human operator only if the computer decides the operator should be told;
- at level 10 the computer does the action if it decides it should be done. The computer informs the human operator only if it decides the operator should be told.

Raja Parasuraman, Sheridan, and Wickens also distinguish four models of human information processing:[3]

1. Sensory processing, which refers to the acquisition and registration of multiple sources of information and includes the positioning and orienting of sensory receptors, sensory processing, initial pre-processing of data prior to full perception, and selective attention. This model can be translated in the function of information acquisition.
2. Perception and/or working memory, which regards conscious perception and manipulation of processed and retrieved information in working memory. It includes cognitive operations such as rehearsal, integration, and inference, but these operations occur prior to the point of decision. This model can be translated in the function of information analysis.
3. Decision making, which means that a decision is based on such cognitive processing. This model can be translated in the function of decision and action selection.
4. Response selection, which involves the implementation of a response or action consistent with the decision choice. This model can be translated in the function of decision and action implementation.

On the four functions discussed above, it is possible to provide an initial categorization for types of tasks in which automation can support the human operator:

1. Information acquisition: the automation of information acquisition can be applied to the sensing and registration of input data.
2. Information analysis: the automation in this function involves cognitive functions such as working memory and inferential processes.
3. Decision and action selection. The decision and action selection involve selection from among decision alternatives.
4. Action implementation, which refers to the actual execution of the action choice.

[3] Parasuraman et al. (2000), pp. 286–297.

This pattern becomes important for designing the different human choice and automated choice games.[4] This is a relevant perspective for thinking about the legal effects of humans related but non-human choices.

2 Ethical Issues

2.1 Ethical Guidelines

Automation can reduce human errors, and also damages, but cannot exclude the latter, especially when they cannot be eliminated.

Dilemma situation cases can arise. The question then becomes choosing which target is best to hit. The classic example is that of the autonomous vehicle that has provided a dramatic, but unavoidable alternative between killing the driver or a pedestrian walking across the street.[5]

Some countries, such as Germany, try to find an answer by creating guidelines for these "dilemma situations." This could be an ethical solution, but what is the consequence if the software is programmed according to these guidelines? What is the impact of that on liability? The Guidelines issued by the German Transport Ministry, thanks to the work of a committee of expert, try to find solutions for the issues of liability. In a Report published in 2017, it stressed that in every driving situation, it must be clearly regulated and apparent who is responsible for the driving task: the human or the computer. This information must be documented.

As Lütge, member of the above-mentioned German ethics committee for automated and connected driving, noted that "As levels of automation progress, with partially automated driving already becoming standard in new cars from a number of manufacturers, the question of ethical and legal standards becomes virulent. For example, while automated and autonomous cars, being equipped with appropriate detection sensors, processors, and intelligent mapping material, have a chance of being much safer than human-driven cars in many regards, situations will arise in which accidents cannot be completely avoided. Such situations will have to be dealt with when programming the software of these vehicles. In several instances, internationally, regulations have been passed, based on legal considerations of road safety, mostly. However, to date, there have been few, if any, cases of a broader ethics code for autonomous or automated driving preceding actual regulation and being based on a broadly composed ethics committee of independent experts."[6]

[4] Ching-Yao (2017), pp. 208–216. Among the first contributions "Hearings on automation and technological change," in *Subcommittee on Economic Stabilization of the Joint Committee on the Economic Report*, (US Congress October 1955) 14–28; Buckingham (1961).

[5] Sara Landini (2020a), pp. 291–312.

[6] Lütge (2017), p. 9; Ethik-Kommission (2017) Automatisiertes und Vernetztes Fahren. https://www.bmvi.de.

On the dilemma situation, the Committee introduced guidelines. Guideline 5 provides that "Automated and connected technology should prevent accidents wherever this is practically possible. Based on the state of the art, the technology must be designed in such a way that critical situations do not arise in the first place. These include dilemma situations, in other words a situation in which an automated vehicle has to decide which of two evils, between which there can be no trade-off, it necessarily has to perform. In this context, the entire spectrum of technological options—for instance from limiting the scope of application to controllable traffic environments, vehicle sensors, and braking performance, signals for persons at risk, right up to preventing hazards by means of intelligent road infrastructure—should be used and continuously evolved. The significant enhancement of road safety is the objective of development and regulation, starting with the design and programming of the vehicles such that they drive in a defensive and anticipatory manner, posing as little risk as possible to vulnerable road users." Thus, prevention is the most important task of automated cars and technology should help. But what about in case of unavoidable damage?

According to the guideline 7 "In hazardous situations that prove to be unavoidable, despite all technological precautions being taken, the protection of human life enjoys top priority in a balancing of legally protected interests. Thus, within the constraints of what is technologically feasible, the systems must be programmed to accept damage to animals or property in a conflict if this means that personal injury can be prevented." Such guideline simply states that damage to humans takes priority respect to animals and property.

The most important dilemmatic decision is the choice between one human life and another. Such decisions, according to guideline 8, " can thus not be clearly standardized, nor can they be programmed such that they are ethically unquestionable. Technological systems must be designed to avoid accidents. However, they cannot be standardized to a complex or intuitive assessment of the impacts of an accident in such a way that they can replace or anticipate the decision of a responsible driver with the moral capacity to make correct judgements. It is true that a human driver would be acting unlawfully if he killed a person in an emergency to save the lives of one or more other persons, but he would not necessarily be acting culpably. Such legal judgements, made in retrospect and taking special circumstances into account, cannot readily be transformed into abstract/general ex ante appraisals and thus also not into corresponding programming activities. For this reason, perhaps more than any other, it would be desirable for an independent public sector agency (for instance, a Federal Bureau for the Investigation of Accidents Involving Automated Transport Systems or a Federal Office for Safety in Automated and Connected Transport) to systematically process the lessons learned."

Thus, in this case, the decision of deliberately sacrificing specific lives should not be taken by a programmer. Moreover, the guideline 9, in case of unavoidable accident situations, any distinction based on personal features (age, gender, physical, or mental constitution) is prohibited.

In the European Union, since the theme is linked to the freedom of movement in the European space, it soon led to the design of guidelines at a European level. As we are going to see in the next paragraph, guidelines for ethical decisions have also been adopted by the European Commission.

2.2 European Ethical Guidelines

The European Commission considers Artificial Intelligence (AI) an area of strategic importance and a key driver of economic development. At the same time, the EC addressed socio-economic, legal, and ethical impacts of the AI.[7] The European Commission expressed in all its communication a European approach to Artificial Intelligence based on three pillars:

- Being ahead of technological developments and encouraging uptake by the public and private sectors. The EU Commission is increasing its annual investments ordered to connect and strengthen AI research centers across Europe. The EU Commission supports the development of an "AI-on-demand platform" that will provide access to relevant AI resources in the EU for all users and supports the development of AI applications in key sectors. On 10 April 2018, 25 European countries signed a Declaration of cooperation on Artificial Intelligence. It builds further on the achievements and investments of the European research and business community in AI.
- Preparing for socio-economic changes brought about by AI. The European Commission will support business-education partnerships to attract and keep more AI talent in Europe; it will set up dedicated training and retraining schemes for professionals; it will foresee changes in the labor market and skills mismatch; it will support digital skills and competences in science, technology, engineering, mathematics (STEM), entrepreneurship and creativity.
- Ensuring an appropriate ethical and legal framework.

On 8 April 2019, the final Ethics Guidelines for Trustworthy Artificial Intelligence[8] prepared by the High-Level European Group on Artificial Intelligence was published. The European Commission will also develop and make available guidance on the interpretation of the Product Liability directive.

They focus on new useful kinds of machine learning approach such as the so-called "reinforcement learning." In this approach, the AI system is free to make its decisions, over time, and at each decision we provide it with a reward signal that tells it whether it was a good or a bad decision. The goal of the system is to maximize the positive reward received. This approach is used, for example, in recommender system (such as the several online recommender systems that suggest users what they might like to buy), or also in marketing.

[7] Bonab et al. (2021); Dafoe et al. (2017), pp. 729–754.

[8] Independent High-Level Expert Group on Artificial Intelligence set up by The European Commission (2019). Ethics Guidelines For Trustworthy AI, Brussels.

Thus, at the end, the group proposes a new definition of AI, which could improve the task and the liability of the manufacturer:

> Artificial intelligence (AI) systems are software (and possibly also hardware) systems designed by humans that, given a complex goal, act in the physical or digital dimension by perceiving their environment through data acquisition, interpreting the collected structured or unstructured data, reasoning on the knowledge, or processing the information, derived from this data and deciding the best action(s) to take to achieve the given goal. AI systems can either use symbolic rules or learn a numeric model, and they can also adapt their behavior by analyzing how the environment is affected by their previous actions.
>
> As a scientific discipline, AI includes several approaches and techniques, such as machine learning (of which deep learning and reinforcement learning are specific examples), machine reasoning (which includes planning, scheduling, knowledge representation and reasoning, search, and optimization), and robotics (which includes control, perception, sensors and actuators, as well as the integration of all other techniques into cyber-physical systems).

The Commission's Communication of 8 April 2019 sets out a human-centric approach, AI is seen as a tool operating in the service of humanity and the public good, aiming to increase individual and collective human well-being. Since people will only be able to confidently and fully reap the benefits of a technology that they can trust, AI's trustworthiness must be ensured. With this Communication, the EU Commission welcomes the publication of Guidelines on AI.

As said, the AI HLEG presented a first draft of the Guidelines in December 2018. Following further deliberations by the group in light of discussions on the European AI Alliance, a stakeholder consultation and meetings with representatives from Member States, the Guidelines were revised and published in April 2019.

Based on fundamental rights and ethical principles, the Guidelines published in 2019 a list of seven key requirements that AI systems should meet to be trustworthy:

- Human agency and oversight
- Technical robustness and safety
- Privacy and data governance
- Transparency, particularly in algorithms
- Diversity, non-discrimination, and fairness
- Societal and environmental well-being
- Accountability

On April 8, 2020, the Committee of Ministers of the Council of Europe (CoE) released a new regulatory framework "the Recommendation on the human rights impacts of algorithmic systems."[9]

The CoE requested Member States to take a precautionary approach to the development and use of algorithmic systems and adopt legislation, policies, and practices that fully respect human rights. The precautionary principle enables decision-makers to adopt precautionary measures when scientific evidence about

[9]Committee of Ministers of the Council of Europe (CoE) (2020) Recommendation on the human rights impacts of algorithmic systems. Brussels.

an environmental or human health hazard is uncertain and their risk is high.[10] The precautionary principle emerged since the 1970s and has since been enshrined in a number of international treaties on the environment, in the Treaty on the Functioning of the European Union and the national legislation of certain Member States. The Precautionary Principle is said to have first appeared in the 1970 German water protection law that introduced Der Vorsorgeprinzip (foresight principle).[11]

Opinions are divided on the method for determining when to apply precautionary measures. The application of the precautionary principle presents many opportunities as well as challenges. The precautionary principle is closely linked to three aspects of governance: risk governance (risk assessment, management, and communication), science-policy interfaces and the link between precaution and innovation.

The CoE's recommendation issued a set of guidelines on addressing the human rights impacts of algorithmic systems. It called on governments to ensure that they do not breach human rights through their own use, development, or procurement of algorithmic systems.

Not only guidelines have been issued at the European Union level. The Commission published its AI package in April 2021, proposing new rules and actions to turn Europe into the global hub for trustworthy AI. This package consists of a Communication on Fostering a European Approach to Artificial Intelligence.

It includes the following actions:

- enabling the development and uptake of AI in the EU
- making the EU the place where AI thrives from the lab to the market
- ensuring that AI works for people and is a force for good in society
- building strategic leadership in high-impact sectors
- the Coordinated Plan with Member States updated. It outlines a vision to accelerate, act, and align priorities with the current European and global AI landscape and bring AI strategy into action in maximizing resources and coordinating investments
- a proposal for an AI Regulation laying down harmonized rules for the EU (Artificial Intelligence Act)

The proposal offers a risk-based approach designing a process that allows to identify potential high risks and develop strategies to mitigate them.

Following the Commission's proposal in April 2021, the regulation will enter into force in the second half of 2022 after a transition period, where standards would be mandated and developed, and the governance structures set up would be operational. During the second half of 2024 regulation could become applicable to operators with the standards ready and the first conformity assessments carried out.

As AI is a fast-evolving technology, the proposal is based on a future-proof approach, allowing rules to adapt to technological change. AI applications should

[10] Bodansky (1991), pp. 4–44.

[11] Boehmer-Christiansen (1994), pp. 38–39.

remain trustworthy even after they have been placed on the market. This requires ongoing quality and risk management by providers.

2.3 *The Central Role of Accountability*

Accountability is perhaps the more complex and more important concept as it can resume all the above-mentioned AI requirements. Generally, accountability means responsibility and therefore it means being responsible, being trustworthy. The conceptual difference between the terms Accountability and Responsibility is based on the concepts of "having to do" and "responding to results."[12]

Generally, we can link the concept of "responsibility" to having to act, while that of "accountability" to giving an account of the action done, to responding to the results obtained.

Responsible is the "operational manager," while accountable is the "ultimate manager."

Basically, those who have "accountability," take risks personally, going beyond the formality of the assignment received and guiding others towards the common result.

On the other hand, those who have "responsibility" are involved based on the organizational role held, professionally carrying out their specialist work.

For a specific project, only one person is accountable, while multiple people can be responsible.

Accountability plays different role in AI management. From the point of view of the governance, accountability includes governance processes and structures. Appropriate governance of AI can help manage risk, demonstrate ethical values, and ensure compliance. Accountability for AI means looking for solid governance including clear goals and objectives for the AI system; well-defined roles, responsibilities, and lines of authority; a multidisciplinary workforce capable of managing AI systems; a broad set of stakeholders; and risk-management processes. Additionally, it requires documented technical specifications of the particular AI system, compliance, and stakeholder access to system design and operation information.[13]

From the point of view of data managing, it is important to have documentation of how data is being used, especially in two different stages of the system: when it is being used to build the underlying model and while the AI system is in actual operation. Good AI oversight includes having documentation of the sources and origins of data used to develop the AI models. The reliability and representativeness of the data needs to be considered, including the potential for bias, inequity, or other

[12] Christoph Lütge (2017).

[13] Chen et al. (2015); Kroll et al. (2016), p. 633; Kim (2017), pp. 166, 189; Sandvig et al. (2014); Surden and Williams (2016), p. 121.

societal concerns. Accountability includes the analysis of AI system's data security and privacy.

From the point of view of performance goals and metrics, accountability requires metrics and the methods used to assess the performance. Management must be able to ensure an AI application meets its intended goals. It is crucial that these performance assessments take place at the broad system level but also focus on the individual components that support and interact with the overall system.

From the point of view of review monitoring plans, AI accountability includes establishing a range of model drift that is acceptable, and sustained monitoring to ensure that the system produces the expected results. Long-term monitoring must also include assessments of whether the operating environment has changed and to what extent conditions support scaling up or expanding the system to other operational settings. Important questions to ask are whether the AI system is still needed to achieve the intended goals, and what metrics are needed to determine when to retire a given system.

In the EU Commission Guidelines "Ethics of connected and automated vehicles Recommendations on road safety, privacy, fairness, explainability and responsibility" accountability play a relevant role.

Accountability can be separated from culpability and liability. According to the EU Guidelines, a fair system of accountability on the duty to explain requires that: (a) relevant formal and informal mechanisms of accountability are created; (b) different actors should be aware in advance of their respective potential need and they should have the possibility to explain and justify the operation of the system and to acknowledge failures; (c) the socio-technical system of which CAVs are a part is not too complex, opaque, or unpredictable and relevant actors have sufficient insight into its functioning and their role in it.

Another form of accountability is the public accountability of manufacturers and deployers, who should ensure that the logic behind sensitive decisions made by CAVs are transparent and explainable to the public.[14]

One of the problems related to accountability is the complexity and interaction of different agents that may create accountability gaps: situations in which it is not clear which of the agents in the responsibility network can be held to explain the behavior of the vehicle.

The regulation of accountability is not a matter of the policymaker only. They should collaborate with researchers and manufacturers and deployers to develop clear and fair criteria for assigning culpability to individual actors or organizations in the event that something goes wrong with CAVs.

[14] Mittelstadt et al. (2019); Miller (2019), pp. 1–38.

3 Liability Issues

3.1 Damage Compensation in Case of Automation

Most of the literature and of the European Institutions attention, dealing with compensation for damage caused in the event of automation, has raised the question of the possible responsibility or co-responsibility of the producer and/or programmer. The most part of the efforts are on the reform of the directive on manufacturer liability and on cybersecurity. Scholars have been also concerned to review the concept of guilt of the owner and/or user of the automated product, for example by raising the level of diligence required.[15]

Without raising any criticism to the aforementioned theories that surely have seized central aspects of the theme, we intend to draw attention to aspects linked to the problem of man-machine interaction and the c.d. self-learning of the machine in the cause of losses highlighting how civil liability is perhaps not sufficient to compensate and above all to prevent damages.

The recommendations to the EU Commission on Civil Law Rules on Robotics (2015/2103(INL) stressed on this point.

The interaction between machines and men (owner, user, producer, programmer, machine's manager, etc.) are so many and complex that it becomes difficult to identify the person (s) responsible.

Even the recourse to solidarity between co-responsible parties fails to resolve the difficulties for the victim to identify the person responsible. In fact, each potentially responsible person will try to prove that the action of the others has excluded his/her own responsibility.

There could be a real risk of extending the time needed to compensate the damage. Moreover, the damage is usually so huge and with domino effects that prevention, rather than compensation, becomes fundamental.

One solution can be found in special legal provision about joint liability. This issue of the presence of a plurality of actors in the wrongdoing does not emerge if we say that all subjects involved are automatically joint tortfeasors. In this case they are liable jointly and then it is their task to find out internally. In any case, this solution needs a legislative intervention to put the liability on all actors independently from the proof of their culpability and of the causal chain in the determination of the wrongdoing.

Some authors propose "to construct a system of strict liability, completely uncoupled from notions of fault for this select group of cases. A strict liability regime cannot be based here on the argument that the vehicles are 'ultra-hazardous' or 'unreasonably risky' for the simple reason that driver-less vehicles are likely to be far less hazardous or risky than the products they replace. Indeed, it is precisely because these machines are so technologically advanced that we expect them not to fail. For these reasons, a true strict liability regime will be needed; one that does not

[15] Patrick Hubbard (2015), p. 1803, Chopra and White (2011); Gless and Seelmann (2016).

resort to a risk-utility test or the re-institution of a negligence standard for the simple fact that those tests will be difficult, if not impossible, for the injured party to overcome." The author also says "Lest there be any doubt, my argument is not based on notions of a 'no-fault' liability system, that is, a system that substitutes mandatory insurance and eliminates access to the judicial system. My proposal is a strict liability regime implemented by the courts. Although the idea of 'no fault' systems took hold in the 1970s and 1980s, and was expected to drive down insurance costs by limiting the transaction costs related to litigation, it is by now apparent that those systems have not worked as envisioned. It is likely, however, that the introduction of driverless cars will shift liability from the 'driver' to the manufacturer, and that shift may trigger a resurgence of interest in 'no fault' insurance regimes."[16]

3.2 *The Proposed Solutions*

As said by EU Commission on Civil Law Rules on Robotics (2015/2103(INL): "the more autonomous robots are, the less they can be considered to be simple tools in the hands of other actors." Thus it is important to reshape civil liability and/or find other mechanisms to prevent and compensate losses when "the cause cannot be traced back to a specific human actor and (when) the acts or omissions of robots which have caused harm could have been avoided."[17]

Tentative to find solution to compensation could be found in some interventions at level of national law and of EC legislation trying to reshape liability in case of motor liability and manufacturer liability.

In the US, different legislation has been enacted about the use of automated vehicles. Some states (California, Florida and Nevada) provide that it is mandatory for drivers of automated vehicles to submit an insurance or a surety bond or to give proof of a self-insurance. Moreover, Nevada's legislation does not require a licensed operator for a "fully autonomous vehicle" if the vehicle can achieve "a minimal risk condition" in the event of a failure.[18] A different solution has been proposed by other states which require a human operator to be present and capable of taking over in an emergency. In German law, a new § 1a StVG (the German law on motor liability) on

[16] Vladeck (2014), p. 146. See also Anderson et al. (2010), http://www.rand.org. We have to say that a strict liability rule on the owner or the user could discourage the use of AI in contrast with the common idea that AI represent an important instrument to reduce risks. In this sense, see Landini (2020b), p. 159. Different would be in the case of strict liability on the producer, which is the entity that obtains profits and can limit the risks. On this matter, see, among the others, Borges (2021), pp. 32–39; Zech (2021), pp. 147–158; Wendehorst (2020), pp. 150–180; Spiecker gen. Döhmann (2012); Spindler (2019); Čerka et al. (2015), pp. 376–389; Fluet (2002), pp. 845–861.

[17] European Parliament, Resolution of 16 February 2017 with recommendations to the Commission on Civil Law Rules on Robotics (2015/2103(INL)).

[18] Aida Joaquin Acosta (2018) "What Governments across the globe are doing to seize the benefits of autonomous vehicles." Cyber Harvard edu. https://cyber.harvard.edu/.

"Motor vehicles with highly or fully automated driving function" has been introduced on 16 June 2017. Under German law, the liability of the car owner as in § 7 StVG, in the case of an autonomous vehicle, remains unaffected anyway, since the owner is liable for all damages that can be referred to the "operation of a motor vehicle." So it is just an additional liability of the motor vehicle driver. For this purpose, the new norm contained in § 1a StVG says that the user must remain receptive to be able to take control immediately.

In fact, under § 1a StVG an automated vehicle: shall be manually overridden or deactivated by the driver at any time; shall recognize the necessity of the vehicle hand control by the driver; shall visually, acoustically, tactually, or otherwise discernibly indicate to the vehicle driver the requirement of the vehicle hand control with sufficient time reserve before the vehicle control is delivered to the vehicle driver.[19]

In Italy, on 28 February 2018, the Minister of Transport and Infrastructure (MIT) issued a decree which permits road testing of automatic guided vehicles. The Decree of 28 February 2018 was implemented taking into account Regulation (EC) No. 377/2014 of the Parliament and of the European Council of 3 April 2014, establishing the program Copernicus and repealing Regulation (EU) No. 911/2010; and having regard to Directive No. 2010/40 / EU of the European Parliament and of the Council of 7 July 2010, on the general framework for dissemination intelligent transport systems in the transport sector and interfaces with other modes of transport. About liability in case of accident, article. 1 letter J of the decree says that "the occupant of the vehicle, which must be always able to take control of the vehicle regardless of the degree of automation of the same, in any moment the need arises, acting on the vehicle controls with absolute precedence over automated systems and which, therefore, is the person responsible for the circulation of the vehicle. When you take it effective guide, in manual mode, assumes the role of driver."

We must also say that insurance in this case should take a relevant role.

In May 2018, the European Commission presented a proposal to amend the motor insurance directive. Under the revamped rules, once adopted by the European Parliament and the Council: victims of motor vehicle accidents will receive the full compensation they are due, even when the insurer is insolvent; drivers who have a previous claims history in another EU country will be treated equally to domestic policyholders, and will potentially benefit from better insurance conditions.

No special provision on automated car is included in the text. We also have to consider the fact that at the end of 2017, Insurance Europe (the European insurance and reinsurance federation) has responded to the European Commission's REFIT consultation on the Motor Insurance Directive (MID), which should be an essential tool in the protection of road traffic accident victims and should be preserved.

On 17 May 2018, the European Commission published a new communication, in which it focuses on the importance of non-personal data sharing while protecting

[19] Greger (2018), pp. 1–5.

cybersecurity and on the importance of fostering vehicle connectivity for automation.

On the safety on the roads and victims' compensation, the European Commission in its communication of May 2018 affirms that the motor insurer can take actions against the manufacturer.

The attention of policymakers needs to also focus on the socio-economic perspective. Some scholars highlight increasing reservations towards autonomous robotic systems in Europe.[20]

3.3 *Product Liability and AI*

Automation also imposes to reshape rules on manufacturer liability especially on the concepts both of defect an of product.[21]

In 2018, the European Commission submitted the fifth report on the application of Directive on product liability. The Commission carried out a formal evaluation of the Directive, assessing whether the Directive continues to be an adequate tool and continues to meet its objectives today—also in the light of new technological developments. The Commission launched a public consultation on the Directive on product liability to collect stakeholders' feedback on the application and performance of the Directive.

In 2019, the Commission issued guidance on the Directive on product liability and a report on the broader implications with regard to the liability and safety frameworks for Artificial Intelligence, the Internet of Things, and robotics. On new technological challenges, the EU Commission's Report considers the following questions: Does it adequately address the challenges of increasingly autonomous devices and cybersecurity? What about sustainability and circular economy? Does the Directive discourage producers from placing innovative products on the market? Does it deter manufacturers from placing faulty and unsafe products on the market? Does it protect injured persons in a changing world?

Some gaps of the Directive on product liability are underlined by its evaluation:

- The application of the Directive is problematic for products on which software and applications from different sources can be installed after purchase, that are connected to the Internet and can perform automated tasks based on algorithms and data analysis, automated tasks based on self-learning algorithms or shared with other users through collaborative platforms.
- In case of damage caused by software there could be a problem of proof: in case of open-source software used, for instance, in the medical field, it could be

[20] Gnambs and Appel (2019), pp. 53–61.

[21] Gallage Alwis (2020).

difficult to prove the damage resulting from a misdiagnosis due to a failure in the software.
- In case of interconnected products correctly attributing liability for defects can be difficult.
- A new concept of production is emerging. New technological developments such as 3D printers, which enable consumers to become manufacturers, could potentially undermine the attribution of the product that caused the damage. Here is the so-called *Prosumerism* where consumers become producer.
- Typical technological damage has to be compensated. Let us think to service failures such as downtime or loss of data.

The Bureau Europeen des Unions de Consommateurs[22] finds out nine ways to ensure that the EU product liability framework remains fit for the digital age and delivers its full potential to consumers:

1. Clarify the objective and scope of the Product Liability Directive that should apply to all tangible and non-tangible goods, including digital services and digital content.
2. The notion of "defect" should be broadened and no longer limited to users' safety expectations: A product should be considered "defective" when it deviates from the reasonable expectations that users may have, which include: the product should be safe, it should be free from cybersecurity failures and other personal security risks and should be GDPR-compliant. Moreover, in case of product using self-learning capacities and taking autonomous decisions should be deemed defective when its unintended behavior has caused harm to the user.
3. The "time at which the product was placed on the market" should no longer be relevant when assessing defects.
4. The Directive should extend the scope of compensable damage to also cover damage to digital assets, including data.
5. The burden of the proof concerning defect and causation should be reversed. It should be enough for the injured party to prove the existence of damage resulting from a product.
6. All professionals involved in the production of digital goods should be held jointly liable if their activities have affected the reasonable expectations which users may have about their products.

Online marketplaces should bear subsidiary liability as "suppliers" under the PLD.

7. Review the exemptions to liability is needed as following: abolish the risk-development defense; Compliance with standards should not be valid grounds for escaping liability.
8. The ten-year limitation period should be extended to 20 years.

---

[22] A. Biard, PRODUCT LIABILITY 2.0 How to make EU rules fit for consumers in the digital age.

9. Introduce greater transparency measures on defective products (e.g., creation of a registry of product liability cases).

Some scholars consider the benefits of a preventive treatment of product liability. Warning becomes a key word.[23] A commonly made distinction is that between warnings and instructions. While warnings indicate (sometimes inherent) risks associated with a product, enhancing the consumer's freedom of choice, instructions indicate how to avoid the risks, and what to do if the risks materialize.

- On 22 November 2021, the EU Commission launched a consultation. The Commission invites interested parties "to express their views on the revision of the Product Liability Directive and on whether other national liability rules still provide legal certainty and consumer protection in an age of smart and AI-based products and services. This is especially important since the safety of these products and services does not depend only on their design and production, but also on software updates, data flows and algorithms. The public consultation covers questions such as which economic operator should be liable for harm. Another important aspect is the upgrade and refurbishment of products and components, something that is becoming more and more important in our transition to a circular economy. The current liability rules are based on two pillars: the Product Liability Directive and non-harmonised national liability rules. The Product Liability Directive protects consumers who suffer injury or property damage from defective products and covers products ranging from garden chairs to medicines, cars and AI-driven products. The non-harmonised national liability rules include various differing liability rules, which cover different types of damage and claims against any liable person."[24]

4 Insurance and Risk in Case of Automated Cars

4.1 More Automation Less Risk

According to a survey conducted in the USA "despite claims to the contrary, self-driving cars currently have a higher rate of accidents than human-driven cars, but the injuries are less severe. On average, there are 9.1 self-driving car accidents per million miles driven, while the same rate is 4.1 crashes per million miles for regular vehicles."[25]

[23] Verheyen (2019), pp. 44–56. See also Pape (2012), p. 257; Lenze (2005), p. 105; Verheyen (2018), pp. 119–140; Taschner (2005), p. 159.

[24] The Commission has launched a public consultation on the rules on compensation for damage caused by defective products. A specific focus will be on the use of Artificial Intelligence (AI) in products and services. https://digital-strategy.ec.europa.eu/.

[25] Clifford Law Office (2021).

Among the dangers inherent with driverless cars are:

- False Sense of Security
- Danger of Fire
- Imperfect Technology
- Cyberattacks
- Complex, Real-Life Driving Conditions
- Lack of Self-Driving Regulations[26]

Is it possible to reduce risk avoiding high level of automation, the so-called highly automated driving (HAD)?

Some studies demonstrate that in case of full automation the risk can decrease. The final result of such studies is that drivers' situational awareness decreases when using HAD. In this HAD context, drivers can engage non driving tasks as reading or sleeping. These non-driving tasks lead to increased reaction time in case of hazardous situations or risky events (hardware, sensor, actuator failures, or front obstacle or crashes, or dense traffic congestion, or adverse conditions). The survey proposes a global risk indicator using local information coming from surrounding vehicles or infrastructures (V2X communication). Finally, the survey first shows the gain of such a global risk indicator comparatively to the local one, and second its impact on the behavior of both the autonomous car and the driver.[27]

How is it possible to manage such risks and transfer it to a party (an insurer) who can manage it?

Given such a situation scenario, it is important to compensate victims. So far, the European Commission has focused particularly on driving automated vehicles and underlined the importance of coordinating the responsibility of the user, of the owner and of the manufacturer.

It could be possible to put strict liability on the user, the owner, or the manufacturer. Is it enough? Could strict liability represent a disincentive to automation?

As said, the EU law considers automation under different perspectives: First, in May 2018, the European Commission presented a proposal to amend the Motor Insurance Directive. The proposal stressed the importance of victim's compensation, but it does not contain any specific norms regarding automated vehicles. Second, some gaps in the EU law have been underlined by the evaluation of the Directive on product liability. It considers the approximation of laws, regulations and administrative provisions of the Member States concerning liability for defective products. The evaluation report also considers typical technological damage that needs to be compensated, for instance, service failures such as downtime or loss of data. Third, on 17 May 2018, the European Commission published a communication where connections between product liability and cyber liability are underlined. Fourth, the European Parliament's Resolution of 16 February 2017 proposes a form of strict liability of users or owners of robots. We can underline three different objectives in

[26] Nikitas (2015).

[27] Sébastien et al. (2019), pp. 390–395.

the above-mentioned interventions: (i) the protection of road victims in general, (ii) the protection of consumers in case of defective product, (iii) cybersecurity.

The solutions till now proposed focus on liability. Of course, civil liability plays a relevant role in compensation and prevention of losses. Are we sure that in case of damage caused by a machine running in full automation civil liability can prevent damage thanks its deterrence function?

Covering the liability of the owner or the producer or acting as delegate of a public funds in compensating damages to victims of AI, insurers can create and update standards and guidelines to "educate" machine with a relevant role in the prevention of damage caused by AI, thanks, for instance, to tools like the "reinforcement learning." The purpose of reinforcement learning is for the agent to learn an efficient policy that maximizes the "reward function."[28] It could be compared with that appear to occur in animal psychology.

In this case, insurance contracts can also play a relevant and social function. Moreover, insurers could, thanks to the data sharing, help in the management of new risk emerging in AI dimension.

5 Distribution

5.1 New Distribution Dynamics of Auto Insurance Products

The increasing level of automation of vehicles and the circulation, in the near future, of totally self-driving cars, thanks also to systems that allow connecting with other cars and infrastructures, will have considerable advantages in terms of road safety, as it will bring about a sustained reduction in accidents.[29] This will undoubtedly be a direct and immediate consequence of automation: the more vehicles are automated, the less the driver's activity will be and, consequently, the margins of inherent human error, with a drastic reduction in accidents due to driver negligence. Moreover, the possibility of guaranteeing connection systems between cars and infrastructure (roads, road signs, both horizontal and vertical, traffic lights, crosswalks, bridges, etc.), determining a constant information flow, will allow achieving a further reduction in accidents. Suffice it to think, for example, of all the accidents caused by poor road surface conditions.

As this reality spreads, it will have as a consequence—on closer inspection—that drivers will no longer be able to be defined as such, since they will perform, at most,

---

[28]Kaelbling et al. (1996), pp. 237–285; Pandian and Noel (2018), pp. 16–29; van Otterlo and Wiering (2012), pp. 3–42.

[29]For additional reading on the various aspects of the issue, see Pongelli (2020), p. 83 et seq; Albanese (2019), p. 995 et seq; Nazzaro (2018), p. 77 et seq; Cerini (2018), p. 401 et seq; Rizzuti (2018), p. 215.

the role of "supervisor" of the computerized vehicle management system.[30] The supervisor-driver, then, will simply step in case the vehicle malfunctions, assuming this is possible. In the face of all this technology, it seems realistic to assume, on the other hand, an increase in risks due to malfunctions of self-driving vehicles or their components, with increasingly likely electrical and computer failures, in addition to classic mechanical problems.

The imagined reality—with the consolidation of connection systems between vehicles and infrastructures—will be based on the massive, unceasing, and immediate circulation of information flows, where the slightest problem with the connection network or a possible slowdown in data transmission becomes a decisive factor for citizens to use roads correctly and safely. While there will be an incredible reduction in road accidents caused by drivers, there will be, in turn, an increase in claims related to vehicle malfunctions. In my opinion, it is not clear whether the latter increase will make up for the disappearance of the first category of accidents, but it is very likely that there will be a significant reduction—in the long term—in motor liability premiums.[31]

This scenario will also most probably affect—at insurance level—the distribution dynamics of products connected with risks deriving from road traffic, where currently—as is known—there is a legal obligation for third-party liability insurance resulting from the use of motor vehicles by a driver.[32] To date, in fact, the motor liability market is almost entirely focused on a retail customers since insurance companies channel their distribution strategies towards hundreds of millions of individual vehicle owners. The cases of motor liability insurance managed directly by the manufacturer or, in any case, by the manufacturer's satellite companies are still an exception (in this sense, suffice it to think of cars rented on a long-term basis or leased).

The transformation on the horizon is far more profound and is not merely a subjective change of policyholder—from the driver-owner to the automaker—but an evolution of the liability system. This will no longer focus on the driver, but on the vehicle as such, as the primary source of road accidents. There will therefore be a significant increase in the demand for insurance cover linked to hypothetical malfunctions of self-driving vehicles. Particularly—with the regulatory framework unchanged—vehicle manufacturers, producers and/or managers of their IT systems, as well as the entities that own or manage the infrastructures will have to seek suitable solutions to protect them from possible claims for damages occurring during road circulation due to electrical or IT system failures. The degree to which all these risks associated with the proper operation of self-driving vehicles and related infrastructure are likely to affect the number of claims in quantitative terms is hard

[30] Rizzuti (2018), p. 216.

[31] Particularly, Nazzaro (2018), p. 83, bases his argument on behavioral pricing models, where insurance players develop new offers using data provided by car manufacturers.

[32] For some initial interesting considerations on the issue, which the doctrine has not yet addressed, see Cerini (2018), p. 401 et seq.

to say. What is certain is that for insurance companies—also in the road traffic sector—large segments of the market will open up for the distribution of non-life products, other than those of traditional third-party motor liability.

Therefore, it seems plausible to assume that insurance companies will be called upon to implement distribution strategies where the core business will no longer consist in placing the policy with the individual consumer-owner of one or two vehicles, but rather in interfacing with professional customers, made up of car manufacturers, producers of computerized automation systems, as well as operators of related infrastructures. Without considering, then, the possibility that lawmakers may introduce compulsory insurance for damage caused by products in road traffic. In the imagined future (which closer than one might think), there will be a whole series of risks, for the time being only marginal compared to vehicular traffic, which will become central to the adequate protection of the actual needs of road users. Just think of cyber risk, understood as the risk associated with the processing of information in the self-driving vehicle's computer system and which can be a decisive factor in whether an accident occurs in case the stored data are violated, tampered with, or deleted. Among other things, the fact that certain types of risks will become "normal" in road traffic will also affect the product oversight governance (POG) models developed by producers and distributors of insurance contracts, given that the change in the demand for cover consequently modifies the reference market based on which the POG is built and periodically modified and updated.[33]

In this perspective, the distribution of insurance products in the road traffic sector will presumably be aimed at placing these in the market inseparably linked to the sale of the vehicle, which will be purchased by the customer together with non-life insurance aimed at providing cover against risks of malfunction, as happens every day when purchasing household appliances of all kinds. These are, therefore, native insurance products as if they were integrated with the vehicle itself to overcome the traditional reluctance of end customers to purchase additional insurance coverage, which is very often considered superfluous. The distribution channels will, therefore, have to be funneled—as already highlighted—towards a type of professional customers, which will act as an intermediary for the placement of its own insurance solutions.

Alternatively, or at least as a supplement, distributors—especially small and medium-sized agents and brokers—will be called upon to make a significant digitalization effort, aimed at exploiting the full potential of IT systems. Since it is impossible to reach large car makers, in fact, the massive use of online platforms capable of reaching consumers-drivers, offering "tailor-made" contracts based on the specific needs also linked to the technical characteristics of the vehicle (so-called "embedded insurance"), will be of crucial importance.

[33] On this aspect, see Cerini (2018), p. 401 et seq.

In short, what emerges is that technological progress will bring about—and it is already doing—profound changes in the road traffic insurance system.[34] What will change will be the very subject of liability, with the centre of gravity shifting from the "human driver" to the "driverless vehicle." In terms of distribution, this will translate into a change in market dynamics, which will have to be oriented towards a type of professional customers, to the advantage of the large players and to the detriment of small and medium-sized distributors who will no longer be able to compete in a market of this type. However, at the same time, technology may be the only tool that can ensure the survival of small agents and brokers, as it will enable them to better reach the vast retail market.

6 Conclusions

The inclusion of Artificial Intelligence systems in the decisions to drive a vehicle highlights: (1) the importance of prevention rather than compensation of harmful events, and (2) the importance of allowing the implementation of algorithms to reduce risks rather than sanction harmful conduct.

On this last aspect, we have seen how the rules of civil liability, although rethought in the logic of the algorithmic decision, are not sufficient and if applied in an inadequate manner they risk being harmful.

Strick liability systems that automatically place the responsibility for algorithmic choices (in the case of level 4–5 automation) on the user or owner of the machine risk discouraging the use of automated driving and at the same fail to prevent the event harmful to repeat itself as the deterrent efficacy of civil liability cannot be explained in the event of automated decisions. Therefore, it becomes important to identify a system that allows you to log errors and create new data to implement algorithms and reduce risks. Hence, as mentioned, the central role of insurers that do not limit themselves to compensate the damage but intervene in risk management. In this case, the insurers could intervene, requiring the insured to update the software, monitoring the claims, and creating new data, customizing the insurance coverage based on the needs of the insured in terms of kilometers traveled, driving style, etc.

These considerations are also relevant to the first aspect noted—prevention. As mentioned, on the prospects for the reform of producer's responsibility, the key word is "warning."

If a manufacturer fails to adequately warn of potential dangers of its products' use, it may be liable for any resulting injuries. A manufacturer, distributor, or retailer can be held liable for a failure to provide adequate warnings on a product if a

[34]For further information on digitalization in insurance distribution, see Marano (2021), p. 143 et seq.; Stoeckli et al. (2016), p. 287 et seq.; and the paper published by Sigma-Swiss re. (2014), pp. 1–37.

consumer suffers an injury as a result. In this case, we use to speak about "liability from failure of warning."

It is a common idea that product warnings and disclaimers are ineffective. Consumers fail to appreciate risk because of: cognitive biases, willful ignorance, or the overwhelming appeal of advertising. And yet, the law has not abandoned reliance on warnings or disclaimers. In any case, the importance of warning in product liability is recognized by courts and policymakers.

While behavioral economics has contributed many important insights in this regard, literary theory provides an additional consideration. Some authors suggest that "by recognizing that warnings and disclaimers are "texts" (in the literary theory sense)—and by taking advantage of recent marketing literature on "reading" advertisements—we might well discover that drafting the warning or disclaimer in a particular way may solve some of consumers' interpretive problems. In other words, to the extent that the conventional wisdom is predicated on a belief that consumers do not have the ability to process these types of contradictory messages, literary theory counters this belief by explaining how readers interpret parody and other similar texts. It also tells us that the most effective way to yield a preferred meaning among readers, given that each reader will bring her own meaning to the table, is to appeal to context and interpretive communities to encourage similar patterns of reading such texts."[35] Perhaps technology can represent a new tool for alerting consumers of the risk. Examples of this already exist, just think of the cell phone that tells the user that too high a volume can damage hearing.

In conclusion, the theme of automation in road traffic has given us the opportunity to reflect on significant changes in the way of thinking about the law that does not reduce it to the binomial rule and sanction. Moreover, Law is increasingly RegTech. The law governs technology, but technology also becomes a regulatory tool.

References

Acosta Aida Joaquin (2018) What Governments across the globe are doing to seize the benefits of autonomous vehicles. Cyber Harvard Edu. https://cyber.harvard.edu/

Albanese A (2019) Le responsabilità civile per i danni da circolazione di veicoli ad elevata automazione. Europa e diritto privato:995 et seq

Anderson M et al. (2010) The US experience with no fault automobile insurance: a retrospective. http://www.rand.org

Biard A. Product liability 2.0 How to make EU rules fit for consumers in the digital age, Bruxelles

Bodansky D (1991) Law: scientific uncertainty and the precautionary principle. Environ Sci Policy Sustain Dev 33(7):4–44

Boehmer-Christiansen S (1994) The precautionary principle in Germany–enabling government. Interpr Precaut Princ 31(1994):38–39

[35] Heymann (2010), pp. 393–415, 2010, William & Mary Law School Research Paper No. 09-50.

Bonab AB, Rudko I, Bellini F (2021) A review and a proposal about socio-economic impacts of artificial intelligence. In: Dima AM, D'Ascenzo F (eds) Business revolution in a digital era. Springer proceedings in business and economics. Springer
Borges G (2021) AI Systems and Product Liability. In Proceedings of the 18th International Conference on Artificial Intelligence and Law, ICAIL 2021, pp 32–39
Buckingham W (1961) Automation: its impact on business and people. Harper & Brothers, New York
Cerini D (2018) Dal decreto smart roads in avanti: ridisegnare responsabilità e soluzioni assicurative. Danno e responsabilità, p 401 et seq.
Čerka P, Grigienė J, Sirbikytė G (2015) Liability for damages caused by artificial intelligence. Comput Law Secur Rev 31(3):376–389
Chen L, Mislove A, Wilson C (2015) Peeking beneath the hood of Uber. IMC 2015
Ching-Yao C (2017) Advancements, prospects, and impacts of automated driving systems. IJTST, 208–216
Chopra S, White LF (2011) A legal theory for autonomous artificial agents. University of Michigan Press, Ann Arbor
Clifford Law Office (2021) The dangers of driverless cars. Natl Law Rev, 5 May 2021
Committee of Ministers of the Council of Europe (CoE) (2020) Recommendation on the human rights impacts of algorithmic systems. Brussels
Dafoe A, Evans O, Grace K, Salvatier J, Zhang B (2017) When will AI exceed human performance? Evidence from AI experts. J Artif Intell Res 62:729–754
Ethik-Kommission Automatisiertes und Vernetztes Fahren. Bericht 2017. https://www.bmvi.de
Fluet C (2002) Liability insurance and moral hazard under the strict liability and negligence. Revue D'économie Politique 112(6):845–861
Gallage Alwis S (2020) Updating the EU product liability directive for the digital era. Droit Technol, 24 February 2020
Gless S, Seelmann K (eds) (2016) Intelligente Agenten und das Recht. Nomos, Baden-Baden
Gnambs T, Appel M (2019) Are robots becoming unpopular? Changes in attitudes towards autonomous robotic systems in Europe. Comput Human Behav 2019:53–61
Greger R (2018) Haftungsfragen beim automatisierten Fahren. Zum Arbeitskreis II des Verkehrsgerichtstags. Neue Zeitschrift fuer Verkehrsrecht 31:1–5
Heymann LA (2010) Reading the product: warnings, disclaimers, and literary theory (June 3, 2010). Yale J Law Human 22:393–415, 2010, William & Mary Law School Research Paper No. 09-50
Independent High-Level Expert Group on Artificial Intelligence Set Up by The European Commission (2019) Ethics Guidelines for Trustworthy AI, Brussels
JCER (1955) Hearings on automation and technological change. In: Subcommittee on Economic Stabilization of the Joint Committee on the Economic Report, US Congress October 1955, pp 14–28
Kaelbling LP, Littman ML, Moore AW (1996) Reinforcement learning: a survey. J Artif Intell Res. 4:237–285
Kim PT (2017) Auditing algorithms for discrimination. Univ Pa Law Rev 166:189
Kroll JA, Barocas S, Felten EW, Reidenberg JR, Robinson DG, Yu H (2016) Accountable algorithms. Univ Pa Law Rev 165:633
Landini S (2020a) Ethical issues, cybersecurity and automated vehicles. In: Noussia K, Marano P (eds) InsurTech: a legal and regulatory view. Springer, pp 291–312
Landini S (2020b) The insurance perspective on prevention and compensation issues relating to damage caused by machines. Ital Law J:159
Lenze S (2005) German product liability law: between European Directives, American Restatements and common sense. In: Fairgrieve D (ed) Product liability in comparative perspective. CUP, p. 105
Lütge C (2017) The German ethics code for automated and connected driving. Philos Technol, p 9
Marano P (2021) Management of distribution risks and digital transformation of insurance distribution—a regulatory gap in the IDD. Risks 9:143 et seq

Miller T (2019) Explanation in artificial intelligence: insights from the social sciences. Artif Intell:1–38

Mittelstadt B, Russell C, Wachter S (2019) Explaining explanations in AI. In Proceedings of the Conference on Fairness, Accountability, and Transparency (FAT* '19). 2019. New York, NY: US

Nazzaro AC (2018) Macchine intelligenti (smart cars): assicurazione e tutela della privacy. Diritto del mercato assicurativo e finanziario, p. 77 et seq.

Nikitas A (2015) Automated cars: a critical review of the potential advantages and disadvantages of driverless technologies. In: First International Workshop on Smart Urban Mobility, 26–27 November 2015

Nof SY (2009) Automation: what it means to us around the world. In: Nof SY (ed) Springer handbook of automation. Springer, Berlin, pp 13–52

Pandian BJ, Noel MM (2018) Control of a bioreactor using a new partially supervised reinforcement learning algorithm. J Process Control 69:16–29

Pape B (2012) Warnings and product liability – lessons learned from cognitive psychology and ergonomics. Eleven International Publishing, p. 257

Parasuraman R, Sheridan Thomas B, Wickens Christopher D (2000) A model for types and levels of human interaction with automation. IEEE Transact Syst Man Cybernet Part A Syst Humans 30(3):286–297

Patrick Hubbard F (2015) Sophisticated Robots: balancing liability, regulation, and innovation. Florida Law Rev 66:1803

Pongelli G (2020) Autonomous vehicles and legal issues on civil liability and motor insurance. Diritto del mercato assicurativo e finanziario, p. 83 et seq.

Rizzuti M (2018) I veicoli automatici, problemi e prospettive. Diritto del mercato assicurativo e finanziario, p. 215

Sandvig C, Hamilton K, Karahalios K, Langbort C (2014) Auditing algorithms: research methods for detecting discrimination on Internet platforms. 64th Annual Meeting of the International Communication Association, May 22

Sébastien D, Dominique G, Jean-Marie B, Sébastien G, Grégoire L, Olivier O, Andry R (2019) Global risk assessment in an autonomous driving context: impact on both the car and the driver. IFAC-PapersOnLine 51(34):390–395

Sigma-Swiss re (2014) Digital distribution in insurance: a quiet revolution. 2:1–37

Simon HA (1979) Models of thought. Yale University Press, New Haven, p 10

Spiecker gen and Döhmann I (2012) Rechtliche Begleitung der Technikentwicklung im Bereich moderner Infrastrukturen und Informationstechnologien. In: Hill H, Schliesky U (eds) Die Vermessung des virtuellen Raums

Spindler G (2019) User liability and strict liability in the internet of things and for Robots. In: Lohsse S, Schulze R, Staudenmeyer D (eds) Liability for artificial intelligence and the internet of things. Nomos, Baden-Baden

Stoeckli E, Dremel C, Uebernickel F (2016) Exploring characteristics and transformational capabilities of InsurTech innovations to understand insurance value creation in a digital world, p. 287 et seq.

Surden H, Williams MA (2016) Technological opacity, predictability, and self-driving cars. Cardozo Law Rev 2016:121

Taschner H (2005) Product liability: basic problems in a comparative law perspective. In: Fairgrieve D (ed) Product liability in comparative perspective, p. 159

Van Otterlo M, Wiering M (2012) Reinforcement learning and Markov decision processes. Reinforcement Learn Adapt Learn Opt 12:3–42

Verheyen T (2018) Full harmonization, consumer protection and products liability: a fresh reading of the case law of the ECJ. Eur Rev Priv Law:119–140

Verheyen T (2019) Modern theories of product warnings and European product liability law. Utrecht Law Rev 15(3):44–56
Vladeck DC (2014) Machines without principals: liability rules and artificial intelligence. Wash Law Rev 89, 117:146
Wendehorst C (2020) Strict liability for AI and other emerging technologies. J Eur Tort Law 11(2): 150–180
Zech H (2021) Liability for AI: public policy considerations. ERA Forum 22:147–158

Legal and Ethical Aspects of Autonomous Vehicles

Kyriaki Noussia, Zuhal Gocmen, and Maria Glynou

Abstract With the acceleration of the pace with which artificial intelligence (AI) penetrates our lives, we will, sooner rather than later, move into a world of automated transportation system which will dominate our everyday life and movements, and which has already been—to a certain extent—introduced in the market. Fully autonomous vehicles (AVs) are expected to be widely used within the next decade or earlier, however, technological, and regulatory obstacles may delay their massive use and public deployment. This chapter discusses the main legal and ethical issues deriving from the use of AVs to showcase the regulators' significant role in providing solutions to such issues. It also makes some propositions for the way forward in relation to issues concerning necessity situations, liability issues and insurance implications, criminal and contractual liability and lastly data implications, before drawing overall conclusions.

Keywords Autonomous vehicles · Automated transportation system · Data and autonomous vehicles

1 Introduction

Parking assist, anti-blocking or lane keep assist systems which have been named as 'Advanced Driver Assistance Systems', are features of an automated transportation world which have already been introduced in the market. One of the final steps in the automation process is the one related to fully autonomous vehicles (AVs) which do

K. Noussia (✉)
University of Reading, School of Law, Reading, UK
e-mail: k.p.noussia@reading.ac.uk

Z. Gocmen
Attorney at Law, Istanbul, Turkey

M. Glynou
Attorney at Law, Athens, Greece

LLM candidate and Research Assistant, LSE, London, UK

K. Noussia, M. Channon (eds.), *The Regulation of Automated and Autonomous Transport*, https://doi.org/10.1007/978-3-031-32356-0_10

not require any human intervention for driving.[1] This technology has been expected to be used widely by the mid-2020s.[2] However, for both technical and security reasons as well as reasons deriving from the lack of political and legal framework, which regulates this sophisticated technology, their wide public circulation may be delayed for an indefinite, but nevertheless hopefully short time. The endeavour for the autonomous vehicle technology to become more utilisable and safer remains. While manufacturers focus on rendering them both profitable, practical, and safe for users, policy and lawmakers strive to find solutions so as to address legal and ethical issues deriving from this technology.

This chapter points out the main legal and ethical issues deriving from the use of AVs and demonstrate the regulators' significant role in providing solutions to those issues. It will then go on to draw some conclusions and make some propositions for the way forward in relation to issues concerning necessity situations (e.g. ability to choose the programming choice of an AV and who should have this authority to do so), liability issues and insurance implications, criminal and lastly data implications.

1.1 A World with AVs

Were one to consider the introduction of AVs, he/she would most probably think of car owners who possess private autonomous cars. However, from the moment AVs appear in the market, this technology will not be used for privately owned vehicles only. Platforms such as UBER, tourism agencies, public buses, transportation companies will all benefit from this driverless technology eventually.[3] For instance, self-driving buses, named as 'Smart Panda Bus', which are manufactured by DeepBlue Technology, have already been in use in China for a couple of years and such technology has been introduced in European countries such as Greece and France as well.[4] It is to be noted that UBER has already started to use driverless cars.

Moreover, in 2018, the Society of Automotive Engineers (SAE) proposed motor vehicle automation systems, which would perform the dynamic driving task partially or on its whole and would be categorised according to their level of automation, which is determined by the features that are engaged. Specifically, the level of automation is put to a scale ranging from level 0, representing no driving automation, to level 5, which stands for full driving automation.[5] Presently, automated

---

[1]Lohmann (2016), p. 335.

[2]Lohmann (2016), p. 335.

[3]Bagloee et al. (2016), p. 288; Hancock et al. (2019), p. 7684; Landini and Noussia (2022), pp. 131–172; Noussia (2020), p. 253; Goodall (2014); Matthaei et al. (2015).

[4]Shepherd (2019). https://www.ft.com/content/72bf3046-bca0-11e9-b350-db00d509634e. Accessed on 1 August 2022.

[5]On-Road Automated Driving Committee (2018) Taxonomy and definitions for terms related to driving automation systems for on-road motor vehicles. *SAE International, Standard J3016*. https://www.sae.org/standards/content/j3016_201806/. Accessed on 1 August 2022.

vehicles of level 0 and level 1 are increasingly in use, which incorporate a system controlling the vehicle's longitudinal and lateral motion. Also a few vehicles of level 2 are on the roads as well, for example Tesla Model S, which has been described as incorporating a semi-autonomous driving system.[6] The difference between SAE levels 0, 1 and 2 lays in that the latter provides control for both the longitudinal and the lateral motion of the vehicle. The driver performs the rest of the dynamic driving task, for instance the object recognition, and intervene only when necessary. As far as SAE level 3 is concerned, the vehicle is expected to perform the dynamic driving task in its entirety, while the driver will be able to intervene only in cases where the system is unable to handle a situation. SAE level 4 will perform the dynamic driving task on its whole without any human interference and lastly SAE level 5, as being the ultimate form of automated vehicles, are anticipated to be able to perform the driving task under any road and weather circumstances.[7]

The fact that AVs will not be owned by individuals only but also companies and governments manifests that the issues arising from the use of AVs will be far reaching. For instance, contract law issues; it can be predicted that these driverless systems will be able to form a contract with the users of their services.[8] Passengers may directly be able to form a contract with a driverless taxi rather than forming a contract with a platform such as UBER in the future.[9] Therefore, a number of questions arise as to the ability of an AI machine to form a contract and subsequently as to liability concerning the results of such a contract. Although those questions related to contract law are significant, they are not the subject of this chapter, and therefore, will not be examined in detail.

1.2 Benefits of AVs

AV technology has been conceived as a disruptive technology by some scholars.[10] Although it has been developing for some time now, there is still space for improvement, since there have been some examples of unfortunate events. For instance, in 2016, a Tesla Model S car was involved in an accident, which occurred and caused the death of a person, due to the car's sensors' failure to distinguish a large white

---

[6]Dorian (2021) 2022 Tesla Model S. https://www.caranddriver.com/tesla/model-s. Accessed on 1 August 2022.

[7]On-Road Automated Driving Committee (2018) Taxonomy and definitions for terms related to driving automation systems for on-road motor vehicles. SAE International, Standard J3016. https://www.sae.org/standards/content/j3016_201806/. Accessed on 1 August 2022. See in general, Automated Driving Systems, A Vision for Safety (2017).

[8]Chinen (2019), pp. 35–36.

[9]See Lieber and Puente (2016), accessed on 1 August 2022; Pakusch et al. (2021), pp. 1–49.

[10]See Morrow et al. (2014); Crayton and Meier (2017), pp. 245–252; Divino (2020), pp. 245–252.

18-wheel truck and trailer crossing the highway.[11] This accident was one of the first incidents, which put the AVs' safety into question. However, it is emphasised that it is significant to distinguish a car on the autopilot mode from a fully automated self-driving car (an AV). AVs, which are the subjects of this chapter, are far more sophisticated technologies than cars with the automated pilot mode, as AVs will have among others laser beams, maps, Internet connection incorporated. AVs will be classified as level 4 while Tesla's autopilot model lays somewhere between level 2 and level 3. Thus, if we compare AVs with driverless cars, AVs will be much safer than driverless cars.

Furthermore, after using AVs, reliable software will most probably replace human drivers, there will not be risks deriving from careless driving, such as issues due to alcohol overdose, stress, or violation of traffic rules anymore. To this end, it has been predicted that accidents will be reduced by 90%.[12] Moreover, this technology will improve mobility for children, the elderly and disabled eliminating or even rendering the need for them to be escorted redundant.[13] Safety distances between cars will be shorter as AVs will be more reliable, interconnected and thus will be able to share information with each other instantly. Subsequently, since it will be easier to distribute AVs more efficiently at the same road, there will be a significantly higher number of vehicles than before, resulting to AVs being safer and faster. Finally, despite their increase, it is anticipated that AVs will reduce environmental pollution.[14]

1.3 Risks

Although AVs will certainly bring numerous benefits, the risks, which are borne by such technologies, should by no means be overlooked; even if the risk of accidents will decrease significantly after AVs start to be deployed, the risk of accidents will not be eliminated completely.[15] During the reasonably long period of transition from cars controlled by humans to AVs, human drivers' behaviour—to the extent that it is unpredictable—will create a challenge for AVs. Even when all cars are autonomously driven, the accident risk will not be reduced to zero,[16] as there will be various circumstances which may cause accidents such as among others software errors, cyber risks, wrong programming choices or mechanical faults, careless

[11] Yadron and Tynan (2016) Tesla driver dies in first fatal crash while using autopilot mode. https://www.theguardian.com/technology/2016/jun/30/tesla-autopilot-death-self-driving-car-elon-musk. Accessed on 1 August 2022.

[12] Lohmann (2016), p. 336.

[13] Harper et al. (2016), pp. 3, 7.

[14] Coca-Vila (2017), p. 60.

[15] Mueller et al. (2020), p. 313.

[16] Coca-Vila (2017), p. 60.

pedestrians, animal and bad weather conditions.[17] Thus, it is significant to determine the rules applicable if and when an AV is involved in an accident to determine the liable party and to ensure compensation of injured parties. Further to this, the way a potentially liable party would escape liability is significant. To illustrate, in the context of product liability, a producer could escape product liability if he/she could sufficiently demonstrate that the state of scientific and technological knowledge was not such as to render the defect discoverable at the time of the product's circulation.[18] In the context of the driver's liability, the driver could escape liability in case he/she demonstrates that the damage is not attributed to his fault.[19] On the other hand, in cases where strict liability regime applies, the party, which is held liable, cannot escape liability.[20]

For the public to use AVs, their benefits should exceed their risks; AVs should be at so high a level that they are considered as safer than the best human driver.[21] As can be observed, AVs will reduce the risk of accidents significantly. However, the public may be concerned about safety while using AVs, as AVs may be vulnerable to hacking and thus it is perceived that they will not reach 100% of safety soon, due to software bugs. Moreover,, when an AV is involved in an accident and causes harm, people may react in a more emotional manner as the harm caused will be the result of a technology which is supposed to benefit them.[22] This is the case even if, statistically, the safety innovation promotes safety for most daily life's aspects and causes harm very rarely. Thus, in relation to accidents caused by AVs, public reaction will also be a pressure point for manufacturers and regulators. At this point, it is significant to keep in mind that there are risks even with current automatic cars. The most common problem has been brakes' defects. Manufacturers have been sued for such cases and paid compensation to the victims. However, the public has been using them to benefit from them. Moreover, to ensure that the public benefits from them, the traffic laws do not prevent the use of automobiles due to the risks, but endeavour to enhance safety consistently.[23] The approach to be taken by regulators should be the same for AVs.

Even if using AVs will reduce the number of accidents, the risk of accidents will remain. In addition to this, AVs will certainly be involved in necessity situations time to time. Thus, they should be pre-programmed to respond to such situations in which they would otherwise inevitably either violate the law or cause damage.[24] These situations have also been described as dilemma situations as in such

[17] Uzair and Hosain (2020), p. 114.

[18] Schellekens (2015), p. 511.

[19] Ibid., p. 510.

[20] Rylands v. Fletcher, [1868] 3 LR HL 330 (Eng.); Rosenberg (2017), p. 218.

[21] Gasser et al. (2016), p. 523; Pammer et al. (2021), p. 257.

[22] Bonnefon et al. (2016), p. 1573.

[23] Gasser et al. (2016), p. 527.

[24] Coca-Vila (2017), p. 60; Contissa et al. (2017), p. 366.

circumstances, AVs might be in front of making difficult decisions by choosing between different but certainly harmful options for different third parties.

Necessity situations have been experienced by human drivers as well, and therefore, this is not an unprecedented danger brought by AVs. Thus, it can be claimed that the only difference between the current Driver Assistance Systems and fully Autonomous Vehicles is 'the underlying automated control'.[25] Moreover, as AVs are advanced technologies, and for the most part, they are expected to make better decisions than humans, they will usually prevent dilemma situations even from occurring and will significantly reduce the number of them. Thus, if we elaborate on the advantages and disadvantages of AVs from this perspective, the former will outweigh the latter. However, even if there will be fewer cases, they will still occur, and therefore, pre-programmed decisions should be based on ethical and legal policies provided by the regulators.

2 Programming for Necessity Situations

'A generic pattern of conduct'[26] is needed to be preset for a conflicting situation that may occur in the future.[27] This poses several concerns; first, the programmer cannot possibly be aware of the choices which due to the necessity situation will be brought in front of an AV. For instance, an AV might be found having to decide between killing an old person or five young people or it might need to choose to save the life of either an innocent third party or a faulty driver. Examples might be countless. Commentators have been trying to come up with scenarios and provide an ethical basis for decisions made by AVs to save the life of one over another in countless unpredictable situations.[28]

Second, if a human is the driver of a car and if the necessity situation is not attributed to him, he will not be liable even if his decision results in harming somebody.[29] However, if a car's behaviour is pre-programmed, and its pre-programmed action resulted in harming somebody, we might not conclude the same as its choices and responses are the result of deliberate programming.[30] Moreover, it is highly likely that the manufacturer will be held responsible in the end.

There are three main scenarios on which scholars have relied their studies and make suggestions for ethical programming of AVs.[31] In all of them, an AV

[25] Jeffcott and Rose (2017), p. 22.

[26] Coca-Vila (2017), p. 60.

[27] Coca-Vila (2017), p. 60.

[28] See Bonnefon et al. (2016), p. 1581; Bigman and Gray (2020), p. 81.

[29] Contissa (2017), p. 368.

[30] Contissa (2017), p. 368.

[31] Contissa (2017), p. 366.

unavoidably harms a person or persons and is found in dilemma of harming pedestrians, occupants or passengers-by.[32] At this point, the first question pertains to 'by whom *the principles that an AV should comply with when deciding in such conditions will be selected: Should the ethics settings be mandatory and used by all cars or should be tailored, meaning that the option to choose their own ethics setting be given to every driver'?*[33] The second question pertains to the manner in which ethical programming should be performed.

2.1 Who Should Decide on Programming?

First, it has been suggested that owners of AVs be entitled to make a decision on programming of their AVs and bear responsibility for their own decisions.[34] Under such view, AVs should contain an 'ethical knob',[35] which enables the users of AVs to switch ethical mode on their AVs. Mainly, the allocation of liability would be the same as for cars driven by humans. However, since the car's behaviour is pre-set, there will not be any difference between an omission, meaning not intervening while the car is proceeding in its direction, and active behaviour, which means changing the direction to avoid pedestrians on the street.[36] This approach might reduce the manufacturers' liability, but might also cause anxiety for the public as their lives might be at stake. However, one could possibly also suggest that we are not currently aware of what a human driver will decide when a necessity situation occurs.

In a study conducted in 2015, it was reported that people considered it preferable that AVs be programmed according to utilitarian approach so as for the number of deaths to be reduced as much as possible.[37] Despite their ethical view, however, they seemed to be inclined to hold a selfish attitude when it comes to sacrificing their own lives. While they require that other vehicles be programmed based on a utilitarian approach, they would most probably feel unsafe while driving a vehicle, which is likely to sacrifice their life for many more lives to be protected. As can be predicted, most users will avoid having their AVs pre-programmed, and therefore, this will cause chaos and even more perils than current cars. Moreover, knowing that others may also avoid having their AVs programmed in advance, this may cause insecurities among the users and reduce the use of AVs. In the authors' view, leaving the programming choice to the users, could be problematic as to issues of liability.

---

[32] Contissa (2017), p. 366.

[33] Contissa (2017), p. 366.

[34] See Kochupillai et al. (2020), pp. 285–299.

[35] Contissa (2017).

[36] Contissa (2017), p. 370.

[37] Bonnefon et al. (2015) Autonomous vehicles need experimental ethics: are we ready for utilitarian cars?. https://pdfs.semanticscholar.org/13d4/56d4c53d7b03b90ba59845a8f61b23b9f6e8.pdf. Accessed on 1 August 2022.

The alternative possibility would be to leave the programming decision to the manufacturers. However, this is also a problematic option for the following reasons; first, most manufacturers may endeavour to sell their products by marketing them as '*tailorable AVs*'. As the users will prefer AVs protecting themselves, manufacturers will not profit from utilitarian AVs, and therefore, they will produce more AVs, which can be adaptable. Consequently, the result will be identical to giving the decision power to the users. Second, leaving the decision to the manufacturers may result in their excessive responsibility towards the accident's victims. If their profits reduce because of producing tailorable self-driving cars, and if they do not consider producing utilitarian cars, since they will not profit them, the use of AVs will end up being prevented.

Lastly, the decision may be left to the regulators. First, if programming choice is subject to regulation, it will be ensured that the way different manufacturers ethically program AVs is consistent. Subsequently, this will provide predictability in terms of behaviour of cars, it will tilt the balance in favour of the customers and provide fairer outcomes.[38] It should be highlighted that it is not conceived as a proper policy to allow manufacturers to commercialise their products by emphasising on the level of safety provided for the occupants of their vehicles even when it is to the detriment of other road users.[39] Second, if the framework is set by the regulators, manufacturers complied with the framework and there is no defect in their product, in a necessity situation manufacturers' responsibility would be limited because 'damages due to control actions would then be just as unavoidable as they would be in the case of the existence of independent risks flowing from the road traffic system', and they do not have any role in the choice of programming.

Because, as we suggest, regulation can mitigate several issues deriving from the use of AVs, a clear and thorough framework concerning the manner of AVs programming should be introduced. To illustrate the above, particular rules are required that will sufficiently address the issue of different levels of programming in conjunction with the way, in which liability will be attributed. There are two reasons providing for such proposal; first, a detailed legislative material would provide certainty for the public. Second, it is imperative that such a decision is not left to the hands of people who may occasionally act selfishly and to the detriment of the wider public. In the latter case, unless adequate regulation is set, just by opting for an AV, which has been programmed based on the utilitarian approach, a user can easily shift the liability due to accidents or defects to the manufacturer.

---

[38] Jeffcott and Rose (2017), p. 23.

[39] Jeffcott and Rose (2017), p. 23.

2.2 How Should the AVs be Programmed by the Regulators?

As elaborated previously, tailorable or 'selfish' self-driving cars, whose programming has not been performed in advance, as per the user's preference, are not considered as an ethically acceptable choice. Some authors suggest that as autonomous vehicles are dangerous, the harm that might be caused by it should be suffered by its occupants. Moreover, in a survey, 1928 participants agreed with this view.[40] However, as was discussed above, it is obvious that a consumer would not purchase a product which would almost definitely harm himself.[41] Probably this was the reason Mercedes Benz has decided to save the occupants of the car in cases where there is a danger caused by it.

The regulation's aim should be to render fatal outcomes illegal considering especially the fact that everyone's life values the same.[42] Thus, to the extent that in the programming stage, the focus is on protecting and/or saving either the occupant, namely the user, or the third party/parties, self-driving cars should by no means be acceptable choices. Therefore, it is suggested that AVs be programmed in such a way to result the least possible harm.

However, the first difficulty could be claimed to be defining 'the most harmless decision' to AVs for each scenario. At this point, increasing the safety of the equipment of AVs such as cameras, radars and sensors should be the first aim to reduce the number of scenarios which may cause harm. Moreover, as AVs will be self-learning and self-deciding vehicles, even the programmers will not be aware of AVs' decisions in a necessity situation in advance.

A second issue is that in a study conducted in 2016 by Bonnefon et al.,[43] it was reported that most of the respondents rejected the idea of regulators following a utilitarian approach as to the use of AVs.[44] Thus, it can be conceived that if it is compulsory to use and rigidly implement 'an impartial (utilitarian) ethical setting' into all AVs, many people may avoid using AVs; even if they may have considerable advantages in comparison to cars driven by humans, especially in relation to safety.[45] Hence, although utilitarian approach seems the most beneficial option for the wider public, this would not be preferable by the public until after it is proved that AVs are accident-free or entail only insignificant risks. In relation to this point, it should also be noted that these issues will no longer constitute a current subject by the time when road safety is improved and when autonomous cars start being used widely.[46] Thus, as it was pointed out above, if AVs' safety becomes pivotal, meaning if the focus is on improving autonomous cars' safety rather than choosing

---

[40] Jeffcott and Rose (2017), p. 23.

[41] Jeffcott and Rose (2017), p. 23.

[42] Noussia (2020), p. 254.

[43] Bonnefon et al. (2016), pp. 1573–1576.

[44] Contissa (2017), p. 366.

[45] Contissa (2017), p. 366.

[46] Jeffcott and Rose (2017), p. 22.

who to save, following the utilitarian approach might not be so easily dismissed by the public.

3 In Between Promoting Innovation and Protecting the Users of AVs

3.1 The Liability of the Users

Given that an overview of the different liability schemes has already been presented in literature,[47] this section aims to examine how each scheme would potentially work in practice. There are examples of existing national laws as well as legislation expected to be introduced, according to which the driver shall be attentive to the road and be able to regain control, when necessary, even when the car is autonomously driven.[48] Otherwise, in the case of an accident caused by AVs, drivers would be held liable due to lack of their attention. This is in line with generally accepted traditional tort liability regime, where drivers owe a duty of care to all other road users.[49] However, in relation to this, it is stated that AVs would not turn out to bring any significant benefit in comparison with conventional vehicles if the 'drivers' are required to pay attention to the road when AVs drive themselves.[50] However, if the drivers' duty to pay attention to the road is to be completely abolished, a fair approach would be to make sure that they are not held liable anymore, and consequently, this would call for changes in the liability regime in most countries as will be described below. Moreover, changes pertaining to the license holders' scheme will be required as the consumers should be protected from such a result caused by the products produced by the manufacturers. It should be emphasised that the explanations that will be made below will rely on the understanding that human users of AVs do not have any obligation to intervene.

First, in most European Countries, such as France, Switzerland and Germany, the license holder of a vehicle, the human driver, is strictly liable for traffic accidents even if they do not have any fault in the occurrence of the accident.[51] Under such a regime, even when a car is autonomously driven, the harm caused by it could be attributed to the license holder of the AV as there is no need to prove any fault on the part of the licence holder. Even if the fairness of this result is disputable in the case of the use of AVs, the benefit is that allocation of liability is ensured under such a

---

[47] Noussia (2020), pp. 255–260.

[48] See German Road Traffic Act (Straßenverkehrsgesetz, StVG) and Survey conducted by the Centre for Connected & Autonomous Vehicles concerning possible amendments to the British Highway Code (2021).

[49] Madden (2005), p. 271.

[50] Goldstein (2016), p. 242.

[51] Lohmann (2016), p. 336.

regime, and the harm caused to the victim is definitely compensated by the license holder or by their insurer.[52] However, as this approach is not fair on the users who do not have any effect on the occurrence of the accident, scholars and regulators endeavour to switch the liability from licence holders to manufacturers.

Second, in some countries,[53] liability of driver, who is not the owner of the vehicle, for traffic accidents depends on their fault. In English Law, road traffic accidents are subject to law of negligence. Particularly, drivers of vehicles have a duty of care towards other drivers, pedestrians, and other road users. The standard of care is 'reasonable care' expected from a careful driver. However, even if drivers have a duty of care, they can only be held liable if they are at fault. Thus, if the driver shows reasonable care expected from him/her, and the accident still occurs, and consequently has no fault in the occurrence of the accident, he/she will not be liable. Notably, the standard applies for learner drivers[54] and drivers who suffered health issues[55] during the accident as well. In the case of *Mansfield v Weetabix Ltd*,[56] the driver was not found liable as he was not informed of the condition endangering his safety at the time of the accident. Even if such a regime applies for vehicles driven by human drivers, , it would be difficult to find liability on the license holder of an autonomously driven vehicle when it is involved in an accident as he does not have any effect in the occurrence of the accident and as there is no possibility to find fault on the license holder.[57] Hence, it will be difficult to ensure (proper) allocation of liability in the case of the use of AVs in a fault-based liability regime as well as that the victim will be compensated.

As can be seen, while in the first regime, it is possible to compensate the harm caused to the victim by rendering the license holder liable, in the second regime, it is not possible to allocate liability by rendering the user responsible for the accident caused by an AV. While the first regime may turn out to be unfair for the users, the second may result in the victims being unfairly treated as there is no possibility for them to compensate their damages. To ensure both fairness for both sides, namely users and accident victims, and allocation of liability, scholars suggest[58] that manufacturers should bear the liability for accidents and harms caused by AVs. Moreover, this approach could be favoured to enhance protection of consumers using AVs.

In the following section, the applicability of the current regime for manufacturers' liability will be examined and its fairness will be critically analysed. Moreover, it would be discussed whether this approach would prevent innovation, and if so, how such a risk would be reduced.

---

[52] Lohmann (2016), p. 336.

[53] See, for example Swiss Law and English Law.

[54] Nettleship v Weston [1971] 2 QB 691.

[55] Roberts v Ramsbottom [1980] 1 WLR 823; Broome v Perkins [1987] RTR 321.

[56] Mansfield v Weetabix Ltd [1998] 1 WLR 1263.

[57] Lohmann (2016), p. 336.

[58] Lohmann and Rusch (2015), p. 352. See in general, Channon (2019), Oliphant (2019).

3.2 Manufacturers as the Risk Bearers

As has been described above, it is anticipated that AVs' introduction on the market will result in the shift of the acceptance that the responsibility for car accidents can only be borne by a human license holder/driver[59] since with the use of AVs, car crashes will be caused by a defect or by defects in AVs themselves rather than the conduct of the driver. Thus, as will be described below, the shift in liability from users to manufacturers is supported by the most scholars. However, opponents are of the view that if an expert system, which can self-learn from the changing environmental features around it and from the inhabitants in that changing environment, and which gains experiences from its own decisions, is uploaded to the AVs the allocation of liability between manufacturers would be more challenging.[60] The benefits and drawbacks of shifting liability from the license holders to the manufacturers will be examined briefly below.

First, it can be considered that the adequate risk bearer is the manufacturer as this will serve as a motive for them to improve their products' safety design.[61] Even if the users may have a degree of effect in reducing risks by indicatively uploading software update, the manufacturer remains the common denominator as being the one who is responsible for the product's safety by designing, constructing, and by instructing the user about the risks. Moreover, the party who earns profits from AVs' placing on the market is the manufacturer and thus the manufacturer is anticipated to be able to recover some profits from consumers by selling the vehicles in higher prices.[62] In this respect, it appears that holding the manufacturers liable rather than the users is a fairer approach.

However, this approach has been criticised by some as well. The criticisms are as follows: First, rendering manufacturers liable may prevent innovation. Scholars are concerned mainly that if manufacturers bear exceeding liability under the current product liability regime which will be examined in the following sections, this would impede manufacturers from producing AVs —the so-called risk of technology chilling.[63] Nevertheless, a counterargument to such criticism could be that, if claims are raised against them, they would have several defences to bring forth such as the 'state of art' defence found in the Consumer Protection Act 1987, the 'contributory

---

[59] Lohmann (2016), p. 336.

[60] Lohmann (2016), p. 336.

[61] Lohmann (2016), p. 338.

[62] Lohmann (2016), p. 338.

[63] Marchant and Lindor (2012), it has been argued that *'liability has the potential to present significant deterrent to the development of autonomous vehicles'* (p. 1322); Marchant and Bazzi (2020), p. 118; U.S. Chamber Institute for Legal Reform (2018) Torts of the Future: Autonomous Vehicles (Special Edition), A Look at Emerging Trends in Liability and Regulation. https://instituteforlegalreform.com/research/torts-of-the-future-autonomous-vehicles/, accessed on 1 August 2022, p. 5; see Kim (2017), pp. 312, 317; Montagnani and Cavallo (2021), p. 212; Viscusi and Moore (1991), pp. 105–150, 163; Huber and Litan (1991), p. 729; Parchomovsky and Stein (2008), pp. 286, 288.

or comparative negligence' defence[64] and 'misuse' defence -both enshrined in the Product Liability Directive,[65] 'assumption of risk' defence,[66] which derives especially from American case law,[67] the 'development risk' defence, which derives particularly from the Product Liability Directive, and the 'economic loss doctrine' which also has been upheld in American case law.[68] Furthermore, unless there is a defect in the product at the time of the sale, it would be unfair to render manufacturers liable for self-learning AVs' decisions, which are unpredictable.

From the users' perspective, the issue is that even if they are set to recover their damages from the manufacturer of their AVs, it may be difficult for them to find the party who is liable as the sophisticated nature of AVs may increase the number of potentially liable parties. Moreover, as will be described below, in practice, finding and proving a defect will be proved to be difficult for the party who is harmed by the actions of an AV.[69]

The question which arises is that even if it is desirable for the users to escape liability for the damages caused by their AVs' autonomous decisions in which they did not intervene, is it really acceptable or even fair to render manufacturers liable under the current product liability regime for the harms caused by self-learning and self-deciding AVs? Thus, in the following section, the current product liability regime in the EU will be critically analysed. Further, whether such regime is suitable for AVs will be discussed.

3.3 A Summary of the Current EU Product Liability Regime

Under this section, the relevant regulations for AVs will be examined, which are the EU Machinery Directive and the EU Product Liability Directive.[70] The relevance of the EU Product Liability framework has been widely accepted as it is considered that

[64] Anderson et al. (2016), p. 153; Gursten (2013) Driverless Cars: Who's Liable in an Accident?. https://www.michiganautolaw.com/blog/2013/05/07/driverless-cars-whos-liable/. Accessed on 1 August 2022. See also Patsakis & Solanas (2013); Scharring et al. (2017).

[65] Gurney (2013), pp. 258, 259, 267, 268.

[66] Ibid., p. 269, 271; Marchant and Lindor (2012), p. 1336.

[67] Keeton (1961), p. 133; see Pokrajac v. Wade Motors, Inc., 266 Wis. 398, 63 N. W. 2d 720 (1954); Moss v. Fortune, 340 S. W. 2d 902 (Tenn. 1960); Tyler v. Dowell, Inc., 274 F. 2d 890, 895 (10th Cir. 1960).

[68] according to which product liability tort claims are precluded when the claimed damages pertain to financial loss and not bodily injury or damage of property other than the product itself. The rationale is that, whereas a consumer should not be charge with the risk of physical injury, he can be fairly charged with the risk that the product would not match his economic expectations, unless the manufacturer agrees that it would; Geistfeld (2015), p. 399; see *Seely v. White Motor Company* 403 P.2d 145 (Cal. 1965); *East River Steamship Corp v. Transamerica Delavel Inc.*, 476 U.S. 858, 866 (U.S. 1986); Bugra (2020), pp. 180–181.

[69] Lohmann (2016), p. 336.

[70] Noussia (2020), p. 255.

liability could be imposed on the manufacturers of AVs, and the human drivers would be freed from liability, if only the EU Product Liability regime applied to AVs. Moreover, it is also claimed that there is already an abundance of precedent according to which manufacturers of automobiles' liability was established for defects in automobiles which result from malfunction and cause injury.[71] Further to this, manufacturers' liability can also arise in the context of AVs since errors in the programming of AVs or system failures will result in malfunctions of AVs causing manufacturers' liability.

Under the current product liability regime in the EU, liability can be imposed on the manufacturer for harm suffered by a person by reason of the product produced by the manufacturer if there is a defect in the product at the time of the sale, which can be classified in one of the following categories, namely design defect, manufacturing defect and warning defect.[72]

Once a defect which falls into one of the categories described above is detected, both negligence and strict liability doctrines have been used in different countries to ensure recovery of injuries suffered while the EU Member States applied the strict liability regime which will be described below. It is necessary to compare the two as the fairness of strict liability regime can only be sufficiently analysed if it is compared to its opposite: It could be argued that the fundamental factor in detecting fault is manufacturer's conduct in the negligence regime, while it is the product itself in the strict liability regime.[73] Moreover, in terms of AVs, it is generally claimed that strict liability is an approach more favourable to consumer as manufacturers' liability would be evaluated by accepting that they possess expertise knowledge for producing an AV while negligence regime evaluates manufacturers' liability by comparing their actions to 'what is reasonable for them to do'.

The discussion on whether or not software can be accepted as a product is outdated now, as with the status of the software as a product has been established with the court decisions. However, the defect in the product shall be proved by the injured party to recover damages which is expected to be adequately difficult for a sophisticated technology as AVs.[74] Thus, a regulation which defines the concept of 'defectiveness' in the context of AVss required. Moreover, the product liability doctrine is only restricted to property damage and physical injury and therefore, do not cover other types of damages such as pure economic harm and privacy violations which are the claims that might arise excessively after the wide use of AVs. Thus, to ensure consumer protection, pure economic harm of the users and harms caused by privacy violations should also be compensated even if not by manufacturers but by a compulsory insurance scheme brought by the regulators.

[71] Goldstein (2016), p. 247.

[72] Goldstein (2016), p. 248.

[73] Goldstein (2016), p. 248.

[74] Goldstein (2016), p. 248.

3.4 Strict Liability

The EU Directive on Product Liability has brought strict liability regime to the EU Member States. Under this regime, a product is accepted as defective if it is not as safe as it is reasonably expected.[75] In the context of highly safety critical AVs, it is predicted that these expectations will be especially intensified as designing AVs in the safest way possible will be significantly important for their social acceptance.

Moreover, under the strict liability regime, manufacturers are held liable without the need of detecting a fault on their behalf if the product is defective and caused injury.[76] However, even if the strict liability regime is conceived by most scholars to be more favourable to the consumer, manufacturers will most probably be excessively burdened by this regime; since even if AVs are designed and produced by manufacturers, they will be designed to self-learn and make decisions by themselves. Thus, some of the decisions made by AVs might not be foreseeable even by manufacturers while several types of risks caused by AVs such as, personal injury, damage to the property can definitely be foreseen.[77] Thus, since AVs are designed to perform unpredictably at times even if the types of risks are foreseeable, the manners, in which such risks should be mitigated will not be clear enough and subsequently it may be harder than one imagined for manufacturers to assess them appropriately.[78] Despite this, the ability to predict the risks will prove to suffice so as for manufacturer's liability to be established under the strict liability regime.[79] This is the reason why it could be conceived as unfair to hold manufacturers liable under the EU Product Liability Regime.

Furthermore, despite the endeavour to make AVs safe, for instance, it will not be possible to ensure that their software be reliable. Moreover, despite the unreliability of the software, the users will still have the right to require a safe product and consequently, require that the manufacturers be held liable if a crash had been caused due to a failure in the AV's software. The only way a manufacturer can escape liability will be to satisfy the state of the art in the designing stage of an AV, to monitor the system by updating it regularly and by fixing bugs. Thus, it could be argued that incorporating safety standards in AVs at a preliminary stage is significant for the boundaries for rendering the manufacturers liable to be set, since it would be fairer for manufacturers to be aware of the safety standards accepted by governments while producing an AV and be informed on the new developments. It should be stated that it is beyond acceptable for manufacturers to go above and beyond what is accepted in terms of safety, as being the innovators. However, falling below the standards set by the regulators should result in their liability. This proposition will be examined in more detail in the following section.

[75] Lohmann (2016), p. 336.

[76] Osmani (2020), p. 62; Noussia (2020), p. 255.

[77] Osmani (2020), p. 62.

[78] Rachum-Twaig (2020), p. 1144.

[79] Rachum-Twaig (2020), p. 1153.

For all the reasons given above, it could be concluded that strict liability scheme would be burdensome for manufacturers, and therefore, as the improvement of this technology would be impeded, the benefits which are likely to be brought by AVs may remain untapped.[80] Thus, regulators should find ways to reduce the manufacturers' liability. Providing a compulsory mutual insurance scheme would be the solution as will be described below.

3.5 Insurance as the Solution

Providing a compulsory mutual insurance scheme would be the solution to the exceeding liability of manufacturers under the current strict liability regime under the EU Product Liability Directive. It is suggested that if an insurance scheme, funded by the governments, users and manufacturers is formed, it would lessen the burden of the manufacturers.

At this point, the UK's Automatic and Electric Vehicles Act 2018 (AEVA 2018)[81] could be provided as an example because it requires a compulsory single insurance policy for AVs so that in cases when an AV is involved in an accident, the insurance company will compensate the damages. However, there are still issues to be investigated under the AEVA 2018.

First, the burden of forming the insurance policy lays only with the users of AVs. This may cause problems as AVs is a novel technology and the risks entailed are not clear yet. Thus, until a point is reached where the risks of AVs are foreseeable, measurable, and reducible by taking certain measures so that premiums do not reach so high levels, it would be too heavy of a burden for the users to pay the premiums and insure their AVs. It is not very clear at the time of the writing of this article what the single insurance policy introduced by the AEVA 2018 means. However, the issue would be more properly addressed if manufacturers and governments contributed to the payment of the premiums along with the users so that they can share the liability for the harms caused to the victim.

Second, the exact terms of such single insurance policy constitute an issue, which is still left unclear. Although the insurer will provide payment when an AV gets involved in an accident, there may be several cases where the insurer may require that an amount of money paid by him/her be retracted by the manufacturer or the user. Particularly, this is exactly the case in Section 5(1) of the AVEA 2018, according to which the insurer may pursue the manufacturer as well as the insurer of the other user of a conventional vehicle for any reimbursement or contribution provided that it can establish liability for the accident under scrutiny. Moreover, it should be mentioned herein that is not just the manufacturer of the vehicle or the other vehicles insurer that may be pursued under Section 5 (1) AEVA 2018 as the

[80] Osmani (2020), p. 63.

[81] See Channon et al. (2019); Marson et al. (2020), pp. 395–416.

provision does not limit itself to them. The Recovery takes place in accordance with the existing product liability laws in force, but it could also be under negligence law. Usually, such disputes between insurers and manufacturers will end up in court, however it is anticipated that targeted processes will be developed between these two sides for recovery claims to be easily and efficiently resolved.[82]

The user's liability may arise against the insurer if the former neglects updating the software, as is provided for in Section 4 of the AVEA 2018, while manufacturer's liability may arise if the product is defective at the time of the product's placement on the market or in cases where the manufacturer omits to take necessary measures to monitor and update the product on a regular basis, as provided for in Schedule 3, para. 5(d) of the CPA 1987.

On manufacturers' responsibility, relying on the traditional product liability can hardly be efficient. There is the need for governments' intervention by regulating manufacturers' responsibilities and by setting the boundaries for their liability. The regulators should ensure the following:

First, the issues as to the defectiveness concept of the current Product Liability Regime have been described above. Defectiveness of an AV should be defined by the regulators by considering the latest and the safest equipment and software available for AVs. Soft law mechanism has been deemed a right choice for overly protective measures that would impede investing in innovative technologies to be avoided.[83] This is significant as most manufacturers are prone to use more economical equipment to ensure that they gain profit from their products. Thus, regulators should set the minimum standards for safety equipment and update those standards where necessary. By and large, it is easier for soft criteria to be issued since more room is left for agencies to act whereas in cases of formal rules those agencies' discretion might otherwise be constraint.[84] As long as manufacturers comply with those standards, they should not be held liable. As has been stated above, while going above and beyond these standards could be acceptable, falling below them should be punished. An example for such a soft law mechanism is provided in the US. The guidelines for AVs are published by National Highway Traffic Safety Administration. The second measure to be taken by the governments is to impose the obligation to manufacturers to monitor and update AVs regularly. If they do not comply with these obligations, the insurers could retract the amount of money paid by themselves from the manufacturers.

---

[82]See also Centre for Connected & Autonomous Vehicles (2017) Pathway to Driverless Cars: Consultation on Proposals to Support Advanced Driver Assistance Systems and Automated Vehicles: Government Response. https://assets.publishing.service.gov.uk/government/uploads/system/uploads/attachment_data/file/536365/driverless-cars-proposals-for-adas-and_avts.pdf. Accessed on 1 August 2022.

[83]Hagemann et al. (2018), p. 98; Hutter (2017), '*Regulators must have access to accurate information so that they have a clear idea of the risks they are regulating*' (pp. 101–114); Owen (2015).

[84]Hagemann et al. (2018), p. 98.

Before discussing users' liability, it must be underlined that manufacturers have an obligation to instruct the users in cases where it is necessary to update their product or if they need to intervene when there is a dangerous situation. If the users' do not adhere to these instructions, the insurance company will be entitled to retract the amount of money paid by themselves from the users. At this point, the instructing obligation of manufacturers would be the same as it is under the current product liability regime. It is claimed that instruction obligation of manufacturers may get complicated if the users are required to intervene in dangerous situations as manufacturers may not be able to foresee every scenario beforehand. At this point, rules stipulating manufacturers' obligations do not have to be explicit in their wording and hence describe every dangerous scenario to the users. They rather impose the users the duty to pay attention to the road if this is what regulation requires for safety. Moreover, it has been held that the users' intervention and requirement of attention should be minimised as much as possible—if ever required—to ensure that they benefit from such technology to the utmost extent. However, if there is the requirement that the users be attentive, intervene and update the software, manufacturers and insurers may rely on the users' actions or inaction with devices such as black boxes. The evidence provided by the black box should be proof of the fact.

4 Criminal Liability: *An Example of Corporate Manslaughter Under the UK Law*

To find someone guilty of a crime, two elements of criminal liability should be established: *actus reus* and *mens rea*. *Actus reus* stands for the action and conduct element of a crime while *mens rea* stands for the mental element. Thus, even if AI-based robots committed the *actus reus* of a crime, the mental element is challenging to be established when AI-based machines are in question.[85] However, since robots are able to acquire knowledge and make decisions accordingly, it could be argued that robots may soon become as intelligent as or even more intelligent than human-beings, and their actions would be unpredictable and uncontrollable by even their manufacturers or their users.[86] In this context, it is suggested that there needs to be a new subject in criminal law, which will be referred to as '*machina sapiens criminalis*'[87] along with humans and legal entities.[88]

There are examples in which legal person status was granted to AI-based machines. First, a humanoid robot, Sophia, was granted citizenship in Saudi Arabia

[85] Osmani (2020), p. 57; Pagallo (2013), p. 51.

[86] Osmani (2020), p. 58; Asaro (2015) Liability Problem for Autonomous Artificial Agents. https://peterasaro.org/writing/Asaro,%20Ethics%20Auto%20Agents,%20AAAI.pdf. Accessed on 1 August 2022.

[87] Osmani (2020), p. 58.

[88] Osmani (2020), p. 58.

in 2017.[89] Second, Japan gave a residency permit to a therapeutic robot, Paro in 2010. Lastly, a resolution entitled *Civil Law Rules on Robotics* is adopted by the European Union.[90] With this resolution, it is foreseen that in case where robots are granted a particular legal status, which is defined as 'electronic persons',[91] this will render them liable for the harms and injuries caused by their autonomous decisions.[92] Consequently, allocation of liability for the harm caused by an autonomous decision that is made independently from any human intervention would be ensured by giving such a legal status to AI-based machines.[93] From this perspective, if autonomous vehicles are granted the status of legal persons, they will be committing an offence when they are involved in accidents and cause harms by violating traffic rules or by acting in a manner, that may be classified as careless driving or dangerous driving pursuant to the current regulations regulating humans' driving behaviour.

However, the purpose of criminal law is to punish criminal conduct. Moreover, the more human features, such as fear, freedom of movement and pain are absent in robot systems,[94] the less meaningful imprisoning a robot would be. Thus, making AI-based machines criminally liable party remains a problematic option. To resolve the issue of punishment, different scholars have shared different views. Asaro has supported that robot systems are built-systems and their schemes are selected by their designers, and therefore, the responsibility for the machines' actions will be borne by designers and programmers of the learning mechanisms of the machines.[95]

Hallevy presented the 'Perpetration-by-Another Liability Model'.[96] According to this model, AI-based machines could be treated as 'innocent agents' as they are machines having no human characteristics and therefore, they cannot be a perpetrator of a crime. Under this model, the programmer of an AI machine, who creates a software with definite instructions resulting in the machine committing a specific offence, is the real perpetrator.[97] Thus, when an autonomous vehicle gets involved in an accident as explained above, the real perpetrator would be the manufacturer. However, there may be more than one manufacturer, and therefore, it would prove difficult to determine the liable party even if this approach is supported

In both these instances, if the responsible party is determined successfully, it will result in the excessive responsibility of both designer and the manufacturer companies of AVs. At this point, it should be stated that as was claimed under Sect. 3, in

[89] Osmani (2020), p. 59; Divino (2020), p. 196.

[90] European Parliament (2017) Civil Law Rules on Robotics. https://www.europarl.europa.eu/doceo/document/TA-8-2017-0051_EN.html#title1. Accessed on 1 August 2022.

[91] Osmani (2020), p. 60.

[92] Osmani (2020), p. 60; Barfield (2018).

[93] Osmani (2020), p. 60.

[94] Osmani (2020), p. 61; Saripan and Putera (2016), p. 827.

[95] Osmani (2020), p. 61; Asaro (2006), p. 10.

[96] Osmani (2020), p. 61.

[97] Osmani (2020), p. 62.

relation to the product liability of manufacturers, both the manufacturers' and the designers' liability for crimes committed by AVs may impede this technology.

Another approach is to treat AI machines in the same way as we treat corporations.[98] As corporations have legal personalities separate from their shareholders and managers, they can be a party to a contract, have rights and own assets. Moreover, they are also accepted to be able to commit crimes by regulators such as corporate manslaughter, as is regulated under the Corporate Manslaughter and Corporate Homicide Act 2007.

Under the Act, 'an organisation to which this section applies is guilty of an offence if the way in which its activities are managed or organised—

(a) causes a person's death, and
(b) amounts to a gross breach of a relevant duty of care owed by the organisation to the deceased'[99]

As can be seen, although a corporation's activities are managed or organised by the humans behind it, the corporation is penalised rather than its managers or people who have control in the corporation. The penalties for such crimes, include unlimited fines, publicity orders and remedial orders against corporations. It could be imagined that if AI-based machines have separate legal personalities and are able to own assets, they may be penalised in the same way for crimes that are regulated in a way that can be committed by AI machines by the governments. While it is open to discussion that an AI-based machine could commit a crime as they do not have any motive to do so, it may be a significant discussion point. For example, it was discussed above that AVs may be owned by platforms such as UBER and may be able to form contracts and may interact with a great deal of people. If it is accepted that the AV in question which provides travelling services for its customers have a legal personality and is not owned by another corporation but has those corporations as its shareholders; and if a fraud committed through the AV, the results would be far-reaching. As a result, it may be regulated that AVs are able to commit fraud, and offences similar to corporate manslaughter might be regulated, such as excessive fees. As those AVs will provide services for the public, they may have assets and could pay fees that may remedy the damages caused to the victims. However, at this point, it should be kept in mind that even if AI-based machines may own assets and have liabilities in the future, they will not have any motive to have more money or commit a crime. Thus, in the end, the party who actually commits a crime or who is actually punished by paying fees due to those crimes will be the so-called shareholders of such AVs.

---

[98] Solaiman (2017).

[99] Corporate Manslaughter and Corporate Homicide Act 2007, s. 1(1).

5 In Between Enhancing Predictivity and Protecting Personal Data of the Users

While previous literature on this issue provides the general regulatory framework,[100] the aim of the present chapter is to go deeper in exploring how the functioning of autonomous vehicles can be aligned with data protection rules, i.e. the General Data Protection Regulation (GDPR). Having this as an introductory reference, and bearing in mind the several considerations prescribed in recent UN and EU instruments, this section will describe the way data tracing and processing should be carried out to fully abide by the law, by giving specific examples.

The use of telematics, for instance dongles and smartphone apps, has brought additional changes to the way in which we drive. Telematics deploy wireless devices and 'black box' technology to relay information back to an organisation. Further, the sophisticated nature of AVs will require the use of a number of different tools incorporated such as cameras, navigations, wireless connection, advanced sensors, and devices that are similar to black boxes. These will be used in the AV technology not only to recognise features of the environment in which the AV functions, but also to observe the actions of the occupants. With the constant use of these tools, it is inevitable that massive amount of data be acquired about the occupants as well as the environment in which they travel. Manufacturers, insurers, governments and other third parties will be interested in those data for the purposes of developing the AV technology, decide on insurance premiums and offer new insurance products and to investigate whether there is any violation of traffic laws and subsequently to profit from those data by means of selling it respectively. The latter, namely the issue of gaining benefit from the data is the most controversial, considering that it goes against the data protection laws which will be examined below. Furthermore, there will also be Vehicle to Vehicle and Vehicle to Instruction connection between AVs to achieve more efficient road traffic.

Thus, it could be argued that such monitoring and connection is eventually vital for ensuring efficiency in travelling experience and for evidence and insurance purposes. However, even if through monitoring and connection efficiency, predictivity and certainty are achieved, the user's as well as the AV occupant's data privacy remains as a question. In relation to this, it should be highlighted that more and more people will be affected by AV's intrusive character, especially if one considers that AV technology is highly likely to be used in public transportation as well as by private companies providing tourism services and travelling platforms such as UBER. Thus, it could be concluded that private companies other than manufacturers of a particular AV and the insurer of that AV's occupant will require to access those data acquired through an AV.

Due to many vehicles' being equipped with advanced sensors, information not only on the vehicle, such as the model, parking location, air bag deployment, speed,

[100] Noussia (2020), pp. 261–263.

braking patterns and collisions but also the driver, such as driving behaviour, habits, age, gender and marital status can all be fed back.[101] The technology has been recently utilised by insurance companies to obtain information on a driver's behaviour to assess the quality of their driving and tailor their premium accordingly. It is likely that this will also be harnessed by autonomous vehicle manufacturers to establish whether it was the driver or vehicle in control at the time of a collision.[102]

As elaborated above, in Section A.1, SAE vehicles of levels 0 to 5 incorporate or will potentially incorporate systems which will be enabling different levels of autonomous driving. Considering that the human interference is decreasing as the SAE vehicles will be performing more and more tasks, in conjunction with the fact that the user will be able to perform other activities, while driving as well, one could easily conclude the following; the higher the level of automation the more the data, which is to be collected, since the system deploys as much information as possible to perform autonomously and the human factor to be deteriorated.

Particularly the fact that personal data are collected and fed back to be processed by insurance or car manufacturing companies, paves the way for the issue of compliance with the GDPR Regulation to be investigated. It should preliminarily be noted that to elaborate on this matter, the applicability of the Regulation should be ensured; specifically, the GDPR applies in cases where a company, which processes personal data, is based in the EU. Alternatively, the Regulation applies when a company, which is established outside the EU, processes personal data in relation to the offering of goods or services to individuals in the EU, or monitors the behaviour of individuals within the EU.[103] In principle, according to the EU data protection rules, the processing of personal information shall be carried out in a fair and lawful manner and is permissible for conclusively enumerated reasons;[104] the process' purpose should be legitimate, sufficiently specified and the data should be limited to that which is necessary so as for the specified purpose to be fulfilled. Furthermore, a company is entitled to process its customers' personal data, when one of the following criteria is met; consent of the individual concerned has been given by an affirmative act or the personal data is needed so as for a contractual obligation with the individual to be fulfilled or the personal data is required for a legal obligation to be performed or the personal data is needed for the vital interests of the individual to be protected or the process of personal data is being carried out in the interest of the public or the personal data process serves the company's legitimate interests, as long as the fundamental rights and freedoms of the individual whose data are processed are not seriously impacted. Concerning the latter, the person's rights shall not override the company's interests, since then the personal data should not be subject to any process.[105] Lastly, it should be highlighted that, pursuant to the Regulation's

[101] Tian et al. (2020), p. 156.

[102] Jeffcott and Rose (2017), p. 23.

[103] Data Protection Under GDPR (2021).

[104] Article 5, GDPR.

[105] Data Protection Under GDPR (2021).

preamble, personal data must be considered and balanced with other fundamental rights, in accordance with the principle of proportionality.

To this end, vehicle data can be considered as personal data to the extent that it is associated with a specific individual.[106] According to the prevailing *Theory of Combined Qualifications*,[107] proponents of which are the European Commission[108] as well as a number of legal scholars,[109] information pertaining to a vehicle, irrespective of its technical nature, is classified as personal data if a customer and subsequently its personal circumstances can be directly or even indirectly identified. Moreover, as underlined in Article 29 Working Party,[110] information either being provided knowingly and actively by the customer or generated by his or her activity can very well be considered as personal data, given that it can be linked with the particular individual.

Notwithstanding the fact that insurers and manufacturers both aim at obtaining data to improve their assessments and mitigate accident risks, there are some data privacy issues to be considered; to illustrate, collection of personal data, for example data on location,[111] for an excessive period is highly likely to also reveal sensitive information of an individual, such as health data. Therefore, in such cases, Article 5 GDPR applies, according to which the insurers or manufacturers shall inform the data subjects about the processing of their personal data, in a concise, transparent, intelligible, and easily accessible form, using clear and plain language, avoiding, for example legalese or vague terms. Additionally, under the Regulation, insurance companies and manufacturers have the obligation to sufficiently inform the customers as to the consequences of such processing as well as to explain the rationale behind automated decision-making mechanism.[112]

---

[106] Storing (2017) What EU legislation says about car data. Legal Memorandum on connected vehicles and data. *Federation International Automobile Reg, 1*. https://www.fiaregion1.com/wp-content/uploads/2017/06/20170516-Legal-Memorandum-on-Personal-Data-in-Connected-Vehicles-www.pdf. Accessed on 1 August 2022.

[107] By virtue of the Mutual Exclusion Theory, which has been adopted by several European Automotive Industry Associations, data should be categorised to non-personal and personal data. Information provided by the customer was considered as personal whereas data associated with the vehicle's technical aspects was classified as non-personal. See VDA (2014), p. 2; ACEA (2015), p. 4; ACEA (2016), p. 2 et seq.; SMMT (2017), p. 21 et seq.

[108] The European Commission (2016) Communication from the Commission to the European Parliament, the Council, the European Economic and Social Committee and the Committee of the Regions, A European Strategy on Cooperative Intelligent Transport Systems, A Milestone towards Cooperative, Connected and Automated Mobility. https://ec.europa.eu/energy/sites/ener/files/documents/1_en_act_part1_v5.pdf. Accessed on 1 August 2022.

[109] Storing (2017).

[110] Article 29 Working Party is composed on behalf of representatives from all European data protection authorities, the European Data Protection Supervisor as well as the European Commission and its opinions serve as guidelines in regard to interpretation and implementation of data privacy provisions.

[111] By virtue of Article 4 GDPR, location is classified as personal data.

[112] van den Boom (2021), p. 301.

Furthermore, considering that data collection and process via automated vehicles can be challenging, the Regulation guides the manufacturer or/and insurer to adopt a three-step approach for compliance with the data privacy rules to be ensured. Particularly, first, a Data Protection Impact Assessment (DPIA) shall be applied, which is considered as mandatory if the processing of personal data is likely to result in a high risk to the rights and freedoms of natural persons[113] or the information is classified in the special categories of data.[114]

The DPIA shall then be followed by data protection by design and data protection by default, both stipulated in Article 25 GDPR, and the core of which is the management of data protection risks.[115] As per the obligation of data protection by design, this should be considered from the outset of the system's design, which in the case at hand is incorporated in the automated vehicle, and continuously be observed throughout the technological life cycle.[116] Moreover, data protection by default aims at ensuring that the right to data protection is adequately safeguarded as a default setting[117] by applying to the amount of personal data collected, the extent of the processing, the period of storage and their accessibility.[118]

Hence, in the case of automated vehicles, personal data processed by the manufacturer or/and the insurer shall be directed towards the specific purpose, which is supposed to be served, under the light of the principle of proportionality enshrined in the Regulation; for instance, data on the driver's location or vehicle's direction shall be processed by the manufacturer to mitigate the risk of an accident and not to be sent over to local restaurants for advertising offers.

5.1 *The Case of Event Data Recorders*

To take it a step further, a closely related issue, namely that of Event Data Recorders (EDR) and their use, must also be discussed. EDRs will certainly play a crucial role in providing evidence in a trial and proving the accident occurrence and its cause. At the same time, the installation of EDRs is highly critical in the light of data privacy

---

[113] Article 35 GDPR.

[114] Article 29 Working Party (2018) Guidelines on consent under Regulation 2016/679, WP259 rev.01, 10 April 2018 https://ec.europa.eu/newsroom/article29/redirection/document/51030. Accessed on 1 August 2022.

[115] European Data Protection Supervisor (2018) Opinion 5/2018, 31 May 2018. https://edps.europa.eu/sites/edp/files/publication/18-05-31_preliminary_opinion_on_privacy_by_design_en_0.pdf. Accessed on 1 August 2022.

[116] Article 29 Working Party (2018) Guidelines on consent under Regulation 2016/679, WP259 rev.01, 10 April 2018 https://ec.europa.eu/newsroom/article29/redirection/document/51030. Accessed on 1 August 2022.

[117] Jasmontaite et al. (2018), p. 182.

[118] Article 25(2) GDPR.

law as well as criminal law and its principle that there is no legal obligation to self-incriminate.[119]

EDRs are not a brand-new technology; in fact, they have been deployed in ships, airplanes and generally vehicles since the 1990s.[120] It is to be noted also that in 2006, the Code for Federal Regulations established requirements for Light Duty Vehicles.[121] Their function is summarised in the collection of all the information, which is necessary to reconstruct and simulate events in case that an accident has taken place. In that way, access to information pertaining to the vehicle is enabled and subsequently a court can determine whether for instance the injury has been caused by the seatbelt's or the airbag's malfunction or the driver's negligence.

Given that the EDRs can be used to provide information as to the private life of the user, for instance the driver's workplace or personal communication, the driving behaviour, and the physical status,[122] several concerns have been expressed on the protection of privacy and personal data, especially considering the possibility of privacy infringement by attackers. In 2019, a first attempt towards addressing such privacy concerns has been made by the informal working group AHEAD (Aggregated Homologation proposal for Event data recorder for Automated Driving), which was formatted to promote the standardisation of the database needed for a future EDR. According to the working group, the goal would be that only the necessary information be recorded so as for the accident's circumstances to be clarified and assessed by the court. Indicatively, the process of recording would be the following: the EDR records data during an event for the first time. Consequently, the data recorded is immediately transmitted to a data trustee via a secure data connection and an over-the-air interface. In that way, if a fire occurs, for example, due to which the EDR is destroyed, the data is safeguarded by the data trustee because of the appropriate network coverage. Once the data is fully and properly received by the data trustee, it could also be deleted from the vehicle.[123]

The aforementioned mechanism has been considered to ensure legal certainty and data privacy,[124] given that accidents can be investigated thoroughly for the involved parties', namely the manufacturer's and driver's, liability to be determined. In addition, a method which enables the circumstances of an accident to be reconstructed accurately, contributes to the creation of anonymous accident scenario

[119]Lohmann (2016), p. 338.

[120]Wiewiórowski (2018) Data Protection and Privacy of the car… as a mobile device, *Type-approval requirements for motor vehicles as regards their general safety and the protection of vehicle occupants and vulnerable road users* https://www.europarl.europa.eu/cmsdata/158076/2018_11_29_wiewiorowski_event_recorders.pdf. Accessed on 1 August 2022.

[121]49 CFR Part 563.

[122]Wiewiórowski (2018), p. 21.

[123]AHEAD (Aggregated Homologation-proposal for Event Recorder Data for Automated Driving) (2019). https://www.thi.de/fileadmin/daten/forschung/CARISSMA/Technologiefelder/Sichere_Elektromobilitaet/Positionspapier_AHEAD.pdf. Accessed on 1 August 2022.

[124]Böhm et al. (2020).

databases, which would be of significant importance as far as road safety and accidents' prevention is concerned.

To this direction, several legislative materials have been issued with the aim of regulating EDRs, as well as addressing the privacy protection implications, which such systems may have. Particularly, in June 2021, the UN Regulation No. 160 was issued, which has been in force as of 30 September 2021 and seems to be, in principle, in line with the proposal of AHEAD mentioned above. This is because its provisions deal with the minimum data collection and storage by the EDRs so as for the most valuable information to be collected and investigated under an effective accident analysis. Moreover, this act applies to the approval of vehicles of categories M1 and N1 with regard to their Event Data Recorder (EDR).

It is to be noted that the UN Regulation is without prejudice to requirements of national or regional laws related to privacy, data protection and personal data processing. Further, what seems to be more important as to the issue of compliance with the GDPR, is the data that is excluded from the scope of EDR technology, under the Regulation. Particularly, VIN (vehicle identification number), associated vehicle details, location or positioning data, information of the driver, and date and time of an event do not fall inside the scope of the data to be collected by EDRs.

On data recording, this shall be triggered in specific circumstances of a sensed technical abnormality; particularly, the EDR will start recording when there has been a change in longitudinal vehicle velocity more than 8 km/h within a 150 ms or less interval or when the lateral vehicle velocity changes to more than 8 km/h within a 150 ms or less interval or non-reversible occupant restraint system has been activated or when the vulnerable road user secondary safety system has been activated.

On the issue of data storage, according to the UN Regulation, the EDR shall record the captured data in the vehicle and this data shall remain in the vehicle at least until they are retrieved in compliance with national or regional legislation. In cases where an EDR non-volatile memory buffer void of previous-event data is not available, the recorded data shall be supplemented or replaced by the current event data, on a first-in first-out basis, or according to different strategies decided by the manufacturer and made available to the relevant parties.

The European Commission underlined the value of data collected by EDRs and, in its proposal for supplement of the UN Regulation, it highlighted that EDR data shall not be used accumulatively with data collected by Data Storage Systems for Automated Driving (DSSAD),[125] which relate to the vehicle control, are stored over a longer period of time and serve to identify if the driver or system was (requested to be) in control of the driving task whenever a specific event occurred.[126] The

[125]The European Commission (2021), Commission Delegated Regulation (EU) Draft supplementing Regulation (EU) 2019/2144 of the European Parliament and of the European Commission (2021), Proposal for Supplement 1 to the UN Regulation No. 160 (Event Data Recorder) and to the 01 series of amendments to UN Regulation No. 160. https://unece.org/sites/default/files/2021-04/GRSG-121-32e.pdf. Accessed on 1 August 2022.

[126]DSSAD system has its origins in the National Highway Traffic Safety Administration's (NHTSA) nonregulatory approach to automated vehicle technology safety and has been introduced

Commission's proposal can be conceived as more than reasonable, since if EDRs and DSSAD are deployed jointly, this could result in excessive personal data being collected. To illustrate, in the framework of EDRs, it would go beyond the scope of investigating the circumstances of a particular car accident to gain information on who was requested or actually driving three months ago, as the case would be with DSSAD data collection.

Lastly, in October of 2021, the European Commission made a significant step towards embracing EDRs by issuing a Commission Delegated Regulation (EU) draft, supplementing Regulation 2019/2144.[127] Under the act, motor vehicles of categories M1, namely vehicles designed and constructed for the carriage of passengers, comprising no more than eight seats in addition to the driver's seat, and having a maximum mass ('technically permissible maximum laden mass') not exceeding 3.5 tonnes, and N1, namely vehicles for the carriage of goods with a maximum mass not exceeding 3.5 tonnes, shall be equipped with event data recorders from 6 July 2022 for new vehicle types and from 7 July 2024 for all new vehicles.

Considering the above, it could be argued that the UN Regulation and the following European materials align with the principle of data minimisation and lawfulness of the processing enshrined in the GDPR, seeing that the data to be collected and processed is exclusively limited to that, which is in fact necessary for a specific purpose that is an effective accident analysis.

In the following section, the issue of data acquisitions as well as its implications pertaining to privacy will be addressed, which is significant for the development of AVs, so as for allocation of liability to be ensured and insurance premiums to be determined more precisely.

6 Big Data

While humans have the inherent feature to forget things, Big Data technology ensures that massive amount of information is acquired and stored in the web.[128] Being able to acquire information about the occupants and the environmental features via the devices provided within an AV will enable government, manufacturers, insurers and other private third parties to profile and classify individuals using this technology which causes privacy concerns. The act of 'data mining' may help those parties to identify 'sensitive' information about the occupants such as,

---

by the Working Party on General Safety Provisions. See 'These data should also contain the status of the ADS and whether the ADS or the human driver was in control of the vehicle leading up to, during, and immediately following a crash'; OICA (2019).

[127] The European Commission (2021).

[128] Vrabec and Domnik (2021), p. 129.

movements, race, political opinions and criminal records.[129] Moreover, if and when data acquired through an AV is cross-referenced, an individual can possibly be identified.[130]

SAE has categorised AV systems in a very detailed manner, and its categorisation has been accepted by the European Union and United States. It categorises vehicles from non-automated to fully automated vehicles. In between non-automated and fully automated vehicles, there are different levels of semi-automated vehicles. While the level of automation rises, the need for human intervention declines, and therefore, dependence 'on computer generated control and computer-generated information increases'.[131] The subject of this chapter is fully AVs. However, as the way in which fully automated AVs acquire data resembles the way which semi-automated vehicles acquire information, the latter will be described as well.

6.1 *How Do AVs Acquire Data?*

AVs will obtain a considerable amount of data which will not be 'homogenous' and therefore, will include different groups of data. Moreover, these data may be obtained from multiple sources as indicated by Brandy[132] such as 'image data external to the vehicle; image data internal to the vehicle; event data records; location/route data from navigation system; data location/route data from V2V/V2I (vehicle to vehicle/vehicle to infrastructure) communication; data covering biometric; biological or health factors; data from in-cabin microphones and entertainment systems; external microphone data; input unit data; and electronic control unit data'.

There will be 'unique vehicle identifier tags'[133] for each AV, and each electronic control system of a particular AV will be associated with that AVs identifier tag. Tags of AVs will be anonymised by using encryption system and will be conveyed within 'the data stream' while the AV is active.[134] It would be possible both to store the data produced by an AV within itself to allow functioning of the system and to convey the data to the Internet. Anonymisation will ensure that the occupants of a particular AV are not exposed in violation of privacy while those occupants are identifiable by cross-referencing the sets of data.

There are three ways in which data may be obtained through an AV:

Stored Data Data restrained in the Event Data Recorder of an AV is named as 'stored data', and it has already been used in the most cars to observe the operational

[129] Brady (2019), p. 590.
[130] Brady (2019), p. 591.
[131] Brady (2019), p. 594.
[132] Brady (2019), p. 595.
[133] Brady (2019), p. 595.
[134] Brady (2019), p. 595.

parameters of a vehicle in between the last 5 to 30 seconds prior to an accident. This is a function that is included in the most modern vehicles even if they are not automated.

The information gathered via Event Data Recorders is significant because they have the possibility to change the behaviour of an AV during an accident as these are self-learning vehicles, Moreover, when an AV is used, the significance of data regarding driver will decrease as driver's intervention to driving process will decrease as well. At this point, it could be argued that privacy issues will arise when the data gathered through an EDR is used to identify a driver for purposes other than identifying the liable party.[135]

Shared Data Data circulated by an AV is defined as shared data.[136] For a fully automated AV, the information gathered is not about the human driver anymore. It is rather about the occupants of a particular AV and is dependent on the functioning of the particular AV.[137] Moreover, as was stressed above, this information could be about the occupants of a privately owned AV or occupants of a public transport vehicle or occupants of an AV that is owned by private companies such as travel agencies. Thus, many individuals may be affected by an invasion of privacy.

Dedicated Short Range Communications ('DSRC')[138] are used by AVs to share the relevant positions, vector, and vicinity of other vehicles. Acquisition of information in real time become possible with Connected Vehicles. Moreover, the act of sharing that information with other connected vehicles nearby fulfils the need to recognise the other road users in cases where a human driver is unable to keep an eye on the road and prevents accidents in the event of the use of an AV.

This kind of data transmission should be limited to serving the purpose of allowing safe functioning of the vehicle. AVs are open to attacks at the point when they share data, and therefore there are cyber risks. Although this technology is used to provide efficiency in the road traffic, the risk of cyberattacks may cause violation of the occupants' privacy as well as dangers to their physical integrity.

Broadcast Data Data shared and received by an AV through telecommunication systems, to/from a Cloud storage location that is provided by the AV's manufacturers, or related traffic authorities is named as Broadcast Data. Transmission of data that is related to the functioning of an AV and its control settings, its position, velocity and vector and occupancy information will be made by itself in real time.

The most significant issue with the Broadcast Data is that the misuse of those data may have vital results as the volumes of data is much greater. While the automation is increased, the input of driver will decrease, and therefore, AVs will rely on

---

[135] Brady (2019), p. 595.

[136] Brady (2019), p. 595.

[137] Brady (2019), p. 596.

[138] Brady (2019), p. 596.

external data more. Moreover, the portion of data that is received and shared to allow functioning of an AV increases. Thus, the risk of the AV occupants' privacy violation is much greater. It has been supported that Broadcast Data constitutes the highest risk for privacy as by using the correct codes it may be possible to intercept and decrypt them.[139]

As the three ways in which AVs acquire and provide data has been described, the legal basis for such an acquisition and processing of data remains to be investigated.

Consent Under the GDPR, consent must be 'freely given, specific, informed and unambiguous'. When AVs are in use, occupants of AVs must give consent so as for their data to be acquired. Consent must be given to manufacturers, insurers and other private companies.

Right To Be Forgotten This right concerns removal of information.[140] This is a newly created right that is brought by the needs of the digital age. A massive amount of information is stored in digital locations. Data about people is acquired and stored by government, companies and by social media platforms. Those data are produced even without people who are the subjects of those data realising that they have produced them. Despite this, those data are stored anyway, and there is a possibility that they can be detrimental to those people. In terms of AVs, this right may prove vital for the occupants of AVs, as for instance, the information as to the environment in which an AV is used by those occupants may be stored and this may result in the profiling of the occupants of AVs.

Purpose Limitation Big Data means 'the analysis of large data sets in order to find new correlations'.[141] Although Big Data is useful for making more precise decisions, it causes some legal issues. To prevent such issues, data protection regulations has been introduced, and therefore, today, if personal data is processed in applications with the use of Big Data technologies, it should be made in accordance with the general principles of data privacy laws. One of those principles is purpose limitation. According to purpose limitation principle, 'personal data must be collected for specified, explicit and legitimate purposes and not further processed in a way incompatible with those purposes'. Purpose limitation may ensure that parties who possess information regarding the occupants of AVs such as insurers, manufacturers and travel agencies will limit their usage of those data with the purposes they have stated when they acquired those data.

Human beings have always collected data and made assumptions about each other. However, the source of information has changed enormously after the start of the use of Big Data. Already in the past, the source of information was mostly the person who the information was about. However, at the moment the information derives from that persons' use of social media network, purchase preferences, use of

[139] Brady (2019), p. 597.
[140] Haga (2017), p. 99.
[141] Forgó et al. (2017), p. 17.

digital applications, etc. The preferences of a person serve to profile and categorise that person and hence make assumptions about him. There is a limitless and never-ending acquisition of information in profiling about people, and the use of AVs will be another way to perform this. Thus, abovementioned data protection rules and principles would be a way to protect occupants of AVs from invasion of their privacy. However, it should be stated that in the process of development of AVs and other AI-machines, new data protection rules and principles may be developed as regulation always follows new developments, and therefore, the applicability and efficiency of those rules are to be seen.

7 Conclusions

Advanced Driver Assistance Systems are features of a present and future automated transportation world within which fully autonomous vehicles (AVs) are expected to be used widely in the coming years. Notwithstanding this development, wide public circulation may be delayed due to legal, ethical and regulatory issues. This chapter has sought to highlight such issues and demonstrate the regulators' significant role in providing solutions as well as discuss necessity situations, liability issues and insurance implications, as well as criminal and contractual liability and data implications. Some conclusion to be derived pertain to the below: as our discussion has elaborated, of the final steps in the automation process is the one related to fully automated AVs, which do not require any human intervention for driving[142] and it is anticipated and expected that such technology will be used widely by the mid-2020s.[143] Despite the hope for the wide public circulation of AVs sooner rather than later, it is pertinent that such technology is used and deployed massively in the wider public sector only when it becomes more utilisable and safer remains. It is in the light of the above realisation and necessity, that while manufacturers focus on rendering AVs both profitable, practical, and safe for users, policy and lawmakers strive to find solutions to address legal and ethical issues deriving from this technology. Presently, automated vehicles of level 0 and level 1 are increasingly in use, which incorporate a system controlling the vehicle's longitudinal and lateral motion. When we reach SAE level 4, the vehicle will perform the dynamic driving task on its whole without any human interference and when we reach SAE level 5, i.e. the ultimate form of AVs, these will be anticipated to be able to perform the driving task under any road and weather circumstances.[144] In conclusion, our discussion has shown that if we compare AVs with driverless cars, AVs will be much safer than driverless cars. However, for the public to use AVs, their benefits should exceed their risks, i.e. they will need to be at so high a level that they are considered as safer

[142] Lohmann (2016), p. 335.

[143] Lohmann (2016), p. 335.

[144] On-Road Automated Driving Committee (2018).

than the best human driver.[145] In the light of the above it has been suggested that owners of AVs be entitled to make a decision on programming of their AVs and bear responsibility for their own decisions,[146] and also AVs should contain an 'ethical knob',[147] to enable the users of AVs to switch ethical mode on their AVs. An alternate solution would be to leave the programming decision to the manufacturers, but as our discussion has shown, such a solution entails problematic areas as well. The last resort as a solution is to rely on the regulators, to guarantee that whichever way different manufacturers ethically program AVs, such programming will be consistent, hence, helping to provide predictability in terms of behaviour of cars, tilting the balance in favour of the customers and providing fairer outcomes.[148] The regulation's aim should be to render fatal outcomes illegal considering especially the fact that everyone's life values the same.[149] Thus, to the extent that in the programming stage, the focus is on protecting and/or saving either the occupant, namely the user, or the third party/parties, self-driving cars should by no means be acceptable choices. Therefore, it is suggested that AVs be programmed in such a way to result the least possible harm. Although currently, liability can be imposed on the manufacturer for harm suffered by a person by reason of the product produced by the manufacturer if there is a defect in the product at the time of the sale, which can be classified in one of the following categories, namely design defect, manufacturing defect and warning defect,[150] it is suggested that such liability will be hard to prove for AVs as it will be difficult to assess adequately the risks involved. On liability, the question which arises is whether it is fair to render manufacturers liable under the current product liability regime for the harms caused by self-learning and self-deciding AVs. To prevent such unfair treatment, any strict liability scheme would be burdensome for manufacturers, and therefore, as the improvement of this technology would be impeded, the benefits which are likely to be brought by AVs may remain untapped.[151] Thus, regulators should find ways to reduce the manufacturers' liability. Insurance could act as a solution and a compulsory mutual insurance scheme would mitigate the chance of an exceeding liability of manufacturers under the current strict liability regime. Telematics can also contribute via monitoring and connection to ensure efficiency in travelling experience and for evidence and insurance purposes. However, the user's as well as the AV occupant's data privacy remains as a question. The use of data entailed in AVs vast deployment has also meant that data issues arise. Given that EDRs are widely used already, some concerns have been expressed on the protection of privacy and personal data, especially considering the possibility of privacy infringement by attackers. The

[145] Gasser et al. (2016), p. 523; Pammer et al. (2021), p. 257.

[146] See Kochupillai et al. (2020), pp. 285–299.

[147] Contissa (2017).

[148] Jeffcott and Rose (2017), p. 23.

[149] Noussia (2020), p. 254.

[150] Goldstein (2016), p. 248.

[151] Osmani (2020), p. 63.

GDPR is a regime that addresses issues relating to the use and collection of personal data. Also, the UN Regulation No. 160 of September 2021 applies without prejudice to requirements of national or regional laws related to privacy, data protection and personal data processing. Big Data is also discussed as the AVs will collect and process such amounts of Big Data and hence a sound regulatory framework is needed. To this end stored data, shared data and broadcast data are addressed.

Cases

Broome v Perkins [1987] RTR 321

East River Steamship Corp v. Transamerica Delavel Inc., 476 U.S. 858, 866 (U.S. 1986)

Mansfield v Weetabix Ltd [1998] 1 WLR 1263

Moss v. Fortune, 340 S. W. 2d 902 (Tenn. 1960)

Nettleship v Weston [1971] 2 QB 691

Pokrajac v. Wade Motors, Inc., 266 Wis. 398, 63 N. W. 2d 720 (1954)

Roberts v Ramsbottom [1980] 1 WLR 823

Seely v. White Motor Company 403 P.2d 145 (Cal. 1965)

Tyler v. Dowell, Inc., 274 F. 2d 890, 895 (10th Cir. 1960)

Regulatory Guidance and Decisions

Council by laying down detailed rules concerning the specific test procedures and technical requirements for the type-approval of motor vehicles with regard to their event data recorder and for the type-approval of those systems as separate technical units and amending Annex II to that Regulation, Available at: https://members.wto.org/crnattachments/2021/TBT/EEC/21_6099_00_e.pdf

Data protection under GDPR (2021), Available at: https://europa.eu/youreurope/business/dealing-with-customers/data-protection/data-protection-gdpr/index_en.htm#shortcut-2

European Commission (2016), Communication from the Commission to the European Parliament, the Council, the European Economic and Social Committee and the Committee of the Regions, A European Strategy on Cooperative Intelligent Transport Systems, A Milestone towards Cooperative, Connected and Automated Mobility, Available at: https://ec.europa.eu/energy/sites/ener/files/documents/1_en_act_part1_v5.pdf

European Commission (2021), Commission Delegated Regulation (EU) Draft supplementing Regulation (EU) 2019/2144 of the European Parliament and of the European Commission (2021), Proposal for Supplement 1 to the UN Regulation No. 160 (Event Data Recorder) and to the 01 series of amendments to UN

Regulation No. 160, Available at: https://unece.org/sites/default/files/2021-04/GRSG-121-32e.pdf

European Data Protection Supervisor (2018), Opinion 5/2018, 31 May 2018, Available at: https://edps.europa.eu/sites/default/files/publication/18-0531_preliminary_opinion_on_privacy_by_design_en_0.pdf

European Parliament (2016). Regulation (EU) 2016/679 of the European Parliament and the Council of 27 April 2016 on the protection of natural persons with regard to the processing of personal data and on the free movement of such data, and repealing Directive 95/46/EC (General Data Protection Regulation), Available at: https://eur-lex.europa.eu/legal-content/EN/TXT/PDF/?uri=CELEX:32016R0679&from=EN

European Parliament (2017). Civil Law Rules on Robotics, Available at: https://www.europarl.europa.eu/doceo/document/TA-8-2017-0051_EN.html#title1

IEEE: IEEE code of ethics. http://www.ieee.org/about/corporate/governance/p7-8.html (2014). Accessed on 16 July 2022

Guidelines on consent under Regulation 2016/679, WP259 rev.01, 10 April 2018, Available at: https://ec.europa.eu/newsroom/article29/redirection/document/51030

References

ACEA European Automobile Manufacturers Association (2015) ACEA Principles of Data Protection in Relation to Connected Vehicles and Services, available at: https://www.acea.auto/files/ACEA_Principles_of_Data_Protection.pdf

ACEA European Automobile Manufacturers Association (2016) ACEA Position Paper Access to Vehicle Data for Third Party Services, available at: https://www.acea.auto/publication/position-paper-access-to-vehicle-data-for-third-party-services/

AHEAD (Aggregated Homologation-proposal for Event Recorder Data for Automated Driving) (2019). Available at: https://www.thi.de/fileadmin/daten/forschung/CARISSMA/Technologiefelder/Sichere_Elektromobilitaet/Positionspapier_AHEAD.pdf

Anderson JM, Nidhi K, Stanley KD, Sorensen P, Samaras C, Oluwatola OA (2016) Autonomous vehicle technology: A guide for policymakers. Rand Corporation

Asaro PM (2006) What would we expect from a Robot ethic. Int Rev Inf Ethics 6(12/2006):9–15

Asaro PM (2015) Liability Problem for Autonomous Artificial Agents. Available at: https://peterasaro.org/writing/Asaro,%20Ethics%20Auto%20Agents,%20AAAI.pdf

Automated Driving Systems, A Vision for Safety (2017). Available at: https://www.nhtsa.gov/sites/nhtsa.gov/files/documents/13069a-ads2.0_090617_v9a_tag.pdf

Bagloee SA, Tavana M, Asadi M, Oliver T (2016) Autonomous vehicles: challenges, opportunities, and future implications for transportation policies. J Mod Transp 24(4):284–303

Barfield W (2018) Liability for autonomous and artificially intelligent Robots. Paladyn J Behav Robot 9(1):193–203. https://doi.org/10.1515/pjbr-2018-0018

Bigman YE, Gray K (2020) Life and death decisions of autonomous vehicles. Nature 579(7797):81

Böhm K, Kubjatko T, Paula D, Schweiger H (2020) New developments on EDR (Event Data Recorder) for automated vehicles. Open Eng 10(1):140–146. https://doi.org/10.1515/eng-2020-0007

Bonnefon JF, Shariff A, Rahwan I (2016) The social dilemma of autonomous vehicles. Science (New York, N.Y.) 352(6293):1573–1576. https://doi.org/10.1126/science.aaf2654

Bonnefon J-F, Shariff A, Rahwan I (2015) Autonomous vehicles need experimental ethics: are we ready for utilitarian cars?, Available at: https://pdfs.semanticscholar.org/13d4/56d4c53d7b03b90ba59845a8f61b23b9f6e8.pdf

Brady M (2019) Data privacy and automated vehicles: navigating the privacy continuum. Monash Univ Law Rev 45(3):589–562

Bugra A (2020) Room for compulsory product liability insurance in the European Union for smart Robots? Reflections on the compelling challenges. In: InsurTech: a legal and regulatory view. Springer, Cham, pp 167–197

Centre for Connected & Autonomous Vehicles (2017) Pathway to Driverless Cars: Consultation on Proposals to Support Advanced Driver Assistance Systems and Automated Vehicles: Government Response, Available at: https://assets.publishing.service.gov.uk/government/uploads/system/uploads/attachment_data/file/536365/driverless-cars-proposals-for-adas-and_avts.pdf

Channon M (2019) Automated and Electric Vehicles Act 2018: an evaluation in light of proactive law and regulatory disconnect. Eur J Law Technol 10(2)

Channon M, McCormick L, Noussia K (2019) The law and autonomous vehicles. Taylor & Francis

Chinen M (2019) Law and autonomous machines. Edward Elgar Publishing

Coca-Vila I (2017) Self-driving cars in dilemmatic situations: an approach based on the theory of justification in criminal law. Crim Law Philos 12(1):56–82. https://doi.org/10.1007/s11572-017-9411-3

Contissa G, Lagioia F, Sartor G (2017) The Ethical Knob: ethically-customisable automated vehicles and the law. Artif Intell Law 25(3):365–378

Crayton TJ, Meier BM (2017) Autonomous vehicles: developing a public health research agenda to frame the future of transportation policy. J Transp Health 6:245–252

Divino SBS (2020) Critical considerations on artificial intelligence liability: E-personality propositions. Rev Electronica Direito Sociedade 8:193

Dorian D (2021) 2022 Tesla Model S, Available at: https://www.caranddriver.com/tesla/model-s

Forgó N, Hänold S, Schütze B (2017) The principle of purpose limitation and big data. In: New technology, big data and the law. Springer, Singapore, pp 17–42

Gasser TM, Seeck A, Smith BW (2016) Framework conditions for the development of driver assistance systems. In: Winner H, Hakuli S, Lotz F, Singer C (eds) Handbook of driver assistance systems. Springer, Cham. https://doi.org/10.1007/978-3-319-12352-3_3

Geistfeld MA (2015) The contractually based economic loss rule in tort law: endangered consumers and the error of east river steamship. DePaul Law Rev 65:393

Goldstein D (2016) Autonomous vehicles will drive themselves-but they won't regulate themselves. Hast Bus Law J 13:241

Goodall NJ (2014) Machine ethics and automated vehicles. In: Meyer G, Beiker S (eds) Road vehicle automation. Springer, Cham

Gurney JK (2013) Sue my car not me: products liability and accidents involving autonomous vehicles. Univ Ill J Law Technol Policy 247

Gursten SM (2013) Driverless cars: who's liable in an accident?. Available at: https://www.michiganautolaw.com/blog/2013/05/07/driverless-cars-whos-liable/

Haga Y (2017) Right to be forgotten: a new privacy right in the era of internet. In: New technology, big data and the law. Springer, Singapore, pp 97–126

Hagemann R, Huddleston Skees J, Thierer A (2018) Soft law for hard problems: the governance of emerging technologies in an uncertain future. Colorado Technol Law J 17:37

Hancock PA, Nourbakhsh I, Stewart J (2019) On the future of transportation in an era of automated and autonomous vehicles. Proc Natl Acad Sci 116(16):7684–7691

Harper CD, Hendrickson CT, Mangones S, Samaras C (2016) Estimating potential increases in travel with autonomous vehicles for the non-driving, elderly and people with travel-restrictive medical conditions. Transport Res Part C Emerg Technol 72:1–9

Huber PW, Litan RE (eds) (1991) The liability maze: the impact of liability law on safety and innovation. Brookings Institution Press, Washington

Hutter MB (2017) A risk regulation perspective on regulatory excellence. In: Coglianese C (ed) Achieving regulatory excellence. Brookings Institution Press, Washington, pp 101–114
Jasmontaite L, Kamara I, Zanfir-Fortuna G, Leucci S (2018) Data protection by design and by default: framing guiding principles into legal obligations in the GDPR. Eur Data Protect Law Rev 4:168
Jeffcott O, Rose I (2017) Driverless cars: ethical and legal dilemmas. J Pers Inj Law 1:19–25
Keeton RE (1961) Assumption of risk in products liability cases. Louisiana Law Rev 22:122
Kim S (2017) Crashed software: assessing product liability for software defects in automated vehicles. Duke Law Technol Rev 16:300
Kochupillai M, Lütge C, Poszler F (2020) Programming away human rights and responsibilities? "The Moral Machine Experiment" and the need for a more "Humane" AV future. NanoEthics 14(3):285–299
Landini S, Noussia K (2022) Big data, privacy, and protection of the user of autonomous vehicles: ethical issues, insurance aspects, and human rights. In: Lima Rego M, Kuschke B (eds) Insurance and human rights, Springer, pp. 131–172
Lieber J, Puente L (2016) Beyond the gig economy: how new technologies are reshaping the future of work. Thumbtack, San Francisco, CA. Available at: http://alanbweaver.com/yonkers/beyond_the_gig_economy.pdf
Lohmann MF (2016) Liability issues concerning self-driving vehicles. Eur J Risk Regul 7(2): 335–340
Lohmann MF, Rusch A (2015) Fahrassistenzsysteme und selbstfahrende Fahrzeuge im Lichte von Haftpflicht und Versicherung. HAVE-Haftung und Versicherung, 4
Madden MS (ed) (2005) Exploring tort law. Cambridge University Press
Marchant G, Bazzi R (2020) Autonomous vehicles and liability: what will juries do. Boston Univ J Sci Technol Law 26(1):67
Marchant GE, Lindor RA (2012) The coming collision between autonomous vehicles and the liability system. Santa Clara Law Rev 52:1321
Marson J, Ferris K, Dickinson J (2020) The automated and electric vehicles act 2018 part 1 and beyond: a critical review. Statute Law Rev 41(3):395–416
Matthaei R, Reschka A, Rieken J, Dierkes F, Ulbrich S, Winkle T, Maurer M (2015) Autonomous driving. In: Winner H, Hakuli S, Lotz F, Singer C (eds) Handbook of driver assistance systems. Springer Reference
Montagnani ML, Cavallo M (2021) Liability and emerging digital technologies: an EU perspective. Notre Dame J Int Comp Law 11(2):208
Morrow WR et al (2014) Key factors influencing autonomous vehicles' energy and environmental outcome. In: Meyer G, Beiker S (eds) Road vehicle automation. Lecture notes in mobility. Springer, Cham. https://doi.org/10.1007/978-3-319-05990-7_12
Mueller AS, Cicchino JB, Zuby DS (2020) What humanlike errors do autonomous vehicles need to avoid to maximize safety? J Saf Res 75:310–318. https://doi.org/10.1016/j.jsr.2020.10.005
Noussia K (2020) Autonomous vehicles: legal considerations and dilemmas. In: InsurTech: a legal and regulatory view. Springer, Cham, pp 253–270
OICA (2019) Event Data Recorder (EDR). Available at: https://unece.org/DAM/trans/doc/2019/wp29grsg/GRSG-116-42e.pdf
Oliphant K (2019) Liability for road accidents caused by driverless cars. Sing Comp Law Rev 190
On-Road Automated Driving Committee (2018) Taxonomy and definitions for terms related to driving automation systems for on-road motor vehicles. SAE International, Standard J3016, Available at: https://www.sae.org/standards/content/j3016_201806/
Osmani N (2020) The complexity of criminal liability of AI systems. Masaryk Univ J Law Technol 14(1):53–82
Owen T (2015) Disruptive power: the crisis of the state in the digital age. Oxford Studies in Digital Poli
Pagallo U (2013) The laws of Robots crimes, contracts, and torts. Springer, Cham

Pakusch C, Boden A, Stein M, Stevens G (2021) The automation of the taxi industry–taxi drivers' expectations and attitudes towards the future of their work. Computer Supported Cooperative Work (CSCW), pp. 1–49
Pammer K, Gauld C, McKerral A, Reeves C (2021) "They have to be better than human drivers!" Motorcyclists' and cyclists' perceptions of autonomous vehicles. Transport Res Part F Traffic Psychol Behav 78:246–258
Parchomovsky G, Stein A (2008) Torts and innovation. Mich Law Rev 107:285–315
Patsakis C, Solanas A (2013) Privacy-aware event data recorders: cryptography meets the automotive industry again. IEEE Commun Magaz 51:123
Rachum-Twaig O (2020) Whose Robot is it anyway?: Liability for artificial-intelligence-based Robots. Univ Ill Law Rev 1141
Rosenberg A (2017) Strict liability: imagining a legal framework for autonomous vehicles. Tulane J Technol Intell Prop 20
Saripan H, Putera NSMFS (2016) Are Robots human? A review of the legal personality model. World Appl Sci J 34(6):824–831
Scharring K, Nash S, Wong D (2017) Connected and Autonomous Vehicles: Position Paper
Schellekens M (2015) Self-driving cars and the chilling effect of liability law. Comput Law Secur Rev 31(4)
SMMT (2017) Motor Industry Facts, https://www.smmt.co.uk/wp-content/uploads/sites/2/SMMT-Motor-Industry-Facts-2017_online_May.pdf
Shepherd C (2019) Chinese AI start-up DeepBlue takes self-driving buses global. Available at: https://www.ft.com/content/72bf3046-bca0-11e9-b350-db00d509634e
Solaiman SM (2017) Legal personality of robots, corporations, idols and chimpanzees: a quest for legitimacy. Artif Intell Law 25(2):155–179
Storing M (2017) What EU legislation says about car data. Legal Memorandum on connected vehicles and data. Federation International Automobile Reg 1. Available at: https://www.fiaregion1.com/wp-content/uploads/2017/06/20170516-Legal-Memorandum-on-Personal-Data-in-Connected-Vehicles-www.pdf
Tian X, Prybutok VR, Mirzaei FH, Dinulescu CC (2020) Millennials acceptance of insurance telematics: an integrative empirical study. Am Bus Rev 23:156
U.S. Chamber Institute for Legal Reform (2018) Torts of the Future: Autonomous Vehicles (Special Edition), A Look at Emerging Trends in Liability and Regulation. Available at: https://instituteforlegalreform.com/wpcontent/uploads/2020/10/Torts_of_the_Future_Repackage_Update051418_Web.pdf
Uzair M, Hosain S (2020) An overview of social, economic, environmental, and safety impacts of intelligent electric vehicles. Islamic Univ J Appl Sci II:114
Van den Boom F (2021) Regulating telematics insurance. In: Marano P, Noussia K (eds) Insurance distribution directive. AIDA Europe research series on insurance law and regulation, vol 3. Springer, Cham, p 301. https://doi.org/10.1007/978-3-030-52738-9_12
Verband der Automobilindustrue (2014) Data Protection Principles for Connected Vehicles. Available at: https://www.pdpjournals.com/docs/99009.pdf
Viscusi WK, Moore MJ (1991) Rationalizing the relationship between product liability and innovation. In: Schuck PH (ed) Tort law and the public interest. Competition, innovation and consumer welfare. Norton, New York, pp 105–150
Vrabec HU, Domnik N (2021) Implementation of the GDPR: Slovenia. Neja, Implementation of the GDPR: Slovenia (Working paper) (September 11, 2021)
Wiewiórowski W (2018) Data Protection and Privacy of the car... as a mobile device, Type-approval requirements for motor vehicles as regards their general safety and the protection of vehicle occupants and vulnerable road users. Available at: https://www.europarl.europa.eu/cmsdata/158076/2018_11_29_wiewiorowski_event_recorders.pdf
Yadron D, Tynan D (2016) Tesla driver dies in first fatal crash while using autopilot mode. Available at: https://www.theguardian.com/technology/2016/jun/30/tesla-autopilot-death-self-driving-car-elon-musk

Cyber Risks: Social, Functional, and Ethical Dimensions

Andrea Signorino Barbat

Abstract This chapter intends to approach the topic of technologies applied to insurance and their risks, in a general way to the reader and reflect on cyber risks in a broad sense. In this line, I will analyze this type of risks methodologically classified in three dimensions that I have called social, functional, and ethical dimension. The social dimension refers to the risks that technologies and the cyber world imply for the consumer—insured in the context of the social network; the functional dimension implies the risks that arise for insurance companies in their internal management; and the ethical dimension, perhaps the most transcendent, refers to the dangers that the cyber world can generate for people in their individuality, as human beings, if the risks involved are not adequately managed, regulated, or contextualized. By introducing risk from the insurance point of view, the conditions to make a risk insurable and a risk's delimitation, to enable the analysis of the dimensions of risks generated by technologies which are insurable, this chapter invites the reader to consider risks implied by the new technologies and the solutions the insurance sector provides.

Keywords Cyber · Risks · Insurance · Dimensions · Social · Functional · Ethics

1 General Approach

To understand these lines, it is essential to begin with an etymological aspect—what does cybernetic mean?

As cybernetics we designate everything related to interdisciplinary computational technology used for the extension of human abilities.

The word cybernetic derives from the Greek "kybernetes," which means "the art of handling a ship." It was later used by Plato in his work The Republic to refer to the "art of directing men" or the "art of governing."

A. Signorino Barbat (✉)
University of Montevideo, Montevideo, Uruguay

K. Noussia, M. Channon (eds.), *The Regulation of Automated and Autonomous Transport*, https://doi.org/10.1007/978-3-031-32356-0_11

The modern concept of cybernetics, computer technology based on human communication, was coined by Norbert Wiener (1894–1964) in his work Cybernetics: or Control and Communication in the Animal and the Machine.[1]

Stafford Beer, a philosopher of organizational and management theory, whom Wiener himself said should be considered the father of management cybernetics, defines cybernetics as "the science of effective organization."[2] According to Professor Beer, cybernetics studies the flows of information surrounding a system, and the way in which this information is used by the system as a value that allows it to control itself: it occurs for both animate and inanimate systems indifferently. Cybernetics is an interdisciplinary science, and is as closely linked to physics as to the study of the brain as to the study of computers, and it also has much to do with the formal languages of science, providing tools with which to describe objectively the behavior of all these systems.

Today, cybernetics is characterized as everything related to computer technology, especially, but not only, to the Internet.

Much is being said at this time around the world about Cyber Risks or CyberRisks, referring to the risks that lurk in cyberspace.

Particularly, the subject has been analyzed with respect to the banking system which suffers many losses for these reasons, but also, the insurance world is strongly talking about these dangers that lurk, inside and outside the insurance company.

In fact, "financial services firms fall victim to cybersecurity attacks far more frequently than businesses in other industries. Security breaches lead to lost revenue for banking institutions, interruptions in operations and loss of both reputation and customers.[3]

Here are just a handful of statistics on the devastating effects of breaches in the financial industry from Fortunly:[4]

---

[1] Wiener (1961).

[2] Beer (1959).

[3] https://www.archonsecure.com/blog/banking-industry-cyber-threats (retrieved on 1 August 2022).

[4] https://fortunly.com/statistics/data-breach-statistics/#gref (retrieved on 1 August 2022). More interesting statistics from Fortunly:

"-e Commerce data breaches account for 80% of payment-card-related investigations. When we look at the number of data breaches over the last decade, it becomes evident just how heavily we now rely on digital finance. Consumers are increasingly shopping online, so much so that the same research from just four years earlier had the exact opposite result. Six years ago, 80% of payment-card related investigations on data breaches were for point-of-sale merchants, while in 2020 that figure is only 20% (Security Metrics);

-71% of all data breaches are financially motivated (Verizon);

-The cost of cyberattacks is highest in the banking industry, reaching $18.3 million annually per company (Accenture);

-Stolen cardholder data remains captured for an average of 127 days (SecurityMetrics);

-Over 500 million users don't realize their device is infected with crypto-mining software (AdGuard);

-A staggering 97% of all records stolen are from the United States (Gemalto);

-The average total cost of a data breach amounts to $3.92 million (IBM);

- The cost of cyberattacks in the banking industry reached $18.3 million annually per company.
- 8 out of 10 US citizens fear that businesses are not able to secure their financial information.
- According to FBI, the amount paid to ransomware scammers has reached nearly $1 billion per year. The report from the Department of Justice also noted that on a global scale, ransomware infects some 100,000 computers each day. As these ransom attacks continue to shift from private citizens to businesses and large corporations, the amount of money these scams make is also increasing.
- 92% of ATMs are vulnerable to hacks.

Financial establishments experience threats from a variety of sources led primarily by mobile applications and web portals. Cyber criminals may steal or manipulate valuable user data and or 'clone' banking apps in order to use them for nefarious purposes."[5]

The subject of cyber risks goes far beyond the action of a hacker and is related to illegal computer activities to subtract, alter, modify, manipulate, disable, or destroy information or assets, such as money, bonds or intangible goods, information of the companies or users affected, using for such purposes electronic means or electronic devices.

To understand its scope, it is necessary to analyze the internal risk, i.e., the risk generated or suffered by the individual and the company itself, and the external risk, i.e., the risk with respect to third parties, the liability that is generated in relation to third party users or third parties linked to the systems.

-Global spending on cybersecurity is expected to surpass $6 trillion by 2021 (Cybersecurity Ventures);

-On average, 4,818 websites per month were compromised with formjacking code in 2018 (Symantec);

-Symantec projected that stealing a mere 10 credit cards from these websites could translate to a $2.2 million gain for the criminals. When you look at it like that, it's no surprise formjacking is becoming more and more popular among cybercriminals. The United States suffered 1,473 cyberattacks in 2019, leading to 164.6 million stolen records (Statista);

-Experts predict spending on cybersecurity training for staff will reach $10 billion by 2027 (Cybersecurity Ventures);

-Emotet malware now accounts for 16% of all financial Trojans (Symantec);

-The number of cryptojacking URLs doubled during the last quarter of 2018. However, in the second half of 2019, the number of cryptojacking attacks fell by 78% (Help Net Security; SonicWall);

-Research on major data breaches performed by Carbon Black points to some alarming trends, with 26% of surveyed financial institutions admitting they had fallen victim to a destructive attack;

-Due to a recent coronavirus outbreak, the European Central Bank has urged financial institutions to take measures in protecting themselves from a possible increase in cyberattacks. Since many employees will be absent from a workplace, it may leave banks more vulnerable, and cybercriminals will certainly try to take advantage of the pandemic;

-Cyber-attack stats show that this trend will not slow down in the near future. At current rates, by the year 2021, one business will fall victim to a ransomware attack every 11 seconds."

[5] https://www.archonsecure.com/blog/banking-industry-cyber-threats (retrieved on 1 August 2022).

In this context, we talk about computer fraud, which is one of the most modern challenges for the protection of individuals and companies from both organized and occasional criminals. Fraud ranges from simple information theft to identity theft, network account theft, extortion, and cyber terrorism, as well as corporate espionage and data liability.

These are organizations but also lone individuals, increasingly sophisticated and endangering both individuals and companies, because today we are all connected. Cyberattacks usually target the flow of data, either by preventing communication between the sender and the receiver, or by intercepting, modifying or inventing data that alters the normal flow of information.

Of course, the insurance business is no stranger to these risks. We are in the era of new technologies, the so-called Insurtech world, i.e., the application of new technologies to the insurance activity opens up a whole world, unthinkable a few years ago, of new ways of doing business. This ranges from tools for the sale of insurance to the technification of the entire underwriting, issuance and even the payment of claims using blockchain technology, smart contracts, algorithms, and others based on artificial intelligence.

But of course, it also opens the door to cyber risks.

Particularly, protecting identity and data and providing security to policyholders is crucial to the insurance industry's loyalty strategies. The principles enshrined in data protection laws—legality, consent, purpose, proportionality, clarity, security, protection, recourse… —must be complied with even when the data is used by a virtual medium.

Generally, such principles are enshrined in data protection laws throughout Latin America. For example, Uruguayan Law No. 18.331 of 28 August 2008 establishes them as follows.[6]

Likewise, the insured must be protected as a consumer in an adhesion contract, but at the same time in a context where the exchange of data is a matter of course in today's society. All this constitutes what I call the social dimension of cyber risks.

However, cyber risks also threaten insurers' own operations, in their internal management, which is what I call the functional dimension of these risks.

In this sense, it is necessary to have an adequate management of technological risk in a broad sense, which ranges from managing human talent to the involvement of senior management.

Finally, the efficient fight against cyber risks depends on nothing less than the security of the company itself and of the policyholders who are the reason for its business.

Lastly, the ethical aspect implied by the new technologies and which generates what I have come to call the ethical dimension of cyber risks.

[6] Revista electrónica de derecho comercial. www.derecho-comercial.com.uy - Ley 18.331 habeas data Uruguay y decreto 414 reglamentario Retreived 1 August 2022.

It is a non-traditional approach to what can be considered a risk of the cyber environment, which has to do with the possible discrimination and moral affectation that technologies based on artificial intelligence can generate to human beings.

However, not all cyber risks may be insured and, in this sense, based on the concept of risk and when does a risk become insurable, we will conclude which risks may be covered or managed by insurance.

2 Risk from an Insurance Point of View

2.1 Definitions of Risk

Doctrine provides us with many definitions of "risk." Scholars—as it usually happens in insurance law, due to the insurance contract's origin related to economic activities, in addition to the contracts' own lawfulness- have given both economic and legal definitions of risk.

-From an economic point of view, we may find, for instance, the following definitions:

According to Manes, insurance is an economic fact and risk is the uncertainty of sustaining an economic loss. Gasperoni claims that risk is the possibility an economically disadvantageous fact occurs;[7]

Garrigues defines risk as the possibility of a random event occurring, thus creating an economic need;[8]

Viterbo, with whom Bruck agrees, states that "risk exists as regards an individual whenever the economic situation is not foreseeable by such individual, or at least the economic situation is not foreseeable with a minimum degree of certainty."[9]

-From a legal point of view, risk as regards insurance has been defined or conceptualized as follows, among other definitions:

Lepargneur defines risk as the possibility or probability of occurrence of a loss. Malagarriga says that risk is any uncertain event which may occur at an indefinite moment and is likely to cause damage.[10]

Fanelli makes a distinction between "non-insured" risk, defined as the possibility—or probability—of occurrence of a certain event—usually harmful, as per the ordinary meaning of the word—and the insurable or insured risk, understood as the possibility of complying with a contractual obligation when a certain event occurs.[11]

---

[7] Manes A. and Gasperoni, N quoted by Halperin and Barbato (1983), pp. 503 and 504.

[8] Garrigues (1982), p. 14.

[9] Viterbo (1941) and Bruck (1930), pp. 53 et seq.

[10] Lepargneur and Malagarriga C quoted by Halperin, Isaac, in op.cit., p. 503.

[11] Fanelli (1973), p. 101.

Sánchez Calero claims that: "Risk is the possibility of a harmful event -or as per an economic terminology, the possibility of an event giving rise to a financial need. Possibility is found in between impossibility, that is, when an event cannot occur, and certainty, that is, when an event will definitely occur at a certain moment. Between these two opposite ends, possibility has several degrees, and each such degree is called probability."[12] A risk is the contingency or proximity of damage; it is an event or change of the outside world, uncertain as to its generation or as to the time of its occurrence. This is what happens with life insurance covering the risk of death: death will definitely occur, but its time of occurrence is unknown.

Insurable risk is also defined as a risk threatening an interest which, in turn, is insurable.

2.2 *Conditions That Make a Risk Insurable*

Not every risk may be insured. A risk must have certain characteristics to be an insurable risk.

On the one hand, a risk must be a potential event. This means that the purpose of the insurance is to compensate a potential damage, cover a need. The insured must have an economic interest in the non-occurrence of a fortuitous event which may occur; it cannot be impossible for such event to occur.[13]

In relation to the above, a risk must be contingent; contingency is exactly what excludes certainty and impossibility. Therefore, a risk must be uncertain. Uncertainty must not necessarily be absolute, but economic; uncertainty as to the time of occurrence is enough. We talk about total uncertainty when the damage may occur or not, as it happens with property damage insurance and relative uncertainty, when uncertainty relates to circumstances of time and manner, as it happens with life or death insurance, where the insured will definitely die, but it is uncertain when will this happen.[14]

The doctrine argues whether the uncertainty must be objective or subjective, that is, if it may be an objectively existing uncertainty, although the parties are not aware if the risk exists or not when executing the contract or if the risk has disappeared during the term of the contract, or if the risk must only be uncertain for each individual in particular. In the latter case, we are talking about a putative risk.[15]

---

[12] Sánchez Calero (1961), p. 454.

[13] Greene (1976), p. 2 et seq.

[14] Greene (1976), p. 5 et seq.

[15] I agree with the stance of objectivity of uncertainty as regards a risk, and with the stance of objectivity of the aggravation of the risk. Such stance seems to be the one adopted by the Uruguayan lawmaker under Section 14 of the Insurance Act, Law No. 19,678. On the other hand, Héctor SOTO, LL.D., supports the stance of subjectivity, claiming "The uncertainty attached to a risk is, more than an objective circumstance, a subjective circumstance referred to the parties' awareness. A risk may be subjectively uncertain for the parties, even though the causes that shall inevitably

In relation to the above, some scholars argue against the randomness of risk—and of the insurance contract in general—considering that the randomness of insurance has vanished, and the contract has become commutative. This, among others arguments basically based on the existence of equivalent consideration—obligations of coverage or security (and only potential compensation) and obligations of payment of premium or gross premium-, is based on the continuous evolution of technical actuarial studies and statistics, causing the disappearance of the uncertainty of risk. Insurers know how often losses occur, and hence total randomness as to the occurrence of the risk can no longer be affirmed.[16]

Indeed, according to Ferri, an insurer's current purpose is no longer that of taking risks, but to compensate risks or distribute them among the participants in the economic life. "Through different insurances, risks are offset, an insurer's contingency is removed and the insurer's profits do not rely so much on whether the event occurs or not, but on the degree of technical perfection of its organization and on the range of action of the insurer's activities."[17]

Halperin reacts to this stance and claims that "we must not resort to a 'legal pythagoreanism' which leads us to think that an insurer's calculations have changed the insurance contract to such an extent of turning what is in essence uncertain into completely certain… Even in these technical cases of spread of risk bearing and calculation of the premium the insurer shall get paid to duly take care of potential losses there is a high degree of uncertainty, and contingency remains present."[18]

A risk must also be fortuitous, that is, intentional misconduct is excluded. It cannot be a voluntary act carried out by the insured; it must be completely independent of the insured's will.

An essential condition of a risk is that it must affect lawful interests or operations, current or future ones. Such interest in certain goods, property or individuals must fall within the scope of the law.[19]

---

determine its occurrence or non-occurrence have already occurred." SOTO, Héctor in an AIDA-Uruguay conference in Montevideo, Uruguay, year 2015. In my stance, however, I accept that there is an exception to a risk's objectivity in liability insurance, where the parties may retroactively agree on *claims made* clauses. Obviously, under the retroactive clause, a risk has already occurred as at the time of inception of the insurance—although it is not known; however, such risk will be covered where the parties have expressly agreed on said clause.

[16]This is the stance of Alejandro Vigil, LL.D. (Cuba) and Ernesto Tilursnik, LL.D. (Brazil), both exposed in insurance conferences I attended in their respective countries. They sustain mainly that there is an equivalent consideration between the parties, since the insurer undertakes an obligation to provide a security in exchange for the payment of a premium, that is, the insurer undertakes to cover the risks, not to indemnify, which is contingent in any case. In Brazil, there is a legal argument; since the Civil Code of the year 2002 mentions an insurer's obligation to provide a security.

[17]Ferri (1972), p. 911.

[18]Halperin and Barbato (1983) op cit 10 (a.1) p. 19.

[19]This is clearly provided for under Section 5 of the Uruguayan Insurance Act, Law No. 19,678, which refers to the lawfulness of operations, property, and risks: "Any insurance, the purpose of

In addition, a risk must exist as at the time of inception of the policy, and it cannot have disappeared at the time of the loss. Otherwise, the interest ceases and premiums become due, that is, the cost or price corresponding to the insurance to be paid by the insured, for the time elapsed. On the other hand, risk shall also cease upon occurrence of the loss.

Finally, a risk must be measurable, appraisable, must have statistical basis and be founded on the law of large numbers, which is essential in insurance techniques.

Its measurable nature is material; thus, the insurer must delimit the risk to obtain certainties as to the dimension of the risk the insurer is taking.

2.2.1 Risk Delimitation in the Insurance Sector

The need for measuring risk requires that to legally determine it, the insurer must follow a risk individualization and delimitation process.

The individualization of risk refers to the description of the risk to establish that it entails a possible event, which causes damage, or an economic loss or of any other kind—in life insurance. The risk must meet certain conditions to be insurable, for example: the risk must be capable of occurring, possible, and uncertain.

This detailed description is made through two channels: legal and conventional.

On the one hand, the law individualizes the risk in a generic way for the purposes of guaranteeing those minimum requirements that it must meet to grant a minimum coverage to the insured, thus preventing the insurer from including in the policy exclusions that alter the risk and hamper the coverage.

If pursuant to what different laws provide in some countries, we consider also insurance laws public policy nature, the insured shall receive a guaranteed minimum coverage.

The conventional description is included in the policy clauses, general and specific conditions and, if that be the case, special conditions. These describe risk in an affirmative or inclusive manner when describing covered and protected risks, and in a negative or excluding manner when specifying non-covered risks and other coverage exclusions in a broad sense.[20]

Risk delimitation enables, once the risk is individualized, to limit the risk as regards its extension or coverage scope.

Risk delimitation is not only provided for under the law, it also arises from the parties' discretion, who establish such delimitation as regards its cause, space, and time. Of course, said delimitation must fall within the scope of the law.[21]

---

which are illicit operations, as well as any insurance covering property unlawfully possessed by the insured or covering the risk of an unlawful business or company, shall be null and void."

[20]The Uruguayan Insurance Act, Law No. 19,678, refers to covered and excluded risks under Sections 14, 15, and 16.

[21]Sánchez Calero, F. Comments on Section 100 of Law No. 50/1980- Sánchez Calero et al. (1999), pp. 1846 and 1847.

It is relevant to classify those clauses delimiting risk, which must not be considered "restrictive" clauses as regards an insured's rights, but "delimiting" clauses as regards an insured risk. This entails important consequences for different issues, such as the need for a special warning to the insured as regards its scope. For example, in Spanish Law on the application of Section 3 of Law No. 50/1980[22] and its scope.

In this sense, we can distinguish between a qualitative delimitation and a quantitative delimitation of risk.[23] We may also include a subjective delimitation, based on the insured's conducts—burden.

Qualitative delimitation includes the delimitation of the cause of the risk, the space, territory, or place where a risk occurs, as well as the delimitation of time or validity period of the coverage.

The delimitation of the cause of the risk is the required relationship between the covered event and the cause of damage; the harm or damage or loss must be caused by the occurrence of the covered risk. This delimitation has a legal dimension.

It may also have a conventional dimension as regards the definition of cause under the policy, which cannot be inconsistent with the provisions of the law.

The delimitation of space or territory is the limitation of the risk to a specific place or territory.

For instance, in the case of fire insurance, the fire must occur in the real estate described or individualized under the policy and in the stated location. In the case of home insurance, where insurance covers theft of its contents, the stolen property must have been included in the insured property.[24]

The qualitative delimitation of time refers to specifying a validity period for coverage of the risk, the term of the contract within which the risk must occur and is, in general, included in the provision of insurance specific conditions.

Such delimitation of time admits certain specialization as regards liability insurance, where it is possible to agree on *claims made* clauses, both retroactively and subsequently to the validity period of the insurance. Obviously, under the retroactive clause, the risk has already occurred as at the time of inception of the insurance—although it is not known; however, such risk will be covered where the parties have agreed on said clause. It is a different situation when the *claims made* clause refers to extension, since actually, the damage occurs during the validity period and the claim during the extended period. Anyway, without said claim, the event causing the loss would not occur; hence, we can claim the loss would not occur.[25]

[22] Section 3 of the Spanish Insurance Act, Law No. 50/1980, requires that clauses restricting an insured's rights must be especially highlighted under the contractual conditions and must be specifically accepted in writing by the insured.

[23] Ríos (2015), pp. 156 et seq.

[24] Rios R op cit, p. 157.

[25] On loss requirements and its complexity, please see my book Signorino (2011) "Seguros de responsabilidad civil." Ed. FCU, chapter II.

The quantitative delimitation of risk refers to estimating the *quantum* of a risk, that is, the extent of an insurer's obligation as insured capital value or income or any other insurer's obligation.

3 Dimensions of Cyber Risks

In relation to the characteristics a risk must have to be insured, and the need for delimitation thereof, we should distinguish which of the several dimensions of cyber risks may be covered by insurance and which cannot be covered.

That is, analyzing which cyber risks elude the technical possibilities of the insurance, especially because they cannot be measured or delimited and, hence, they cannot be insurable.

3.1 Cyber Risks and Their Social Dimension

When we mention cyber risks, we are automatically presented with several legal aspects, in my broader vision of a social nature, among which stand out the protection of users' data, their identity, and their security.

If we apply this to the insurance sector, it is clear that the multiple uses latest technologies offer to provide improvements in the procedures of insurers, may involve risks to third parties and to the insurance companies themselves.

As for the internal damage to the company itself, we will refer to this when talking about the functional dimension of cyber risks.

As for the risk to third parties, in fact if we refer to the insured is not a third party with respect to the insurance company, it is its client with whom it has a contract concluded with nothing less than trusting that the insurer will cover, paradoxically, its risks.

For this reason, it is better in this sense to speak of risks external to the insurance company, which of course also bring into play the contractual civil liability towards its clients.

In this sense, many of the risks arise from the users themselves.

On the one hand, there is a great lack of knowledge about whether the Internet is safe or not. This confusion among consumers makes the fight against cybercrime more difficult.

On the other hand, user support in "technological" education and its risks is very dissimilar in different countries, which results in a greater or lesser aversion to cyber risks and therefore a greater or lesser prevention against them, something important because sometimes the only thing that can save us from a cyberattack is prevention.

A recent report by cxLoyalty (ex AffinionGroup),[26] a benchmark in this type of research, shows that the highest levels of concern are in Brazil, with 87%, and the USA, with 75%. In Europe, France, Spain, Italy, and the United Kingdom show levels of concern ranging from 60% to 70%. In contrast, the Nordic countries have relatively lower levels of concern with only 40% of respondents in Sweden and 42% in Finland saying they are worried about cybercrime.

The difference in concern levels could be attributed to the fact that Brazil is consistently among the countries with the highest levels of cybercrime, particularly in relation to *botnets*,[27] banking fraud and financial malware.

Botnet is the generic name for any group of infected PCs remotely controlled by an attacker. Generally, a hacker or a group of hackers creates a botnet using malware that infects a large number of machines. The computers are part of the botnet, called "bots" or "zombies." There is no minimum number of computers to create a botnet. Small botnets can include hundreds of infected PCs, while larger botnets use millions of computers. Examples of recent botnets include Conficker, Zeus, Waledac, Mariposa, and Kelihos.

The botnet is often understood as a single entity, but the creators of this malware sell it to anyone who will pay for it. For this reason, there are dozens of separate botnets using the same malware and operating at the same time.

Users first became aware of this malware in 2000 when a teenager from Canada launched a series of denial-of-service attacks against popular websites. The young man, whose nickname was Mafiaboy, attacked Yahoo, ETrade, Dell, eBay, Amazon, among others, for several days, overloading websites until servers crashed. Mafia Boy, or his real name Michael Cale, did not use a botnet for his misdeed, but security experts warned after this episode that botnets (large networks of computers infected with a variety of malware) and DDoS attacks were a major threat to the stability and integrity of the Internet. Of course, the professionals were not wrong at all.

In addition, Brazil also has a significant number of fake accounts on social networks, a problem that has recently manifested itself in an alarming way with the discovery, within a span of days, of 12.6 million of fake interactions on Facebook in the run-up to the last national elections.

Concern about cybercrime has also increased over time: respondents in the report, across all countries surveyed, say they are more concerned about all types of cybercrime, with identity theft being the most worrying. Moreover, concern about this type of cybercrime has grown the most in the last twelve months.

To put it in perspective, when comparing cybercrime to more "traditional" crimes such as theft or robbery, the figures generally suggest that consumers are now more concerned about cybercrime.

The user-consumer's personal experience of cybercrime undoubtedly plays a key role, increasing concern and keeping awareness levels high across all types.

---

[26] AFFINIONGROUP- CYBERCRIME REPORT - https://www.ciberseguridadpyme.es (retrieved on 1 August 2022).

[27] https://www.kaspersky.es/blog/que-es-un-botnet/755 (retrieved on 1 August 2022).

Many people have been victims or know someone who has been. Awareness of this type of crime is very high, with identity theft, along with hacking, being the most intensely perceived cyber threat.

However, the concern exists but few know how to deal with it and this generates significant social uncertainty.

The research revealed that, despite high levels of concern, there is a lack of understanding about how to stay safe from threats:[28]

- one-third of the world's population—35%—mistakenly believe that a public Wi-Fi network has to be, by law, secure;
- more than half—54%—are unsure or do not know that https:// means a website is secure;
- another third—33%—are unaware that using the same password on different accounts increases the risks of cyberattack.

Despite their fears of identity theft and other forms of cybercrime, many people have not taken more than basic measures to protect themselves online.

Only 16% have a system in place to protect them from identity theft. Of these, about 39% have it because it was included in another product or service. Interestingly, 38% consider its inclusion the main reason for choosing their current provider.

The research also revealed that respondents with a high interest in technology do not necessarily have a greater understanding or confidence regarding the risks. It appears that confusion about security transcends demographic groups and affects even those who demonstrate a high interest in technology.

Additionally, users often do not know what to do when a cyberattack happens to them, especially identity theft which, as we saw, is what they are most concerned about.

The most usual cyber risks are the following:[29]

- Ransomware

It seeks to infiltrate systems to encode or damage them. Systems may be infected when demanding ransom with bitcoins. It is usually concealed in applications or software used daily; for example, email attachments, updates, links in advertisements, among others. Currently, Ransomware of Things (RoT) is common, affecting devices connected to the Internet.

- Information leakage

These cyber risks may imply civil, administrative, and criminal sanctions. They cause companies to miss business opportunities and can also seriously harm a company's reputation. They imply stealing some device or accessing systems such

[28] Affiniongroup- Cybercrime Report - https://www.ciberseguridadpyme.es (retrieved on 1 August 2022).

[29] Instituto Nacional De Compañias De Seguros De España (INESE) (2019).

as databases. Malware, rootkit, social engineering, or backdoor are used for these purposes.

– Phishing

The identity theft of a person or a webpage seeks to steal confidential information, such as access to bank accounts. The most common example is an email informing us that we must enter our password in a link of questionable credibility due to certain issue. At a personal level, it is used when knowledge of certain information of a specific individual to access his or her accounts is demonstrated.

– Advanced Persistent Threat (APT)

These cyber risks attack a specific company to infiltrate its technological infrastructure, thus obtaining sensitive information or damaging processes. This is a customized attack conducted by groups allegedly related to governments.

– DDoS attack

A company's information systems are artificially collapsed to prevent other users from using the same. By saturating these services, the attacker subsequently seeks to obtain an economic remuneration to return to normal operation.

These are some of the most usual cyberattacks which may affect consumers. Broadly speaking, they are based on website malfunction, information extraction and request of actions which shall cause losses. Insurance coverage may be a useful tool to be protected against these attacks and to help prevent the same.

We are therefore facing a great opportunity for insurance companies in our case, which can make a big difference in their products if they include services, supplies, cyber protection solutions, which are increasingly in demand among consumers.

Thus, insurers are trying various strategies to respond to policyholders in the face of these risks. These range from recurring warnings to providing prevention services, rather than compensation, in the pre-damage phase. It is also key to inform clients, through guides, of the best practices with respect to the handling of insurance tools.

Likewise, the international insurance market itself has generated protection through the so-called "Cyber Insurance" even as micro-coverages for families and SMEs (small and medium-sized enterprises).

The coverage offered is liability insurance, due to damage caused to third parties in case of a privacy or security breach of a system to which a third party connects, even for the actions of technology providers. In this sense, we may talk about multimedia, professional, privacy and data security coverage.

The other kind of coverage granted by cyber insurance is loss coverage, such as loss of benefits or profits and expenses to manage crises, reputational risks, among others.

Another usual coverage is the coverage of fines and sanctions imposed in case of breach of personal data protection laws and regulations, which seems to exceed the extensive doctrinal discussion historically provoked by the coverage of fines by insurance laws.

Cyber insurance is often mentioned as a tool to increase cyber resilience—capacity to overcome traumatic situations—, as a risk transfer mechanism and as a useful assessment tool to accompany and assist in the risk calculations of companies. These insurances are marketed as a stand-alone product or as part of a package of other coverages, especially liability, operational all-risk or home insurance, if intended for the family.

But cyber insurance has their own difficulties to overcome to be a reliable insurance.

On the one hand, risk is difficult to measure; cyber risk is a global problem that is constantly evolving as the world becomes increasingly connected, and boy, is it ever. Also, the lack of sufficient data on cyber incidents and the low level of knowledge and experience of cyber risk are obstacles that have yet to be addressed.[30]

However, these are obstacles inherent to new risks and I understand they must be assumed with a degree of boldness, especially if we think of the preventive aspects that this type of insurance can bring to individuals and companies.

Considering the foregoing, it is evident that insurance is a valid response to address the social dimension of cyber risks and defend consumers and businesspersons against these serious hazards involved in the use of technology.

3.2 *Cyber Risks and Their Functional Dimension*

I refer at this point to the risks that may affect the functionality of the insurance company itself and how they should be prevented.

Wc arc in Latin Amcrica in thc cra of Corporatc Governancc which implies the management of business risks, inspired in essence by the "Insurance Core Principles, Standards, Guidance and Assessment Methodology" of the International Association of Insurance Supervisors, IAIS (www.iaisweb.org), particularly core principles 7 and 8:

PBS 7 Corporate Governance
The supervisor requires insurers to establish and implement a corporate governance framework that provides stable and prudent management and supervision of the insurer's business, and that adequately recognizes and protects the interests of policyholders.

[30] The issue that risks must be measurable is an issue that goes beyond cyber risks and applies to all new risks. This is clearly stated by Professor GREENE in his work already cited in chapter 3 on Probability, Risk and Insurance, pp. 27 et seq.

PBS 8 Risk Management and Internal Controls[31]
The supervisor requires the insurer, as part of its overall corporate governance framework, to have effective systems of risk management and internal controls, including effective risk management, compliance, actuarial and internal audit functions.

Other than IAIS' principles, the other fundamental principle is that of a company's sustainability as part of management comprehensive strategy pursuing general interests.

Sustainability is a modern term used to refer to corporate or business social responsibility and its scope and regulations are broad and have an impact on organizations' practices and management and on corporate governance. Sustainability demands a social interest in business activities, as well as a special composition of the administration body, which is essential in Corporate Governance, involving independent advisors, entrusted with specific duties for the sake of the collective social interest, instead of the company' own interest.[32]

Following this path, even service suppliers are influenced, since companies demand them to meet sustainability requirements. That is, companies influence the supplier's behavior by trying to direct and monitor their actions towards the observance and integration of environmental, social, and financial governance (ESG) criteria.

Thus, under the aura of sustainability, an insurance company manages its technology not only to avoid hazards, but also following environmental care criteria, with green criteria—because let us not forget that technology also pollutes, while demanding technology providers with whom the insurance company is involved to act accordingly.

In Uruguay, in application of the core principles of the IAIS, the Superintendency of Financial Services, the body that controls insurance activity, has established minimum management standards for insurance companies based on the integral evaluation and the so-called CERT methodology.

The Superintendency of Financial Services has defined that the supervision process must be comprehensive, proactive, risk-focused and on a consolidated basis.

One of the tools that supervision has to fulfill its tasks is the Comprehensive Evaluation, work carried out in situ at the insurance company. The purpose of the Comprehensive Assessment is to evaluate the quality of the entities' management and, if weaknesses are detected, to assess their impact on the entity's ability to maintain prudential solvency levels in the short, medium, and long term.

To synthesize the results of the evaluation, a methodology called CERT has been defined, where C stands for Corporate Governance, E for Financial Economic Evaluation, R for Risks and T for Technology.

---

[31] "Insurance Core Principles, Standards, Guidance and Assessment Methodology" of the International Association of Insurance Supervisors, IAIS (www.iaisweb.org). Retrieved on 1 August 2022.

[32] Perales Viscasilla (2021), pp. 1, 7, 11.

The objective of the CERT is to synthesize the evaluation by component and in general, of three aspects:

- whether there is a weakness in one of the components that requires priority attention by the institution;
- the stage of resolution of the weakness;
- the potential impact of the weakness found on the institution's ability to maintain prudential solvency levels in the short term.

To apply the CERT methodology to an institution, supervisors will analyze the following components:

C - Corporate Governance: is the system through which institutions are directed, monitored, and controlled. It is the form through which institutions organize themselves to carry out the administration and control of their management.

It is made up of the following structures

- management of the institution (the Board of Directors or equivalent hierarchical authority),
- management (Senior Management, hierarchical officers) and control (Audit Committee, Board of Directors).
- control (Audit Committee, Internal Audit and External Audit, among others).

It implies a set of practices adopted to carry out the daily management, monitoring, and control of the business, within the framework of the applicable laws and regulations.

Corporate governance must also ensure the adherence of the institution's officers to these practices and achieve efficiency in the practices, which entails, among other aspects, an adequate integral risk management system.

E - Financial Economic Evaluation: the financial economic situation of the institution is analyzed with emphasis on the level and quality of the institution's assets and its capacity to support the risks assumed and provide protection to policyholders and beneficiaries.

R - Risks: the institution's risk management system and the institution's ability to identify, control, measure and monitor the following risks:[33]

- Insurance Risk: Insurance Risk can be defined as the possibility of an entity's equity being affected by the adverse modification of the value of commitments undertaken under the insurance, due to the unsuitability of the pricing hypotheses and constitution of technical reserves.
- Market Risk: Market Risk can be defined as the possibility of losses as regards on and off-balance sheet positions arising from an adverse movement of market variables. The following are identified as market risks:
- Interest rate risk: Interest rate risk is composed of the following risks:

[33]Regulations of the Superintendency of Financial Services www.bcu.gub.uy (retrieved on 1 August 2022).

Interest rate risk of the investment portfolio: Is the risk associated with potential losses in the market value of the investment portfolio arising from an adverse movement in interest rates.

Structural interest rate risk: This risk encompasses the entire balance sheet, including off-balance positions. It is the potential risk of an entity's equity being affected due to movements in interest rates.

- Exchange rate risk: Exchange rate risk can be defined as the possibility of an entity's equity being adversely affected by unfavorable movements in exchange rates between currencies as regards on and off-balance sheet positions.
- Adjustment risk: Adjustment risk is the risk of an entity's equity being adversely affected by movements in exchange rates of units of account in local currency on a long-term horizon.
- Other market risks: Other market risks can be defined as the possibility of an entity's equity being affected by adverse movements in the price of shares, goods and/or real estate.

Liquidity Risk: Liquidity Risk refers to the possibility that an entity has not enough liquid assets to face undertaken obligations. Liquidity risk depends on two dimensions defined as funding liquidity risk (Debt) and market liquidity risk (Credit) and the correlation existing between them.

Funding liquidity risk: It includes the institution's failure to manage the decrease or unexpected change in funding sources. This may usually cause the early liquidation of a portion of its assets.[34]

Market liquidity risk: It comes from difficulties arising from changes in market conditions which affect the prompt liquidation of assets with a minimum loss in value.

- Operational Risk: Operational Risk can be defined as the possibility of an entity's equity being affected by losses arising from unsuitable or faulty processes, personnel, or internal systems, or by external events.

It also includes compliance risk, that is, the possibility of an entity being affected by breach of laws, regulations, standards and industry best practices or ethical standards.

- Money Laundering and Terrorism Financing Risk: Money Laundering and Terrorism Financing Risk can be defined as the possibility of loss or damage suffered by an entity when used, directly or through its operations, as an instrument to launder money and/or channel resources to conduct terrorist activities, or concealing money arising from such activities.[35]

[34] Regulations of the Superintendency of Financial Services www.bcu.gub.uy (retrieved on 1 August 2022).

[35] Regulations of the Superintendency of Financial Services www.bcu.gub.uy (retrieved on 1 August 2022).

– Reputational Risk: Reputational Risk is the possibility of an entity's equity being affected by a negative public opinion. It affects the institution's ability to establish new relationships or services, or to maintain already existing relationships. This risk may expose the institution to lawsuits, financial losses, or a reduction of its client base.

T - Technology: is the management of technological risks and reliability and effectiveness of information systems as management tools.

From the methodological point of view, Corporate Governance should be viewed as the central core of the analysis, with which the other components of the system are interrelated.

For greater transparency on the application of the new system and with the idea of providing guidance to insurers on what is expected of them, a series of minimum management standards associated with the four components of the CERT methodology has been drawn up.

From the supervisor's point of view, it is understood that **non-compliance with a standard** constitutes a weakness that must be treated with **priority attention** by the entity.

Insurance companies adopt different schemes and structures to carry out their management, considering the nature, size and complexity of their operations and their risk profile.

The supervisor carries out its supervision and evaluation procedures taking these elements into account.

The standards constitute management practices that the supervisor expects to find in the supervised entities.

As we can see, technology and its management constitute nothing less than one of the four pillars of the CERT methodology, which does not identify it simply as a risk but as a true standard that the company must manage as a central part of its business.

The standards for the evaluation of Information Technology (IT) areas are based on the set of principles known as CobiT, particularly those related to the Acquisition and Implementation domain.

IT (information technology) management must have the ability to identify needs and to develop, acquire, install, and maintain appropriate IT solutions in accordance with the needs of the entity.

It is clear, therefore, that under these guidelines, technological risks must be addressed in a broad sense, covering both risks directly related to IT aspects and those related to human resources, since human error is often the cause, or the gateway, of a cyberattack.

That is, not only the management of aspects directly related to the security of computer systems such as the permanent updating of systems, the proper classification of personal, financial and commercial information, the execution of periodic back-up and its custody in places other than the main location, the eventual restriction of access to public networks, the development of incident response plans, but also the proper management of human resources with access segmentation and password usage rules, ensuring confidentiality, achieving the training of the auditing areas in incident detection, among others.

In addition, given what has been said above about risk management as part of corporate governance, it is clear that senior management must understand data protection and cybersecurity as a central, key element of the business.

It must understand that the source of risk is not only external, that cybercrime can originate in any position, be aware of the applicable regulations and the consequences of non-compliance and create a special and specific risk management policy that is constantly applied, documented, updated, and made known to the organization, allocating resources for its implementation.

In short, in the age of technology and corporate governance, the insurance company should address the prevention and minimization of cyber risk with a specialized and priority risk management at a general level.

As it arises from the above analysis, insurance would not be an effective tool to cover the functional dimension of cyber risks. The focus to handle risks implied in this dimension should be to use expert technology management as a quality standard to be met by insurance companies.

Notwithstanding the above, some of the risks managed by companies may have arisen from a technology-related event and may be covered by cyber insurance. This is the case, for instance, of reputational risk when the damage caused to the company's image has occurred due to a technology-related event, such as computer fraud or data security breach or user privacy breach.

3.3 Cyber Risks and Their Ethical Dimension

Lastly, it is worth analyzing what I call the ethical dimension of cyber risks, this time seen as another type of risk that the cyber environment or space creates for human beings.

As expressed in an article by the Real Instituto El Cano[36] "in the field of cybersecurity, AI (Artificial Intelligence) brings significant improvements through algorithmic analysis applied to large amounts of information, inferring results based on context and learning acquired from previous situations. The capabilities of AI, its algorithms, can be applied in a similar way - both - by those who create insecurity in advanced societies and - as well as - by those who create insecurity in advanced societies - by those who create insecurity in advanced societies and - as well as - by

[36] https://www.realinstitutoelcano.org/analisis/implicaciones-sobre-el-uso-de-la-inteligencia-artificial-en-el-campo-de-la-ciberseguridad/ Author: Alonso Lecuit (2019). ARI 50//2019 (retrieved on 1 August 2022).

those who create insecurity in advanced societies and - like - by those who protect them."[37]

"The direct confrontation between AI algorithms and their escalation may lead to a point where human intervention could be relegated to the background. The response to this situation promotes an international debate on the need to regulate the characteristics and use of AI, mainly from an ethical point of view, but also from the regulatory and control point of view of its use, without limiting the benefits brought by AI innovation to society."[38]

"Although there is unanimity on the need for international regulation, it is particularly difficult to establish a roadmap setting out the steps to be taken. Inaction, i.e., letting market forces set the rules of the game, would lead to the disprotection of fundamental security rights of individuals and nations, similar to that currently experienced globally in the area of privacy. This approaches the potential impact of AI, the effects of its malicious employment and the need for controls and countermeasures that lack a global regulatory framework."[39]

This concern has not escaped the European Union authorities and so it is that in April 2019 there has been the "Communication from the Commission to the European Parliament, the Council, the European Economic and Social Committee and the Committee of the Regions" on "Building Trust in Human-Centered Artificial Intelligence."[40]

The document states that artificial intelligence (AI) has the potential to transform our world for the better: it can improve healthcare, reduce energy consumption, make vehicles safer and enable farmers to use water and resources more efficiently. AI can be used to predict climate and environmental change, improve financial risk management, and provide the tools to manufacture, with less waste, products tailored to our needs.

AI can also help detect fraud and cybersecurity threats and enable law enforcement agencies to fight crime more effectively. AI can benefit society and the economy as a whole. It is a strategic technology that is being developed and used at a rapid pace around the world.

However, it also brings new challenges for the future of work and raises legal and ethical questions.

---

[37] https://www.realinstitutoelcano.org/analisis/implicaciones-sobre-el-uso-de-la-inteligencia-artificial-en-el-campo-de-la-ciberseguridad/ Author: Alonso Lecuit (PDF ARI 50/2019), p. 1.

[38] https://www.realinstitutoelcano.org/analisis/implicaciones-sobre-el-uso-de-la-inteligencia-artificial-enel-campo-de-la-ciberseguridad/ Author: ALONSO LECUIT, Javier (PDF ARI 50/2019), p. 1 (retrieved on 1 August 2022).

[39] https://www.realinstitutoelcano.org/analisis/implicaciones-sobre-el-uso-de-la-inteligencia-artificial-en-el-campo-de-la-ciberseguridad/ALONSO LECUIT (PDF ARI 50//2019) pp. 1 et seq.

[40] https://ec.europa.eu/transparency/regdoc/rep/1/2019/ES/COM-2019-168-F1-ES-MAIN-PART-1.PDF. Brussels 8.4.2019.

To address these challenges and make the most of the opportunities offered by AI, in April 2019 the Commission published a European strategy. [41]

The strategy places the person at the center of AI development-it is human-centered AI.

It adopts a three-pronged approach to boost the EU's technological and industrial capacity and drive the adoption of AI in all areas of the economy, prepare for socio-economic transformations, and ensure that an appropriate ethical and legal framework is in place.

The European AI Strategy and the plan coordinated under it make it clear that trust is a prerequisite for ensuring a human-centered approach to AI: AI is not an end in itself, but a means that must serve people with the ultimate goal of increasing their well-being.

To this end, the reliability of AI must be guaranteed. The values on which our societies are based must be fully integrated into the evolution of AI. The values of respect for human dignity, freedom, democracy, equality, the rule of law, and respect for human rights, including the rights of persons belonging to minorities, must be respected.

These values are common to societies not only in the European Union but can be extrapolated to all of Latin America, where pluralism, non-discrimination, tolerance, justice, solidarity, and equality prevail. In addition to respect for fundamental rights, the individual, civil, political, economic, and social rights that are the foundation of our societies.

In its report, the Commission supports the following essential requirements for a reliable IA. It encourages stakeholders to apply them and to check the list that puts them into practice to create the right environment of trust for the successful development and use of AI.

The seven essential requirements are as follows: human agency and oversight; technical robustness and safety; privacy and data governance; transparency; diversity, non-discrimination and fairness; societal and environmental well-being; and accountability.

Human Agency and Oversight AI systems must help people make better and more informed choices based on their objectives. They should act as enablers of a flourishing and equitable society, supporting human intervention and fundamental rights, and not diminish, limit, or disorient human autonomy. The overall well-being of the user must be paramount in the functionality of the system.

Human oversight helps ensure that an AI system does not undermine human autonomy or cause other adverse effects. Depending on the specific AI system and its scope of application, appropriate degrees of control measures, including adaptability, accuracy, and explanation of AI systems, must be ensured. Oversight should

[41] https://ec.europa.eu/transparency/regdoc/rep/1/2019/ES/COM-2019-168-F1-ES-MAIN-PART-1.PDF. Brussels 8.4.2019.

be achieved through governance mechanisms, such as the human-in-the-loop approach, human-on-the-loop supervision, or human-in-command.

-Powers in accordance with their mandates. Other things being equal, the less oversight a human can exercise over an AI system, the more extensive the testing will have to be and the stricter the governance will have to be.

Technical Robustness and Safety AI reliability requires that algorithms be sufficiently safe, reliable, and robust to resolve errors or inconsistencies during all phases of the AI system's life cycle and adequately cope with erroneous results.

AI systems must be reliable, secure enough to be resilient, both in the face of overt attacks and more subtle attempts to manipulate data or the algorithms themselves, and must ensure a contingency plan in case of problems.

Their decisions must be accurate, at the very least, reflect their level of correctness, and their results reproducible. In addition, AI systems must integrate safety and security mechanisms from the design stage to ensure that they are verifiably safe at every stage, with the physical and psychological safety of all those involved being paramount. This includes minimization and, where possible, reversibility of unintended consequences or errors in system operation.

Processes should be put in place to clarify and assess the potential risks associated with the use of AI systems in various application areas.

Privacy and Data Governance Privacy and data protection should be ensured at all stages of the AI system lifecycle.

Digital records of human behavior may allow AI systems to infer not only people's preferences, age, and gender, but also their sexual orientation or religious or political views. For individuals to have confidence in data processing, it must be ensured that they have full control over their own data, and that data concerning them will not be used to harm or discriminate against them.

In addition to safeguarding privacy and personal data, requirements must be met in terms of ensuring the quality of AI systems. The quality of the data sets used is paramount to the performance of AI systems when data is collected, may reflect social biases, or contain inaccuracies or errors. This must be resolved before training an AI system with a dataset.

Also, the integrity of the data must be ensured. The processes and datasets used must be tested and documented at every stage, such as planning, training, testing and deployment. This must also apply to AI systems that have not been developed in-house, but have been procured externally.

Finally, access to data must be adequately regulated and controlled.

Transparency The traceability of AI systems must be ensured; it is important to record and document both the decisions made by the systems and the entire process—including a description of the data collection and labeling, and a description of the algorithm used—that led to the decisions.

In this regard, the explanation of the algorithmic decision-making process, tailored to the individuals concerned, should be provided to the extent possible.

Ongoing research should be pursued to develop explanatory mechanisms. In addition, explanations of the extent to which an AI system influences and shapes the organizational decision-making process, the system design choices, as well as the rationale for its deployment—thus ensuring not only data and system transparency, but also business model transparency—must be available.

Finally, it is important to adequately communicate the capabilities and limitations of the AI system to the various stakeholders concerned in a manner appropriate to the case at hand.

Moreover, AI systems must be identifiable as such, ensuring that users know that they are interacting with an AI system and which persons are responsible for it.

Diversity, Non-Discrimination, and Fairness Data sets used by AI systems-both for training and operation-can be affected by the inclusion of unintentional historical biases, by incompleteness, or by poor governance models. Persistence in these biases could result in (in)direct discrimination. Damage may also result from intentional exploitation of consumer biases or unfair competition.[42]

Moreover, the way in which AI systems are developed—for example, the way in which the programming code of an algorithm is written—may also be biased. These problems need to be addressed early in the development of the system.

Establishing diversified design teams and creating mechanisms to ensure participation, particularly of citizens, in AI development can also help resolve these problems. Stakeholders who may be directly or indirectly affected by the system throughout its life cycle should be consulted.

AI systems should consider the full range of human capabilities, abilities and needs and ensure accessibility through a universal design approach to strive for equal access for people with disabilities.

Social and Environmental Well-Being For AI to be reliable, its impact on the environment and on other sentient beings must be considered. Ideally, all human beings, including future generations, should benefit from biodiversity and a livable environment.

The sustainability and ecological responsibility of AI systems must therefore be promoted.

Moreover, the impact of AI systems should be considered not only from an individual perspective, but also from the perspective of society as a whole. Special attention should be paid to the use of AI systems, particularly in situations related to the democratic process, including opinion formation, political decision-making or in the electoral context.

The social impact of AI should also be considered. While AI systems can be used to improve social skills, they can also contribute to their deterioration.

[42] Junqueira (2020), p. 228.

-Accountability mechanisms should be put in place to ensure the responsibility and accountability of AI systems and their results, both before and after implementation.

The ability to audit AI systems is critical, as the evaluation of AI systems by internal and external auditors, and the availability of evaluation reports, contributes greatly to the reliability of the technology.

The possibility of external audits must be ensured especially in applications that affect fundamental rights, e.g., safety-critical applications.

Potential negative impacts of AI systems must be identified, assessed, documented, and minimized. The use of impact assessments facilitates this process. These assessments should be commensurate with the magnitude of the risks posed by AI systems. Trade-offs between requirements—which are often unavoidable—must be addressed in a rational and methodological manner, and considered.[43]

Finally, when unfair adverse effects occur, accessible mechanisms must be in place to ensure adequate redress.

Additionally, in the European Union, in April 2021, a draft resolution was generated to be applied by the members, which seeks to regulate the implementation of systems based on artificial intelligence. Indeed, the European Commission proposes a "Regulation of the European Parliament and the Council . . .for the harmonization of rules on Artificial Intelligence (Artificial Intelligence Act)" that amends some other laws of the European Union.[44]

According to their Explanatory Memorandum, the reasons for and objectives of the proposal for a Regulation laying down harmonized rules on artificial intelligence (Artificial Intelligence Act) are as follows. "Artificial Intelligence (AI) is a fast evolving family of technologies that can bring a wide array of economic and societal benefits across the entire spectrum of industries and social activities. By improving prediction, optimising operations and resource allocation, and personalising service delivery, the use of artificial intelligence can support socially and environmentally beneficial outcomes and provide key competitive advantages to companies and the European economy. Such action is especially needed in high-impact sectors, including climate change, environment and health, the public sector, finance, mobility, home affairs and agriculture. However, the same elements and techniques that power the socio-economic benefits of AI can also bring about new risks or negative consequences for individuals or the society. In light of the speed of technological change and possible challenges, the EU is committed to strive for a balanced approach. It is in the Union interest to preserve the EU's technological leadership and to ensure that Europeans can benefit from new technologies developed and functioning according to Union values, fundamental rights and principles."[45]

---

[43] Junquiera Thiago, op.cit p. 230 et seq.

[44] https://eur-lex.europa.eu/legal-content/EN/TXT/?uri=CELEX%3A52021PC0206 (retrieved on 1 August 2022).

[45] Additionally, according to the Explanatory Memorandum "This proposal delivers on the political commitment by President von der Leyen, who announced in her political guidelines for the 2019-

The Commission puts forward the proposed regulatory framework on artificial intelligence with the following specific objectives:

- ensure that AI systems placed on the Union market and used are safe and respect existing law on fundamental rights and Union values;
- ensure legal certainty to facilitate investment and innovation in AI;
- enhance governance and effective enforcement of existing law on fundamental rights and safety requirements applicable to AI systems;
- facilitate the development of a single market for lawful, safe, and trustworthy AI applications and prevent market fragmentation.

"To achieve those objectives, this proposal presents a balanced and proportionate horizontal regulatory approach to AI that is limited to the minimum necessary requirements to address the risks and problems linked to AI, without unduly constraining or hindering technological development or otherwise disproportionately increasing the cost of placing AI solutions on the market. The proposal sets a robust and flexible legal framework. On the one hand, it is comprehensive and future-proof in its fundamental regulatory choices, including the principle-based requirements that AI systems should comply with. On the other hand, it puts in place a proportionate regulatory system centred on a well-defined risk-based regulatory approach that does not create unnecessary restrictions to trade, whereby legal intervention is tailored to those concrete situations where there is a justified cause for concern or where such concern can reasonably be anticipated in the near future. At the same time, the legal framework includes flexible mechanisms that enable it to

---

2024 Commission 'A Union that strives for more' 1, that the Commission would put forward legislation for a coordinated European approach on the human and ethical implications of AI. Following on that announcement, on 19 February 2020 the Commission published the White Paper on AI - A European approach to excellence and trust. The White Paper sets out policy options on how to achieve the twin objective of promoting the uptake of AI and of addressing the risks associated with certain uses of such technology. This proposal aims to implement the second objective for the development of an ecosystem of trust by proposing a legal framework for trustworthy AI. The proposal is based on EU values and fundamental rights and aims to give people and other users the confidence to embrace AI-based solutions, while encouraging businesses to develop them. AI should be a tool for people and be a force for good in society with the ultimate aim of increasing human well-being. Rules for AI available in the Union market or otherwise affecting people in the Union should therefore be human centric, so that people can trust that the technology is used in a way that is safe and compliant with the law, including the respect of fundamental rights. Following the publication of the White Paper, the Commission launched a broad stakeholder consultation, which was met with a great interest by a large number of stakeholders who were largely supportive of regulatory intervention to address the challenges and concerns raised by the increasing use of AI. The proposal also responds to explicit requests from the European Parliament (EP) and the European Council, which have repeatedly expressed calls for legislative action to ensure a well-functioning internal market for artificial intelligence systems ('AI systems') where both benefits and risks of AI are adequately addressed at Union level. It supports the objective of the Union being a global leader in the development of secure, trustworthy and ethical artificial intelligence as stated by the European Council and ensures the protection of ethical principles as specifically requested by the European Parliament."

be dynamically adapted as the technology evolves and new concerning situations emerge."[46]

The regulation seeks to avoid possible discrimination that algorithms, included in the systems may generate, and in turn preserve essential rights of the individual, such as privacy, security, the right to complain, among several others.

This, especially in what the project defines as "High Risk" systems. The regulation[47] highlights the definition and regulation of "High Risk" AI systems characterized by the opacity of the information provided and its source, the complexity of the system, the dependence on data and autonomous learning.

The proposal lays down a solid risk methodology to define "high-risk" AI systems that pose significant risks to the health and safety or fundamental rights of persons. Those AI systems will have to comply with a set of horizontal mandatory requirements for trustworthy AI and follow conformity assessment procedures before those systems can be placed on the Union market. Predictable, proportionate, and clear obligations are also placed on providers and users of those systems to ensure safety and respect of existing legislation protecting fundamental rights throughout the whole AI systems' lifecycle. For some specific AI systems, only minimum transparency obligations are proposed, particularly when chatbots or "deep fakes" are used.

These systems are characterized by, among other features, the primacy of "machine learning" algorithms, i.e., they "learn by themselves" and respond based on the successive collection and exchange of massive data or Big Data.

Title III contains specific rules for AI systems that create a high risk to the health and safety or fundamental rights of natural persons. In line with a risk-based approach, those high-risk AI systems are permitted on the European market subject to compliance with certain mandatory requirements and an ex-ante conformity assessment. The classification of an AI system as high-risk is based on the intended purpose of the AI system, in line with existing product safety legislation. Therefore, the classification as high-risk does not only depend on the function performed by the

[46] https://eur-lex.europa.eu/legal-content/EN/TXT/?uri=CELEX%3A52021PC0206-

Explanatory Memorandum, added: "The proposal sets harmonised rules for the development, placement on the market and use of AI systems in the Union following a proportionate risk-based approach. It proposes a single future-proof definition of AI. Certain particularly harmful AI practices are prohibited as contravening Union values, while specific restrictions and safeguards are proposed in relation to certain uses of remote biometric identification systems for the purpose of law enforcement.

The proposed rules will be enforced through a governance system at Member States level, building on already existing structures, and a cooperation mechanism at Union level with the establishment of a European Artificial Intelligence Board. Additional measures are also proposed to support innovation, in particular through AI regulatory sandboxes and other measures to reduce the regulatory burden and to support Small and Medium-Sized Enterprises ('SMEs') and start-ups." Retrieved on 1 August 2022.

[47] https://eur-lex.europa.eu/legal-content/EN/TXT/?uri=CELEX%3A52021PC0206.

Explanatory Memorandum. Other Elements, 5.2.3. High-Risk AI Systems (TITLE III). Retrieved on 1 August 2022.

AI system, but also on the specific purpose and modalities for which that system is used.

Chapter 1 of Title III sets the classification rules and identifies two main categories of high-risk AI systems:

AI systems intended to be used as safety component of products that are subject to third party ex-ante conformity assessment;

other stand-alone AI systems with mainly fundamental rights implications that are explicitly listed in Annex III.

This list of high-risk AI systems in Annex III contains a limited number of AI systems whose risks have already materialized or are likely to materialize in the near future. To ensure that the regulation can be adjusted to emerging uses and applications of AI, the Commission may expand the list of high-risk AI systems used within certain pre-defined areas, by applying a set of criteria and risk assessment methodology.

Chapter 2 sets out the legal requirements for high-risk AI systems in relation to data and data governance, documentation and recording keeping, transparency and provision of information to users, human oversight, robustness, accuracy, and security. The proposed minimum requirements are already state-of-the-art for many diligent operators and the result of two years of preparatory work, derived from the Ethics Guidelines of the HLEG, piloted by more than 350 organizations. They are also largely consistent with other international recommendations and principles, which ensures that the proposed AI framework is compatible with those adopted by the EU's international trade partners. The precise technical solutions to achieve compliance with those requirements may be provided by standards or by other technical specifications or otherwise be developed in accordance with general engineering or scientific knowledge at the discretion of the provider of the AI system. This flexibility is particularly important, because it allows providers of AI systems to choose the way to meet their requirements, taking into account the state-of-the-art and technological and scientific progress in this field.

Chapter 3 places a clear set of horizontal obligations on providers of high-risk AI systems. Proportionate obligations are also placed on users and other participants across the AI value chain (e.g., importers, distributors, authorized representatives).

Chapter 4 sets the framework for notified bodies to be involved as independent third parties in conformity assessment procedures, while Chapter 5 explains in detail the conformity assessment procedures to be followed for each type of high-risk AI system. The conformity assessment approach aims to minimize the burden for economic operators as well as for notified bodies, whose capacity needs to be progressively ramped up over time. AI systems intended to be used as safety components of products that are regulated under the New Legislative Framework legislation (e.g., machinery, toys, medical devices, etc.) will be subject to the same ex-ante and ex-post compliance and enforcement mechanisms of the products of which they are a component. The key difference is that the ex-ante and ex-post mechanisms will ensure compliance not only with the requirements established by sectorial legislation, but also with the requirements established by this regulation.[48]

On the other hand, the regulation seeks to defend the fundamental rights of the human being, basically: human dignity, respect for privacy and data, non-discrimination, equality of gender and race, freedom of expression, the rights of children—of the disabled, protection of the environment, and the guarantee of the right of defense and of remedy or correction. It also limits the right to trade and the use of the technique, if this is necessary to respect the right of the consumer and to protect personal data.

As can be seen, the concern for the ethical aspects of new technologies and the cybernetic environment merit beginning to speak of an ethical dimension of cyber risks.

This ethical dimension of cyber risks cannot be solved by insurance. It is solved, mitigated, or prevented in general by social and government policies which properly manage these risks.

4 Conclusions

We are all hostages of technology, but at the same time we enjoy it and live it.

This of course brings lights and shadows, cyberspace can create false illusions, unreal personalities, human, professional and even business profiles that do not correspond to reality. It also takes us away from the spontaneity of face to face and can lead to misunderstandings and lack of genuine and deep communication.

[48] Additionally, according to the Explanatory Memorandum. Other Elements, 5.2.3. High-Risk AI Systems (TITLE III) "As regards stand-alone high-risk AI systems that are referred to in Annex III, a new compliance and enforcement system will be established. This follows the model of the New Legislative Framework legislation implemented through internal control checks by the providers with the exception of remote biometric identification systems that would be subject to third party conformity assessment. A comprehensive ex-ante conformity assessment through internal checks, combined with a strong ex-post enforcement, could be an effective and reasonable solution for those systems, given the early phase of the regulatory intervention and the fact the AI sector is very innovative and expertise for auditing is only now being accumulated. An assessment through internal checks for 'stand-alone' high-risk AI systems would require a full, effective and properly documented ex ante compliance with all requirements of the regulation and compliance with robust quality and risk management systems and post-market monitoring. After the provider has performed the relevant conformity assessment, it should register those stand-alone high-risk AI systems in an EU database that will be managed by the Commission to increase public transparency and oversight and strengthen ex post supervision by competent authorities. By contrast, for reasons of consistency with the existing product safety legislation, the conformity assessments of AI systems that are safety components of products will follow a system with third party conformity assessment procedures already established under the relevant sectoral product safety legislation. New ex ante re-assessments of the conformity will be needed in case of substantial modifications to the AI systems (and notably changes which go beyond what is pre-determined by the provider in its technical documentation and checked at the moment of the ex-ante conformity assessment)."

This can make us live in a parallel, fantastic, and unreal world, if we do not know how to handle it properly, with responsibility at all levels and if the aspects that support the application and uses of artificial intelligence, the basis of this new technological reality, are not regulated.

But on the other hand, among many other virtues, the communication of people between different points of the planet, even all at the same time, as if they were talking at a coffee table, is of an immeasurable value?

Undoubtedly, if the technological, cybernetic world did not exist, if it did not coexist with us, these risks would not exist, but how can we refuse the progress in so many positive aspects that they bring?

Again, as in so many issues in life, balance should be the key.

Accomplishing due balance in the design, application and use of technologies is the only thing which can provide peace to human beings and society as a whole, highlighting their strengths and not their weaknesses, within the attractive, exciting, and dizzying maelstrom of this cybernetic world.

Insurance is an unquestionable tool to manage insurable risks. The dimensions that cyber risks may reach, in accordance with this chapter, cannot be nor must be completely covered by insurance. Insurance may only cover those hazards which, according to their technical basis, may be covered following solvency conditions.

Some of these risks are infinite or hard to measure or delimit and, hence, they elude the technical parameters on which the insurance is based and which precisely provide its comprehensive and responsible coverage.

Therefore, and considering the foregoing, the dimension of cyber risks which may be completely covered by insurance is the social dimension, which is precisely the dimension mainly affecting insurance consumers, as well as businesspersons in general.

Thus, insurance remains a strategic tool for due management of risks affecting human beings and the community in general.

References

Affiniongroup-Cybercrime Report. https://www.ciberseguridadpyme.es. Retrieved on 1/08/2022

Alonso Lecuit J (2019) Inteligencia Artificial - ARI 50/2019- Real Instituto El Cano Publication. www.realinstitutoelcano.org/wps/portal/rielcano_es/. Retrieved on 1/08/2022

Beer S (1959) Cybernetics and management. English Universities Press

Bruck E (1930) Das privat-verisicherungsrecht. Lepzig

European Union-Communication from the Commission to the European Parliament, the Council, the European Economic and Social Committee and the Committee of the Regions-Building trust in human-centered artificial intelligence. https://ec.europa.eu/transparency/regdoc/rep/1/2019/ES/COM-2019-168-F1-ES-MAIN-PART-1.PDF. Brussels 8.4.2019. Retrieved on 1/08/2022

Fanelli G (1973) Le assicurazioni, Vol. 1. Dott. A. Giuffrè-Editore, Milano

Ferri G (1972) Manuale de Diritto Commerciale. 3° edizione. UTE Torineses. Torino

Garrigues J (1982) Contrato de seguro terrestre. J Garrigues, Madrid

Greene M (1976) Risk and Insurance. Ed. MAPFRE, Madrid

Halperin I and Barbato (1983) Seguros. Exposición crítica de las leyes No. 17,418 y 20,091. Depalma

Instituto Nacional De Compañías De Seguros De España (INESE) (2019) Ciber riesgos y ciber seguros. Ed. INESE

International Association of Insurance Supervisors-Insurance core principles, standards, guidance and assessment methodology - IAIS -www.iaisweb.org. Retrieved on 1/08/2022

Junqueira T (2020) Tratamento de dados Pessoais e Discriminação Algorítmica nos seguros. Thomson Reuters, São Paolo

Perales Viscasilla MdP (2021) Retos y tendencias actuales en sosteniblidad y gobierno corporativo: una mirada tras el Covid-19. Revista Española de seguros, SEAIDA No. 185–186

Proposal for a Regulation of the European Parliament and of the Council - Laying down harmonised rules on Artificial Intelligence (Artificial Intelligence Act) and amending certain Union legislative acts. https://eur-lex.europa.eu/legal-content/EN/TXT/?uri=CELEX%3A52021PC0206. Retrieved on 1/08/2022

Regulations of the Superintendency of Financial Services. www.bcu.gub.uy. Retrieved on 1/08/2022

Ríos R (2015) El contrato de seguro. Comentarios al Título VIII, Libro II del Código de Comercio; Ed. La Ley, Santiago de Chile

Sánchez Calero F (1961) Curso de Contrato de seguro privado. Ediciones Nauta, Bilbao

Sánchez Calero F, Director-Tirado Suárez FJ, Tapia Hermida A, Fernández Rozas JC, Fuentes Calderón V (1999) Ley de contrato de seguros. Comentarios a la Ley 50/1980 de 8 de octubre y modificativas. Ed. Aranzardi, Madrid

Signorino A (2011) Seguros de responsabilidad civil. Ed. FCU, Montevideo

Viterbo C (1941) Teoría económica del seguro a prima fija. Revista De Economía y Estadística V 3 No. 4

Wiener N (1961) Cybernetics: or control and communication in the animal and the machine, 2nd edn. Hermann &Cie&Camb. Mass. MIT Press, p. 1948

Road Traffic Law and Application to Automated Vehicles

Matthew Channon

Abstract Current road traffic law in the UK is focussed the on persons driving or using the vehicle. With the introduction of vehicles of higher automation, the responsibility of vehicle drivers is likely to change. The application of road traffic law to individuals within automated vehicles and whether reform is needed is a key issue. Moreover, what exactly an individual within a partially automated vehicle should be able to do is also a key issue. This chapter analyses these key issues and further analyses whether responsibility could fall elsewhere when the Automated Driving System is engaged. This chapter focusses on the current road traffic law in the UK and Victoria, an Australian State, and discusses UK reform proposals from the Law Commission of England and Wales and Scottish Law Commission as well as reform proposals in Australia from the National Transport Commission. The chapter further considers whether responsibility could eventually rest with the automated vehicle itself with the potential for legal personality.

Keywords Automated vehicles · Road traffic law · Road Traffic Act 1988 · UK and Victoria

1 Introduction

Automated vehicles have the potential to transform society with huge impact on everyday lives.[1] The UK road traffic law and its application to automated vehicles has received recent attention, and the Law Commission of England and Wales, and Scottish Law Commission (Law Commissions) have proposed reforms.[2] Moreover,

[1] See, e.g. Bissell et al. (2020), pp. 116–134.

[2] There have been three Law Commission of England and Wales and Scottish Law Commission Consultations and a final report. See Law Commissions (2018), 'Automated Vehicles: A joint preliminary consultation paper' (LC 240), Law Commissions (2020), 'Automated Vehicles:

M. Channon (✉)
University of Exeter, Law School, Exeter, UK
e-mail: m.channon2@exeter.ac.uk

K. Noussia, M. Channon (eds.), *The Regulation of Automated and Autonomous Transport*, https://doi.org/10.1007/978-3-031-32356-0_12

in Australia the National Transport Commission (NTC) has further examined the application of Australian road traffic law to automated vehicles.[3] Research has also been undertaken. For example, Van Wees[4] focusses on international traffic law, although does mention the similar proposals made by the Law Commissions and NTC. This chapter goes beyond previous literature by comparing specific laws from Victoria and the UK and their application to automated vehicles. It further discusses and compares some of the most recent recommendations from the Law Commissions and NTC.

This chapter will discuss current road traffic law in the UK and Victoria as well its focus. We will then explore the application of this focus and the application of road traffic law to lower and higher automation levels in both jurisdictions as well as comparing reform proposals in relation to automated vehicles. The Australian Road traffic regime can differ between states and territories, analysing every territory would be unviable due to word constraints, this chapter instead focusses on one state, Victoria. As reform is being proposed at national level in Australia, this chapter will focus on reform at a national level, although with occasional reference to Victoria. This chapter will therefore provide a comparison between the UK and Victoria approaches to road traffic offences and application to automated vehicles. Moreover, the chapter will further compare national reform proposals. The reason for choosing Australia as a comparator is due to recent reform discussions in Australia, which can be compared to the UK. Finally, this chapter discusses whether automated vehicles could themselves be held responsible and punished; this will involve discussion of legal personality.

2 Current Road Traffic Law and Focus

This chapter begins by discussing the general legislative regimes, as well as specific road traffic law for conventional vehicles in the UK and Victoria. Additionally, the function of these laws will be discussed, and this will allow later discussion of the application of these laws and functions to automated vehicles.

Consultation paper 3' (LC 252), Law Commissions (2022b) 'Automated Vehicles: Joint Report' (LC 404).

[3]National Transport Commission (2020), 'A national in service safety law for automated vehicles', National Transport Commission (2022) 'The regulatory framework for automated vehicles', National Transport Commission (2017) 'Changing driving laws to support automated vehicles'.

[4]Kiliaan Van Wees (2020).

2.1 General Legislative Regimes and Offences

Road traffic law is contained largely in legislation[5] and has '*swelled over the past century*'.[6] It is also pertinent to mention that the Highway Code contains an array of rules on motor vehicles.[7] Failure to comply with provisions of the Highway Code is not itself an offence (Section 38, *Road Traffic Act (RTA) 1988*) although can be relied upon, for example in dangerous or careless driving offences.[8] Some updates have been made to the Highway Code in light of automated vehicles[9] after the UK Department for Transport (DfT) issued Calls for Evidence for updates to the Highway Code.[10] Updates to the Highway Code are discussed later in this chapter. In Australia, road traffic offences are determined at state and territory level. Like the UK, Victorian road offences are contained in legislation.[11] It is also worth noting the Australian Road Rules, which are federal model laws with no legal effect and which '*form the basis of road rules of each Australian state and territory*'.[12] Notably, each state and territory can diverge from road rules, and therefore road rules can differ between states and territories. In Victoria, the road rules can be found in the *Road Safety Road Rules 2017.*

Several offences can be committed via motor vehicle with similarities between the UK and Victoria. It would be difficult to discuss all of these in this chapter, although we mention a few offences here. In the UK, some offences relate to the motor vehicle being driven carelessly or dangerously, as well as injuries or death caused by the driving of a motor vehicle carelessly or dangerously.[13] As well as

---

[5] See, e.g the Road Traffic Act 1988.

[6] Cunningham (2008), p. 169.

[7] See the UK Department for Transport (2015) < https://www.gov.uk/guidance/the-highway-code> accessed on 27 July 2022.

[8] We discuss these offences below.

[9] See Department for Transport (2022), Highway Code <https://www.gov.uk/guidance/the-highway-code/introduction> accessed on 23 August 2022.

[10] Department for Transport (2020), 'Safe use of Automated Lane Keeping System on GB motorways: call for evidence' <https://assets.publishing.service.gov.uk/government/uploads/system/uploads/attachment_data/file/980644/Safe-Use-of-Automated-Lane-Keeping-System-ALKS-Call-for-Evidence-FINAL-accessible.pdf> accessed on 11 September 2022, and Department for Transport (2022) 'Consultation Outcome: Rules on safe use of automated vehicles on GB roads' <https://www.gov.uk/government/consultations/safe-use-rules-for-automated-vehicles-av/rules-on-safe-use-of-automated-vehicles-on-gb-roads#consultation-on-the-rules-on-use-for-automated-lane-keeping-systems> accessed on 11 September 2022.

[11] See, e.g. the *Road Safety Act 1986.*

[12] National Transport Commission <https://www.ntc.gov.au/laws-and-regulations/australian-road-rules#:~:text=The%20model%20Australian%20Road%20Rules,%2C%20pedestrians%2C%20passengers%20and%20others> accessed on 27 July 2022.

[13] Offences in the *Road Traffic Act 1988* include causing death (Section 1, *RTA 1988*) or serious injury (Section 1A, *RTA 1988*) by dangerous driving. The UK *Road Safety Act 2006* added new offences to the statute book involving causing death by careless or inconsiderate driving (Section 20, *Road Safety Act* 2006 inserting Section 2B, *RTA 1988*).

offences contained in statute, manslaughter (gross negligence or unlawful act), an offence at common law, can also be committed via motor vehicle. In Victoria, offences include dangerous driving (Section 64, *RSA 1986*) and careless driving (Section 65, *RSA 1986*). Moreover, offences include careless driving causing death or serious injury,[14] and causing death by culpable driving.[15] Moreover, offences exist in both UK and Victoria concerning driving (or being in charge of) a vehicle while under the influence of alcohol or drugs.[16] Furthermore, there are requirements for the driver to have a driving licence in both the UK and Victoria.[17] Additionally, in both the UK and Victoria, it is an offence for a passenger to not wear a seatbelt[18] with additional requirements for the driver to ensure that children are wearing a seatbelt.[19] Further, the use of a hand-held mobile phone while driving a motor vehicle or supervising the driving of a motor vehicle is an offence in the UK under Section 41D of the *RTA 1988*.[20] Similarly, Section *300* of the Victoria *Road Safety Road Rules 2017* makes it an offence to use a mobile phone while the vehicle is moving or stationary (not parked).[21]

It is worth mentioning here that criminal offences in the UK can be committed through putting unsafe vehicles on the market through the *General Product Safety Regulations*.[22] There is also the *Corporate Manslaughter and Corporate Homicide*

[14] Section 319, *Crimes Act 1958*.

[15] Section 318, *Crimes Act 1958*.

[16] For the UK, see Section 4, *RTA 1988* for 'Driving, or being in charge, when under influence of drink or drugs'. For Victoria, see, e.g. Section 49 (1) of the *Road Safety Act 1986* which states '(1) A person is guilty of an offence if he or she— (a) drives a motor vehicle or is in charge of a motor vehicle while under the influence of intoxicating liquor or of any drug to such an extent as to be incapable of having proper control of the motor vehicle…'.

[17] In the UK, see Section 87 (1), *RTA 1988*; in Victoria, see Section 18, *Road Safety Act 1986*.

[18] Section 14 (3), *RTA 1988*, there are some exceptions under Section 14 (1), *RTA 1988*.

[19] Section 15, *RTA 1988*.

[20] See Regulation 110 of the *Road Vehicles (Construction and Use) Regulations 1986*. Note that in 2018 this was amended by *Regulation 2 of the Road Vehicles (Construction and Use) (Amendment) Regulations 2018* which inserted Regulation 110 (5A), to allow the use of a mobile phone when remotely parking a vehicle. Also, see changes to the Highway Code rules 150, 160 and 239. Also, see prior to these changes, the DfT (2017) 'Remote Control Parking and Motorway Assist: Proposals for Amending Regulations and the Highway Code' <https://assets.publishing.service.gov.uk/government/uploads/system/uploads/attachment_data/file/669442/remote-control-parking-motorway-assist-proposals-for-amending-regulation-and-highway-code.pdf> accessed on 25 August 2022, and DfT (2018) 'Remote control parking and motorway assist: proposals for amending regulations and the Highway Code: government Response' <https://assets.publishing.service.gov.uk/government/uploads/system/uploads/attachment_data/file/707095/ccav-consultion-response.pdf> 25 August 2022. It is important to note that the Law Commissions are consulting on changes in relation to remote driving. See Law Commissions (2022a) 'Remote Driving: Issues Paper'.

[21] Note that there are some exceptions to this, see, e.g. Section 300 (1) (a), *Road Safety Road Rules 2017*.

[22] Regulation 5 (1) *General Product Safety Regulations 2005*; Regulation 20 (1) *General Product Safety Regulations 2005*.

Act 2007.[23] Section 1 (1) of the Act states that '*An organisation to which this section applies is guilty of an offence if the way in which its activities are managed or organised— (a)causes a person's death, and (b)amounts to a gross breach of a relevant duty of care owed by the organisation to the deceased*'. The Act provides further detail of what is classed as a 'gross breach' in Section 1 (4).[24] In Australia, we can also see the *Criminal Code Act 1995, which, 'allows for corporate bodies to be found liable for harms caused by operational-level employees acting within the scope of their employment'*.[25] However, it is notable that '*this reform only extended to Commonwealth crimes, while most offences, including manslaughter, were the legislative responsibility of individual states and territories'*.[26] It is important to note that Victoria has not passed enabling legislation. Notably, however, there are offences that can be committed, with corporate bodies able to be liable for 'mainstream criminal law offences, including homicide and manslaughter offences' without legislation specifically in force for corporate manslaughter.[27]

2.2 Function

The function of road traffic law in both the UK and Australia concerns road safety. Offences arguably provide a deterrence against practices, which cause problems for road safety.[28] Specific and general deterrence can be distinguished here. Watson et al. note that '*Behavioural countermeasures have typically aimed to reduce road trauma by deterring road users from committing illegal road behaviours through the threat of detection and punishment (i.e., general deterrence) and through the administration of appropriate sanctions to those who are detected (i.e., specific deterrence)*'.[29] Whether the law has a deterrent effect on practices, has been discussed prevalently in literature.[30] It is not just the law but also the punishment

[23] This was discussed in Channon et al. (2019).

[24] Section 1 (4) of the Act states, '*a breach of a duty of care by an organisation is a "gross" breach if the conduct alleged to amount to a breach of duty falls far below what can reasonably be expected of the organisation in the circumstances'*. Also, see, e.g. Channon et al. (2019), p. 95, who noted challenges concerning the '*gross*' element.

[25] Amond (2013), p. 36.

[26] Amond (2013), p. 36.

[27] Amond (2013), p. 38.

[28] Cunningham (2008).

[29] Watson et al. (2015), p. 27.

[30] See Elvik and Christensen (2007); see also Cunningham (2007).

received[31] and enforcement,[32] which are important when providing a deterrent. It is worth noting in relation to automated vehicles that punishing the individual in the driving seat while the ADS is engaged will not provide a deterrent concerning the driving of the vehicle, as the individual in the driving seat will not have any control over that vehicle in which to be deterred. Moreover, it is notable that deterrence is not the only aim of road traffic law, but also, for example the need to ensure justice.[33] While there would likely be calls for justice if a vehicle caused serious injury or death with ADS engaged, justice would not be achieved through finding the individual in the driving seat responsible. One could apply the '*Desert Constraint*',[34] which is, '*the claim that an offender may not, in justice, be punished in excess of one's desert, which is understood mainly in terms of the culpability one incurs by virtue of one's conduct. The main effect of the desert constraint is to rule out punishments that go beyond what is proportionate to one's culpability'*.[35] Applying this to the individual who is in the driving seat with the ADS engaged, it is evident that punishment of that individual would be disproportionate to (the absence) of culpability. Moreover, the punishment of a human driver could '*discourage public use and acceptance*' of vehicles.[36] Public acceptance does seem to be a key factor on whether autonomous driving will be '*a realistic part of future transportation*'.[37] It is also worth repeating that offences not related to the driving of the vehicle (for example seatbelt offences) would likely be appropriate and could aide safety. Moreover, it can also be noted that technology itself could prevent some road traffic offences, as we will discuss in the next part of this chapter.

3 Can Automated Vehicle Technology Prevent Road Traffic Offences?

Before discussing the application of offences to automated vehicles, we can note that there are a number of offences, which could be preventable through technology. For example, vehicles could prevent driving while intoxicated, either through the vehicle

[31] There is a significant amount of research on specific punishment and deterrence in a road traffic context. See, e.g. Easten and Piper (2015) who look at whether Fixed Penalty Notices would deter careless drivers.

[32] We could mention the article from Nick Reed et al. (2021), p. 780, which mentioned that '*Even where gross breaches of the standard expected of competent and careful human drivers occur, if this is not observed by others or by authorities, prosecution is unlikely*'.

[33] Cunningham (2007), pp. 297–298.

[34] See Ryan Abbott (2020), p. 118. Moreover, see, e.g. Weinrib's article on desert constraint. See Weinreb (1986), pp. 47–80.

[35] Ryan Abbott (2020), p. 118.

[36] Van Wees (2020), p. 34.

[37] See Nastjuka et al. (2020) 2 citing Panagiotopoulos and Dimitrakopoulos (2018); Xu et al. (2018).

not having any controls to allow human control or using breathalysers within vehicles. In fact, the European Union has made it compulsory for all new vehicles to have '*alcohol interlock facilitation installation*' (as part of a number of '*advanced vehicle systems*').[38] Moreover, vehicles could be programmed to not exceed the speed limit, although as noted by the Law Commissions, there are ways in which speeding offences may occur, such as failure to update software.[39] Furthermore as noted by Chesterman '*a more utopian vision is that driverless cars may be so superior to human drivers that there is no need to provide for criminal responsibility at all*',[40] although Chesterman further notes that this vision seems '*unrealistic*'.[41] This is an interesting point. There is the potential for responsibility to become unnecessary due to the introduction of fully automated vehicles. It is likely that there will be changes on offences through the introduction of automated vehicles, with some offences becoming irrelevant where the vehicle has full control, e.g. the requirement of a driving licence. Some offences as they are currently worded may need to be changed (this will be discussed later in this chapter). However, there are some offences which are likely to continue. For example, the wearing of seatbelts.

4 Automated Vehicles: Lower Levels of Autonomy

Before discussing the application of offences in road traffic law, it is important to mention the Society of Automotive Engineer (SAE) Levels.[42] The SAE Levels highlight the differences between the automation of vehicles and range from level zero, a vehicle, which can contain warnings, and 'momentary assistance',[43] to level five, which is a fully automated vehicle where the vehicle performs all aspects of the

[38]See REGULATION (EU) 2019/2144 OF THE EUROPEAN PARLIAMENT AND OF THE COUNCIL of 27 November 2019 on type-approval requirements for motor vehicles and their trailers, and systems, components and separate technical units intended for such vehicles, as regards their general safety and the protection of vehicle occupants and vulnerable road users, amending Regulation (EU) 2018/858 of the European Parliament and of the Council and repealing Regulations (EC) No. 78/2009, (EC) No. 79/2009 and (EC) No. 661/2009 of the European Parliament and of the Council and Commission Regulations (EC) No. 631/2009, (EU) No. 406/2010, (EU) No. 672/2010, (EU) No. 1003/2010, (EU) No. 1005/2010, (EU) No. 1008/2010, (EU) No. 1009/2010, (EU) No. 19/2011, (EU) No. 109/2011, (EU) No. 458/2011, (EU) No. 65/2012, (EU) No. 130/2012, (EU) No. 347/2012, (EU) No. 351/2012, (EU) No. 1230/2012 and (EU) 2015/166, this came into force on 6 July 2022. In the UK, see Norbury and Webster (2021), p. 7.

[39]Law Commissions (2022b) 'Automated Vehicles: Joint Report' (LC 404).

[40]Chesterman (2021), p. 41.

[41]Ibid.

[42]Society of Automotive Engineers (2021), 'Taxonomy and Definitions for Terms Related to Driving Automation Systems for On-Road Motor Vehicles' J3016 202104.

[43]Ibid.

Dynamic Driving Task (DDT).[44] Overall control of vehicles at levels one and two remain with the person in the driving seat, who continues to be responsible for the vehicle. Advanced drivers assistance systems (ADAS) assist the driver but do not perform all aspects of the dynamic driving task (DDT).[45] Examples of ADAS can include, for example cruise control and automatic emergency braking.[46] Level three vehicles involve both driver and vehicle control[47] when the vehicle is in control this will allow the driver to disengage from the driving task, although the driver will need to regain control of the vehicle if requested by the vehicle.[48] Automated Lane Keeping Systems (ALKS) are level three vehicles, approved by the United Nations,[49] which '*keeps the vehicle within its lane for travelling speed of 60 km/h or less by controlling the lateral and longitudinal movements of the vehicle for extended periods without the need for further driver input*'.[50] In instances, which are outside of the vehicles' operational design domain (ODD), i.e. a situation that the ALKS is not designed to cope with,[51] the vehicle will then issue a transition demand and the driver will then need to take control of the vehicle. If the driver does not take control of the vehicle, the vehicle will perform a minimal risk manoeuvre (MRM)[52] and this can include the potential for the vehicle to stop in lane. As we will discuss later, there are therefore questions on what the 'driver' could do when the ALKS is engaged, such as whether they could use the infotainment system. Level four vehicles are highly automated. As noted by the Law Commissions, 'The main difference between SAE Level 3 and SAE Level 4 is that, following a failed prompt to intervene, a Level 4 vehicle can achieve a "minimal risk condition". At Level 4, responding to the prompt is not safety-critical, while at Level 3 it is'.[53]

For ADAS, most offences in the UK and Victoria will continue to apply to the driver of the vehicle such as careless and dangerous driving, with the vehicle

[44] 'In all situations and conditions that a human driver could' Law Commissions (2018) 13. Society of Automotive Engineers (2021), 'Taxonomy and Definitions for Terms Related to Driving Automation Systems for On-Road Motor Vehicles' J3016 202104.

[45] As defined by the BSI, Dynamic Driving Task is the 'Real-time operational and tactical functions required to operate a vehicle safely in on-road traffic' BSI Group (2020) https://www.bsigroup.com/en-GB/CAV/cav-vocabulary/dynamic-driving-task/ accessed on 26 July 2022.

[46] See, e.g. BSI Group (2020) https://www.bsigroup.com/en-GB/CAV/cav-vocabulary/advanced-driver-assistance-system/ accessed on 26 July 2022.

[47] Law Commissions (2018) [3.16].

[48] Society of Automotive Engineers (2021), 'Taxonomy and Definitions for Terms Related to Driving Automation Systems for On-Road Motor Vehicles' J3016_202104.

[49] See United Nations Regulation 157 (2021), 'Uniform provisions concerning the approval of vehicles with regard to Automated Lane Keeping Systems',

[50] UN Regulation No. 157 (2021) at 2.1.

[51] UN Regulation No. 157 (2021) at 5.4.2.3.

[52] As noted by the DfT (2020), 'Safe Use of Automated Lane Keeping Systems', 9, an MRM is 'A procedure aimed at minimising risks in traffic, which is automatically performed by the system after a transition demand without driver response or in the case of a sever ALKS or vehicle failure'.

[53] Law Commissions (2020) [4.33].

assisting the driver. As noted by the Venturer Report '*As long as human drivers remain in control of vehicles, these existing offences will continue to have relevance, notwithstanding growing prevalence of driver assistance features*'.[54] Mobile phone offences, for example will continue to apply, due to the driver remaining in control of the vehicle. Evidently, the relevant driving licence will continue to be required, as the individual is still required to control the vehicle safely.

5 Automated Vehicles: Higher Levels of Autonomy

It is evident that when the ADS is in control, certain offences will not apply in both the UK and Victoria. In some instances, there may be differences concerning offences applying at different automation levels. Where the individual is driving the vehicle (at any level) and the ADS is not engaged, offences are likely to continue to apply in both the UK and Victoria.[55] This is an important point; as illustrated previously in this chapter, the focus of road traffic law is road safety, with offences, for example aiming to provide deterrent. It is important to note that when the ADS is operating, certain offences, would, as written, be unlikely to apply to the person in the driving seat. For example, dangerous and careless driving offences utilise '*person*' in both the UK and Victoria. The Australian NTC noted that current transport law in Australia '*does not envisage a situation where an ADS, rather than a human driver, is in control of the dynamic driving task*'.[56] This was supported in a 2017 paper from Victoria.[57]

As can therefore be seen, human centred provisions would be unlikely to apply satisfactorily to automated vehicles. With the above noted, what if there is some human involvement? For level three vehicles, it is questionable whether the offences of dangerous or careless driving could apply if the vehicle requests driver control without response and the vehicle stops in lane. In the UK, a vehicle stopping in lane on a motorway is currently an offence under Regulation 7 of the *Motorway Traffic (England & Wales) Regulations 1982* (1982 Regulations).[58] Moreover, in the UK DfT's ALKS consultation, a vehicle which is fitted with ALKS and which stops in-lane after no response from a transition demand, was discussed.[59] The DfT have

[54]Venturer Project (2018) <https://www.venturer-cars.com/wp-content/uploads/2018/06/Year-3-Legal-and-Insurance-Report.pdf> accessed on 6 September 2022, 49.

[55]See, e.g. from the Australian National Transport Commission (2022).

[56]Australian National Transport Commission (2017), p. 15. See also Lilla Thiele-Evans, Blake Pepper, Zeleznikow et al. (2021), pp. 79–100, 95.

[57]VicRoads (2016), 'Future Directions Paper: How Victoria will continue to support the development of automated vehicles', <https://apo.org.au/node/71905> accessed on 28 August 2022.

[58]It is worth noting that in Victoria, the Section 177, *Road Safety Road Rules* 2017 prohibits a driver from stopping on a freeway (unless in 'an emergency stopping lane').

[59]Department for Transport Centre for Connected and Autonomous Vehicles (2020).

proposed to limit exceptions from the 1982 Regulations,[60] to provide an incentive for the driver to ensure that they resume control of the vehicle.[61] Whether this will provide such incentive is unclear, and drivers will need to be aware of their responsibilities. In relation to intoxication offences, it is worth noting that an offence can be committed through being '*in charge*' of a motor vehicle '*under the influence of drink or drugs*'.[62] The Court in the *DPP v Watkins*[63] noted that there are '*two classes of case*':

(1) 'If the defendant is the owner or lawful possessor of the vehicle or has recently driven it, he will have been in charge of it, and the question for the court will be whether he is still in charge or whether he has relinquished his charge... he would not be so if in all the circumstances he has ceased to be in actual control and there is no realistic possibility of his resuming actual control while unfit...'.[64]
(2) 'If the defendant is not the owner, the lawful possessor or recent driver but is sitting in the vehicle or is otherwise involved with it, the question for the court is...whether he has assumed being in charge of it. In this class of case the defendant will be in charge if, whilst unfit, he is voluntarily in de facto control of the vehicle or if...he might be expected imminently to assume control...'.[65]

For ALKS systems, the individual in the vehicle would fall within both class 1 and class 2. Where an individual is sat in the '*drivers*' seat of a fully automated vehicle and with no means of being able to control the vehicle, it is likely that this would not fall within being '*in charge*' of the vehicle. We can also mention here reforms in Victoria on automated vehicle trialling, as the RSA 1986 also uses the term '*in charge*' and which was clarified in the *Road Safety (Amendment (Automated Vehicles) Act 2018*.[66] Section 3AA(ba) of the *RSA 1986* now provides an offence concerning the intoxicated supervision in Victoria of an automated vehicle in automated mode.[67]

---

[60] Department for Transport Centre for Connected and Autonomous Vehicles (2020), p. 30.

[61] Department for Transport Centre for Connected and Autonomous Vehicles (2020), p. 30.

[62] Section 4 (2), *RTA 1988.*

[63] [1989] Q.B. 821.

[64] Ibid, 831.

[65] Ibid, 831.

[66] Section 4, *Road Safety (Amendment (Automated Vehicles) Act 2018.*

[67] Section 3AA(ba), *RSA 1986* reads 'a person who is a vehicle supervisor of an automated vehicle for which an ADS permit is in force and which is operating in automated mode at any time while the person is assigned by the ADS permit holder to perform duties as a vehicle supervisor in relation to the vehicle'. It is also important to note new criminal offences created in Victoria in relation to automated vehicles involving those driving or 'being in charge' of an automated vehicle without an ADS permit (Section 33I, *Road Safety Act 1986*) or 'in breach of an ADS permit condition' (Section 33J, *Road Safety Act 1986*).

Vehicles at level three SAE pose questions concerning what the person in the driver's seat can legally do when the Automated Driving System (ADS) is engaged, due to expectations of taking vehicle control within a limited timeframe. The DfT have introduced changes to the Highway Code, which state, inter alia, '*If a self-driving vehicle needs to hand control back to the driver, it will give you enough warning to do this safely. You MUST always be able and ready to take control and do it when the vehicle prompts you. For example, you should stay in the driving seat and stay awake. When you have taken back control or turned off the self-driving function, you are responsible for all aspects of driving*'.[68] Moreover recent changes state that the infotainment system can be used when the ADS is engaged but mobile phones cannot.[69] Many of the responses to the DfT's '*Call for Evidence*' prior to these changes, noted that use of infotainment system was safe when the ADS was engaged, although concerns were evident around driver distraction if the infotainment system was used.[70] The potential use of mobile phones is controversial, with concerns that the use of a mobile phone would provide too much of a distraction and potentially increase the risk of the driver not taking control of the vehicle in time.[71] It is important to state, as mentioned in the Responses to the DfT's ALKS Consultation, that changes to the Highway Code are communicated widely; an '*awareness campaign*' to highlight that the rules on use of ALKS systems could be beneficial to ensure that the 'driver' is aware of what they are able to do when the ADS is engaged, and their responsibilities during a transition demand.[72]

We can discuss the UK proposals from the Law Commissions concerning the introduction of a *'user in charge'* (UIC) while the ADS is engaged.[73] The Law Commissions proposed that '*while the ADS is engaged, users-in-charge would have immunity from any dynamic driving offences or civil penalties*'.[74] Moreover, when the ADS is in control, offences, which do not arise from the dynamic driving task,

---

[68]Department for Transport (2022), *Highway Code* <https://www.gov.uk/guidance/the-highway-code/introduction> accessed on 23 August 2022.

[69]Department for Transport (2022), *Highway Code* <https://www.gov.uk/guidance/the-highway-code/introduction> accessed on 23 August 2022.

[70]DFT ALKS Summary of Responses and Next Steps (2021), 28.

[71]It is worth recognising that the research on this topic is mixed. See, e.g. Sherrie Anne Kaye et al. (2021) who stated that there were not significantly longer takeover times (for younger drivers), compared with Erikson and Stanton (2017) (cited in Sherrie Anne Kaye et al. 2021) who found a longer take over time. However, differences between the two studies were noted in the Sherrie Anne Kaye et al. (2021), 26 article.

[72]See DfT ALKS Summary of Responses and Next Steps (2021), 35.

[73]The Law Commissions noted the challenges surrounding distinguishing Levels 3 and 4 of the SAE levels, Law Commissions (2022) [3.38] citing Law Commissions (2020) 43. Consequently the Law Commissions noted that their recommendations 'do not attempt to distinguish between Level 3 and Level 4 systems'. Law Commissions (2022) [3.40]. Citing Society of Automotive Engineers (2021), 'Taxonomy and Definitions for Terms Related to Driving Automation Systems for On-Road Motor Vehicles' J3016 202104.

[74]Law Commissions (2022) [8.7].

would continue to be committed, such as ensuring that children wear seatbelts.[75] Furthermore, it is important to note that it would be an offence to be a passenger in a vehicle with no UIC.[76] Moreover, the UIC would be required to be '*qualified and fit*' to drive.[77] There are clear reasons this, as noted by Van Wees '*when ODD restrictions are more categorical, a human driver is still needed, for example to drive the vehicle to the motorway when the system is confined to these types of roads (and to take back control when the exit is reached)'*.[78] Notably, however, the Law Commissions stated that some vehicles would not need UICs, because of the so-called 'no user-in-charge' (NUIC), where a licenced NUIC operator would oversee the vehicle remotely.[79] A NUIC operator would not commit driving offences, as the NUIC would not be classed as a driver; however, certain offences would apply such as failing to have insurance as this applies to the vehicle '*user*', a term wider than '*driver*'.[80] It is important to note that the UK government, in response to the Law Commissions' proposals, stated its support for the introduction of both a UIC and NUIC,[81] agreeing, for example with the need for the UIC to be licenced. Moreover, the UK government agreed with the Law Commissions proposals in relation to the responsibility of UIC's post handover.[82]

In Australia, the NTC have stated that automated vehicles will require a '*fallback ready user*'[83] who '*remain(s) sufficiently vigilant to respond to ADS requests, mechanical failure or emergency vehicles and regain control of the vehicle without undue delay when required*'.[84] The '*fallback ready user*' may turn their attention away from thc driving task '*but can only perform secondary activities with appropriate reaction times*'.[85] It is not completely clear as to what these tasks would

[75] Law Commissions (2022) [8.103].

[76] Law Commissions (2022). We can also see proposals by the Centre for Data, Ethics and Innovation in relation to UICs, e.g. that 'sufficient time' is provided to the UIC 'to retain situational awareness'. Moreover, 'ensuring safe operation' if the UIC fails to resume control of the vehicle. See Centre for Data, Ethics and Innovation (2022), 'Responsible Innovation in Self-Driving Vehicles' <https://www.gov.uk/government/publications/responsible-innovation-in-self-driving-vehicles/responsible-innovation-in-self-driving-vehicles> accessed on 5 September 2022, 22.

[77] Law Commissions (2022), [2.45].

[78] See Van Wees (2020), pp. 33–34.

[79] Law Commissions (2022).

[80] Law Commissions (2022), citing Lord Hodge *in R&S Pilling (t/a Phoenix Engineering) v UK Insurance Ltd* [2019] UKSC 16 and Section 143 (1) (a), *RTA 1988*. Moreover, certain offences could apply to the NUIC if they are the '*owner*' or '*registered keeper*' of the vehicle.

[81] See the UK Government (2022), 'Connected & Automated Mobility 2025: Realising the benefits of self-driving vehicles in the UK' (CP 712) <https://assets.publishing.service.gov.uk/government/uploads/system/uploads/attachment_data/file/1099173/cam-2025-realising-benefits-self-driving-vehicles.pdf> accessed on 12 September 2022, 124.

[82] Ibid, 124.

[83] Note that this applies to vehicles at level three SAE. See, NTC (2020), 23. Also discussion of 'fall-back ready users' in Law Commissions (2022) 45-47'.

[84] NTC (2022), p. 43.

[85] NTC (2017), p. 66.

be. Moreover the '*fallback ready user*' must be fully licenced and fit to drive.[86] A significant notable difference between the Law Commissions and NTC concern whether the person in the vehicle needs to respond to anything other than a transition demand.[87] Differently to stated above, the Law Commissions noted that 'Vehicles should not be authorised as self-driving if an individual is expected to respond to objects or events in the external environment (such as low impact collisions and emergency vehicles) in the absence of a transition demand'.[88] This is a very important issue, which concerns the responsibility of the individual in the vehicle and whether they would need to respond to anything other than a transition demand. This would likely have an impact on what that individual could do in the vehicle when the ADS is engaged. Moreover, the NTC have noted that '*the fallback-ready user obligations will be in state and territory road transport laws*'.[89] Importantly the NTC will be guiding states and territories to ensure that rules between states are consistent with each other, this should prevent confusion on potentially different responsibilities or prohibitions between each state and territory.[90] Overall, while there are similarities between the Law Commission and NTC, there are differences in the expectations of the user. It is important to provide clarity to users of vehicles who may be required to take control. Key terminology does need to be clarified to allow the user to know the expectation of them when the ADS is engaged.

This chapter has, so far, explored the liability of person(s) in the vehicle without exploring whether liability could exist elsewhere when the ADS is engaged. Before discussing this, it is worth noting, in relation to both jurisdictions, that some offences may well continue to apply when the vehicle is driving itself and there is an individual in the vehicle, e.g. seatbelt offences. It is also important to note when considering future liability (civil or criminal), that ensuring continued innovation is a consideration, with a potential '*chilling effect*' on innovation,[91] and such chilling effect could lead to the delay in the introduction of vehicles, which are safer than is currently available, due to concerns over potential future liability.[92] Notably, there are alternative approaches rather than traditional punishments, as noted by Chesterman '*monetary and custodial punishments may be less appropriate than seeing enforcement as part of a feedback loop to train the system'*.[93] Chesterman cited the Law Commissions who have proposed the introduction of an Authorised Self Driving Entity (ASDE) which '*puts the ADS (automated driving system)*

---

[86] NTC (2021), p. 22.

[87] NTC (2021). Cited by the Law Commission (2022), 46.

[88] Law Commission (2022) [3.67].

[89] NTC (2022), p. 43.

[90] NTC (2022), p. 43.

[91] See work, e.g. by Schellekens (2015) which focusses more on product liability which was also cited in Channon (2019). See also Law Commissions (2020), p. 243, who note the potential 'stifling effect' of 'attaching criminal liability to wrongdoing by ADSEs'.

[92] Schellekens (2015).

[93] Chesterman (2021), p. 226.

forward for legal categorisation as self-driving and is legally responsible for how the ADS performs dynamic control'.[94] This, for example could be the manufacturer of the vehicle. As noted by the Law Commissions, where a problem exists 'regulatory sanctions' could be applied to the ASDE, which would signal a shift from the current approach of criminal sanctions.[95] Regulatory sanctions include fines, warnings, and compliance orders.[96] This does not mean that criminal sanctions would be removed completely, several offences were proposed by the Law Commissions involving the ASDE's 'duty of candour'.[97] It seems that the responses to the Law Commissions' proposals were favourable concerning the introduction of regulatory sanctions and a move away from criminal liability.[98] It is also important to consider the application of the functions of current road traffic rules to regulatory sanctions, e.g. would regulatory sanctions aide road safety? This would need to be considered further with the potential 'chilling effect' of liability also considered. Moreover, on rationales of road traffic law, it is notable that the Law Commissions have proposed some criminal punishment for '*senior managers*' within ASDEs if offences are committed[99] with their '*consent or connivance*' and '*aggravated*' offences where death or injury results from this.[100] This is aimed at the rationale above in terms of justice.[101] Some future clarification is needed concerning the application of Manslaughter offences to ASDEs with the Law Commissions not proposing to '*exempt AV developers*'[102] in relation to these offences, but also noting that they would '*not wish to see these offences being used widely against ADSEs'*.[103] The extent that, and circumstances where, these offences should be used, is not clear and clarification is certainly needed here. The Law Commissions noted an 'accountability gap', which needed to be filled as, for example manslaughter offences are applicable in relation to death and not serious injury.[104] The Law Commissions proposals seek to fill this

---

[94] Law Commissions (2020, v). It is further worth noting some of the additional proposals in relation to ASDEs from the Centre for Data, Ethics and Innovation (2022). For example, the need for 'explainability', the CDEI noted that 'the ASDE should design the AV so that it is possible to construct an explanation of the key decisions made by the AV when it is undertaking the Dynamic Driving Task (DDT) for a bounded test scenario'.

[95] Law Commissions (2020), p. 175.

[96] Law Commissions (2020), p. 185.

[97] Law Commissions (2022), pp. 274–275.

[98] Law Commissions 'Automated Vehicles: Summary of responses to Consultation Paper 3 and next steps' (2021).

[99] See, e.g. as stated in Recommendation 65, Law Commissions (2022) 274, 'When putting forward a vehicle for authorisation as self-driving, it should be an offence for the ASDE to (1) fail to provide information to the regulator; or (2) provide information to the regulator that is false or misleading in a material particular where that information is relevant to the evaluation of the safety of the vehicle'.

[100] Law Commissions (2022), p. 222.

[101] Law Commissions (2022).

[102] Law Commissions (2020). [3.61].

[103] Ibid.

[104] Law Commissions (2020), p. 249.

gap. The UK government in its response to the Law Commission's proposals noted its support of the proposal for the introduction of an ASDE, as well as the potential for regulatory sanctions (discussed above) and for the introduction of a 'duty of candour'.[105]

In Australia, the NTC have recommended the introduction of a national 'automated vehicle safety law' (AVSL).[106] Moreover, '*Automated Driving System Entities*' (ADSE) have been proposed, who will '*self-certify the safety of the ADS and take responsibility for it over its life*'.[107] This is clearly very similar to the proposals provided by the Law Commissions in the UK; and in fact, the Law Commissions cited some of the Australian recommendations for an ADSE when making its proposals.[108] The NTC also proposed the introduction of a '*general safety duty*'[109] and it is important to note that while this duty is expressed in '*general terms*' there are 'some minimum prescriptive requirements'.[110] Burke[111] has noted several advantages of the introduction of a general safety duty, for example Burke stated that *'Instead of focusing on a myriad of prescriptive requirements, ASDEs can focus their attention on the safety risks'*.[112] Where the ADS is engaged, there is an attempt to move away from, *'relying on the current system of infringements and other relevant sanctions*'.[113] Moreover, executive officers of the ADSE will have a 'due diligence' obligation to ensure that the ADSE meets its general safety duty with consequences for breaching this obligation.[114] The in-service regulator will also have several options, such as warnings and improvement notices.[115] Interestingly and different to the UK approach, the NTC has proposed three categories of offences when determining the penalties under the AVSL.[116] Overall, therefore it is evident that there are differences between the UK and Australian approaches particularly on penalties, although it is notable that there are significant similarities as noted above. The effectiveness of these approaches in meeting the functions of road safety law are unclear, it is evident though that ensuring justice is a focus in the reforms. Moreover, the deterrent effect of the reforms are not currently clear, and will depend on other factors (as noted previously in this chapter).

---

[105] The UK Government (2022), p. 125.

[106] NTC (2020).

[107] National Transport Commission (2022), p. 10.

[108] See Law Commissions (2020).

[109] National Transport Commission (2021), p. 26.

[110] National Transport Commission (2021), p. 27.

[111] Burke (2022). More from Marcus Burke can, e.g. be found here https://www.ntc.gbring-new-challenges-data-access-and-insurance accessed on 10 August 2022.

[112] Burke (2022), p. 45.

[113] National Transport Commission (2022), p. 59.

[114] National Transport Commission (2022), p. 64.

[115] National Transport Commission (2022), p. 79.

[116] National Transport Commission (2022), pp. 64–65.

6 Future Liability of the Vehicle Itself?

In this part of the chapter, it is briefly questioned whether it is possible that the vehicle itself could be legally responsible for a collision and therefore itself receiving punishment. This would also involve questions concerning whether vehicles should be provided with legal personality.[117] Chesterman analysed the rationale behind criminal law and its application to vehicles, noting that a fine would affect behaviour of a vehicle if the programming '*sought to maximize economic gain without regard for the underlying criminal law itself*'.[118] Moreover, Gless et al. note that '*it is difficult, at present, to imagine sanctions*. . .*that would fulfil the same purposes as criminal sanctions imposed on human beings*'.[119] It is difficult to see how a fine, imprisonment or disqualification of a vehicle itself could at the moment meet the function of road traffic law, and the vehicle would unlikely be deterred through the introduction of punishments. Regardless, Abbott considers that punishing AI now would '*be an overreaction to all but hypothetical concerns*'.[120] This author agrees that this is likely to be a question for the future where the vehicle is separate from any form of human control. Currently, there is little question of criminal liability for vehicles, with the driver responsible for conventional vehicles and vehicles in which the ADS is not engaged. For vehicles with the ADS engaged, there are, as discussed above, others who can take responsibility for the vehicle such as the proposed ASDEs. Current proposals providing the introduction of the ASDE who will be responsible for vehicles seems to be a more feasible approach and would better meet the focus of road traffic law. This is not to discount the possibility of legal personality provided to the automated vehicle in the future, however.

7 Conclusion

To conclude, this chapter has explored road traffic offences in the UK and Victoria and application to automated vehicles. While there are some differences, it is evident that there are significant similarities between road traffic offences in the UK and Victoria. The focus of road traffic offences is road safety and this will be a key rationale in terms of automated vehicles. For ADAS and conventional vehicles, responsibility currently rests with the driver and this is unlikely to change. For vehicles at level 3 questions concern tasks that individuals could undertake when they have disengaged from the driving task, and it is notable that a number of offences are likely to continue (such as intoxicated driving). The Law Commissions have proposed the introduction of a UIC, when the ADS is engaged, and a NUIC

[117] See Ryan Abbott (2020) 127 on AI.
[118] Chesterman (2021), p. 125.
[119] Gless et al. (2016), pp. 423–424.
[120] Ryan Abbott (2020), p. 133.

where the vehicle can drive without a UIC. The Law Commissions propose that the UIC would have immunity from offences related to the driving task when the ADS is engaged, although some offences would continue to apply such as where the UIC is not 'fit and qualified' to drive. Australia's approach requires that the 'fallback ready user' is fit and qualified to drive, although with significant differences to the Law Commissions proposal concerning what the person in the vehicle should be expected to do, there is seemingly additional complexity with Australia's approach.

This chapter has also explored liability where the ADS is engaged. Of course, innovation is a consideration when determining this. There are significant similarities between the Law Commission and NTC approaches through the introduction of an entity, some criminal offences are proposed, although there seems to be some move towards regulatory sanctions. The NTC approach uses three categories of offences. The effectiveness of both offences is uncertain. Moreover, so is the application of corporate manslaughter in the UK and clarification here would be welcome. Finally, legal personality and criminal responsibility on the automated vehicle itself is currently not realistic nor necessary although could be a consideration in the future.

Websites

BSI Group (2020) https://www.bsigroup.com/en-GB/CAV/cav-vocabulary/dynamic-driving-task/ accessed on 26 July 2022

BSI Group (2020) https://www.bsigroup.com/en-GB/CAV/cav-vocabulary/advanced-driver-assistance-system/ accessed on 26 July 2022

National Transport Commission <https://www.ntc.gov.au/laws-and-regulations/australian-road-rules#:~:text=The%20model%20Australian%20Road%20Rules,%2C%20pedestrians%2C%20passengers%20and%20others> accessed on 27 July 2022

National Transport Commission, https://www.ntc.gov.au/news/introduction-automated-vehicles-bring-new-challenges-data-access-and-insurance accessed on 10 August 2022

Venturer Project (2018) < https://www.venturer-cars.com/wp-content/uploads/2018/06/Year-3-Legal-and-Insurance-Report.pdf> 49, accessed on 12 August 2022

Vic Roads https://www.vicroads.vic.gov.au/safety-and-road-rules/vehicle-safety/automated-and-connected-vehicles/testing-of-automated-vehicles accessed on 10 August 2022

References

Abbott R (2020) The reasonable Robot. Cambridge University Press

Amond P (2013) Corporate manslaughter and regulatory reform. Palgrave MacMillan

Bissell D, Birtchnell T, Elliott A, Hsu EL (2020) Autonomous automobilities: the social impacts of driverless vehicles. Curr Sociol 68(1):116–134
Burke M (2022) Regulation of in-service safety risks of automated vehicle. In: Meyer G, Beiker S (eds) Road vehicle automation 8 lecture notes in mobility. Springer
Centre for Data, Ethics and Innovation (2022) Responsible Innovation in Self-Driving Vehicles. <https://www.gov.uk/government/publications/responsible-innovation-in-self-driving-vehicles/responsible-innovation-in-self-driving-vehicles> accessed on 5 September 2022
Channon M (2019) Automated and electric vehicles act 2018: an evaluation in light of proactive law and regulatory disconnect. Eur J Law Technol 10(2)
Channon M, McCormick L, Noussia K (2019) The law and autonomous vehicles. Routledge
Chesterman, (2021) *We, the Robots: regulating artificial intelligence and the limits of the law.* Cambridge University Press
Cunningham S (2007) Punishing drivers who kill: putting road safety first? Legal Stud 27(2): 288–311
Cunningham S (2008) Driving offences: law, policy and practice. Routledge
Department for Transport (2017) Remote Control Parking and Motorway Assist: Proposals for Amending Regulations and the Highway Code. <https://assets.publishing.service.gov.uk/government/uploads/system/uploads/attachment_data/file/669442/remote-control-parking-motorway-assist-proposals-for-amending-regulation-and-highway-code.pdf> accessed on 25 August 2022
Department for Transport (2018) Remote control parking and motorway assist: proposals for amending regulations and the Highway Code: government Response. <https://assets.publishing.service.gov.uk/government/uploads/system/uploads/attachment_data/file/707095/ccav-consultion-response.pdf> 25 August 2022
Easten S, Piper C (2015) Fixed penalties for careless driving: the delusion of deterrence. Contemp Iss Law 13(3):175–192
Elvik R, Christensen P (2007) The deterrent effect of increasing fixed penalties for traffic offences: the Norwegien experience. J Saf Res 38(6):689
Gless S, Silverman E, Weigend T (2016) If Robots cause harm, who is to blame? Self-driving cars and criminal liability. New Crim Law Rev Int and Interdiscipl J 19(3):412–436
Kaye S-A, Lewis I, Demmel S, Trespalacios OO, Griffin W (2021) Young drivers' takeover time in a conditional automated vehicle: The effects of hand-held mobile phone use and future intentions to use automated vehicles. Transport Res Part F Traffic Psychol Behav 78:16–29
Law Commissions (2018) Automated Vehicles: A joint preliminary consultation paper. (LC 240)
Law Commissions (2019) Automated Vehicles: Analysis of Responses to the Preliminary Consultation Paper. (LC 240)
Law Commissions (2020) Automated Vehicles: Consultation paper 3. (LC 252)
Law Commissions (2022a) Remote Driving: Issues Paper. <https://s3-eu-west-2.amazonaws.com/lawcom-prod-storage-11jsxou24uy7q/uploads/2022/06/Remote-driving-LC-Issues-paper.pdf> accessed on 25 August 2022
Law Commissions (2022b) Automated Vehicles: Joint Report. (LC 404)
Nastjuka I, Herrenkinda B, Marroneb M, Brendela AB, Kolbea LM (2020) What drives the acceptance of autonomous driving? An investigation of acceptance factors from an end-user's perspective' Technological Forecasting & Social Change 161:120319
National Transport Commission (2017) Changing driving laws to support automated vehicles
National Transport Commission (2020) A national in service safety law for automated vehicles
National Transport Commission (2022) The regulatory framework for automated vehicles
Frank Norbury, Evan Webster (2021) LOCKING OUT THE DRINK DRIVER: Using alcohol interlocks to reduce drink driving in the UK. Parliamentary Advisory Council for Transport Safety Report
Reed N, Leiman T, Palade P, Martens M, Kester L (2021) Ethics of automated vehicles: breaking traffic rules for road safety. Ethics Inf Technol 23:777–789

Schellekens M (2015) Self-driving cars and the chilling effect of liability law. Comput Law Secur Rev 31:506–517

UK Government (2022) Connected & Automated Mobility 2025: Realising the benefits of self-driving vehicles in the UK. https://www.gov.uk/government/publications/responsible-innovation-in-self-driving-vehicles/responsible-innovation-in-self-driving-vehicles

Van Wees K (2020) Technology in the driver's seat: legal obstacles and regulatory gaps in road traffic law. In: Van Uytsel S, Vasconcellos Vargas D (eds) Autonomous vehicles: business technology and law (Perspectives in law, business and innovation). Springer

VicRoads (2016) Future Directions Paper: How Victoria will continue to support the development of automated vehicles. <https://apo.org.au/node/71905> accessed on 28 August 2022

Barry Watson, Vic Siskind, Angela Watson, J.J. Fleiter (2015) Assessing specific deterrence effects of increased speeding penalties using four measures of recidivism. Accid Anal Prevent 84: 27-37

Weinreb LL (1986) Desert, punishment, and criminal responsibility. Law Contemp Probl, pp 47–80

Zeleznikow J, Foster N, Sourdin T (2021) Regulatory approaches to managing artificial intelligence systems in autonomous vehicles in Australia. Int J Law Inf Technol 29(2):79–100

MIX
Papier aus verantwortungsvollen Quellen
Paper from responsible sources
FSC® C105338

If you have any concerns about our products,
you can contact us on
ProductSafety@springernature.com

In case Publisher is established outside the EU,
the EU authorized representative is:
Springer Nature Customer Service Center GmbH
Europaplatz 3, 69115 Heidelberg, Germany

Printed by Libri Plureos GmbH
in Hamburg, Germany